McDougal Littell Science

Electricity and Magnetism

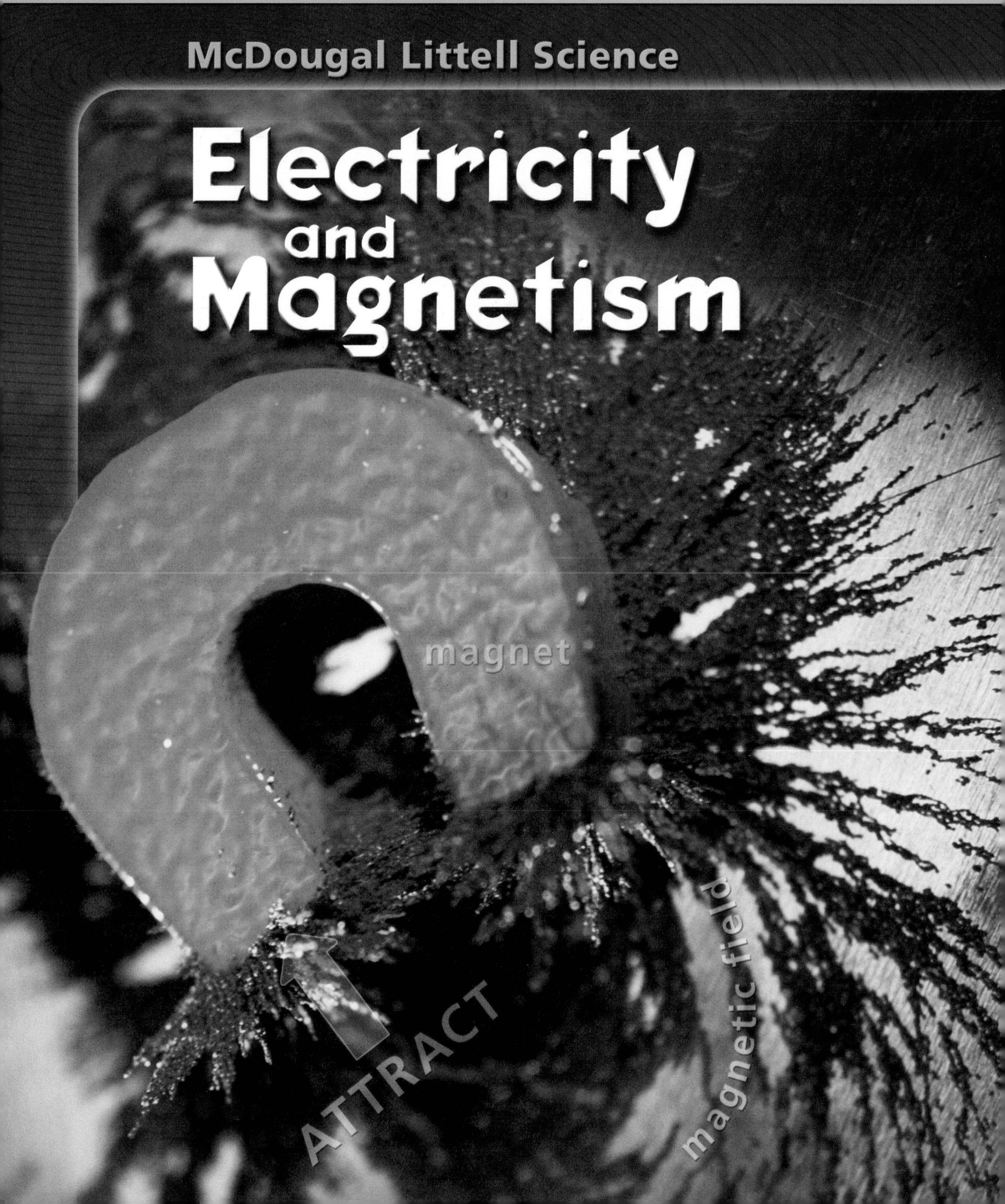

magnet

ATTRACT

magnetic field

Credits
5B Illustration by Stephen Durke; **5C** Illustration by Dan Stukenschneider;
39C © Donna Cox and Robert Patterson/National Center for Supercomputing
Applications, University of Illinois, Urbana; **75B** Photograph by Sharon Hoogstraten.

Acknowledgements
Excerpts and adaptations from National Science Education Standards by the National
Academy of Sciences. Copyright © 1996 by the National Academy of Sciences. Reprinted
with permission from the National Academies Press, Washington, D.C.

ISBN: 0-618-33441-6   1 2 3 4 5 6 7 8 VJM 08 07 06 05 04

Internet Web Site: http://www.mcdougallittell.com

McDougal Littell Science

Effective Science Instruction Tailored for Middle School Learners

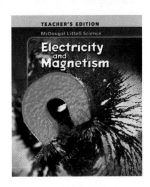

Electricity and Magnetism Teacher's Edition Contents

Consultants and Reviewers

Science Consultants

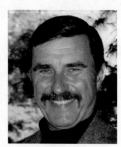

Chief Science Consultant

James Trefil, Ph.D. is the Clarence J. Robinson Professor of Physics at George Mason University. He is the author or co-author of more than 25 books, including *Science Matters* and *The Nature of Science.* Dr. Trefil is a member of the American Association for the Advancement of Science's Committee on the Public Understanding of Science and Technology. He is also a fellow of the World Economic Forum and a frequent contributor to *Smithsonian* magazine.

Rita Ann Calvo, Ph.D. is Senior Lecturer in Molecular Biology and Genetics at Cornell University, where for 12 years she also directed the Cornell Institute for Biology Teachers. Dr. Calvo is the 1999 recipient of the College and University Teaching Award from the National Association of Biology Teachers.

Kenneth Cutler, M.S. is the Education Coordinator for the Julius L. Chambers Biomedical Biotechnology Research Institute at North Carolina Central University. A former middle school and high school science teacher, he received a 1999 Presidential Award for Excellence in Science Teaching.

Instructional Design Consultants

Douglas Carnine, Ph.D. is Professor of Education and Director of the National Center for Improving the Tools of Educators at the University of Oregon. He is the author of seven books and over 100 other scholarly publications, primarily in the areas of instructional design and effective instructional strategies and tools for diverse learners. Dr. Carnine also serves as a member of the National Institute for Literacy Advisory Board.

Linda Carnine, Ph.D. consults with school districts on curriculum development and effective instruction for students struggling academically. A former teacher and school administrator, Dr. Carnine also co-authored a popular remedial reading program.

Donald Steely, Ph.D. serves as principal investigator at the Oregon Center for Applied Science (ORCAS) on federal grants for science and language arts programs. His background also includes teaching and authoring of print and multimedia programs in science, mathematics, history, and spelling.

Sam Miller, Ph.D. is a middle school science teacher and the Teacher Development Liaison for the Eugene, Oregon, Public Schools. He is the author of curricula for teaching science, mathematics, computer skills, and language arts.

Vicky Vachon, Ph.D. consults with school districts throughout the United States and Canada on improving overall academic achievement with a focus on literacy. She is also co-author of a widely used program for remedial readers.

Content Reviewers

John Beaver, Ph.D.
Ecology
Professor, Director of Science Education Center
College of Education and Human Services
Western Illinois University
Macomb, IL

Donald J. DeCoste, Ph.D.
Matter and Energy, Chemical Interactions
Chemistry Instructor
University of Illinois
Urbana-Champaign, IL

Dorothy Ann Fallows, Ph.D., MSc
Diversity of Living Things, Microbiology
Partners in Health
Boston, MA

Michael Foote, Ph.D.
The Changing Earth, Life Over Time
Associate Professor
Department of the Geophysical Sciences
The University of Chicago
Chicago, IL

Lucy Fortson, Ph.D.
Space Science
Director of Astronomy
Adler Planetarium and Astronomy Museum
Chicago, IL

Elizabeth Godrick, Ph.D.
Human Biology
Professor, CAS Biology
Boston University
Boston, MA

Isabelle Sacramento Grilo, M.S.
The Changing Earth
Lecturer, Department of the Geological Sciences
Montana State University
Bozeman, MT

David Harbster, MSc
Diversity of Living Things
Professor of Biology
Paradise Valley Community College
Phoenix, AZ

Richard D. Norris, Ph.D.
Earth's Waters
Professor of Paleobiology
Scripps Institution of Oceanography
University of California, San Diego
La Jolla, CA

Donald B. Peck, M.S.
*Motion and Forces; Waves, Sound, and Light;
    Electricity and Magnetism*
Director of the Center for Science Education (retired)
Fairleigh Dickinson University
Madison, NJ

Javier Penalosa, Ph.D.
Diversity of Living Things, Plants
Associate Professor, Biology Department
Buffalo State College
Buffalo, NY

Raymond T. Pierrehumbert, Ph.D.
Earth's Atmosphere
Professor in Geophysical Sciences (Atmospheric Science)
The University of Chicago
Chicago, IL

Brian J. Skinner, Ph.D.
Earth's Surface
Eugene Higgins Professor of Geology and Geophysics
Yale University
New Haven, CT

Nancy E. Spaulding, M.S.
Earth's Surface, The Changing Earth, Earth's Waters
Earth Science Teacher (retired)
Elmira Free Academy
Elmira, NY

Steven S. Zumdahl, Ph.D.
Matter and Energy, Chemical Interactions
Professor Emeritus of Chemistry
University of Illinois
Urbana-Champaign, IL

Susan L. Zumdahl, M.S.
Matter and Energy, Chemical Interactions
Chemistry Education Specialist
University of Illinois
Urbana-Champaign, IL

Safety Consultant

Juliana Texley, Ph.D.
Former K–12 Science Teacher and School Superintendent
Boca Raton, FL

English Language Advisor

Judy Lewis, M.A.
Director, State and Federal Programs for reading proficiency
and high risk populations
Rancho Cordova, CA

Research-Based Solutions for Your Classroom

The distinguished program consultant team and a thorough, research-based planning and development process assure that *McDougal Littell Science* supports all students in learning science concepts, acquiring inquiry skills, and thinking scientifically.

Standards-Based Instruction

Concepts and skills were selected based on careful analysis of national and state standards.

• National Science Education Standards

• Project 2061 Benchmarks for Science Literacy

• Comprehensive database of state science standards

Standards and Benchmarks

Each chapter in **Electricity and Magnetism** covers some of the learning goals that are described in the *National Science Education Standards* (NSES) and the Project 2061 *Benchmarks for Science Literacy*. Selected content and skill standards are shown below in shortened form. The following National Science Education Standards are covered on pages xii–xxvii, in Frontiers in Science, and in Timelines in Science, as well as in chapter features and laboratory investigations: Understandings About Scientific Inquiry (A.9), Understandings About Science and Technology (E.6), Science and Technology in Society (F.5), Science as a Human Endeavor (G.1), Nature of Science (G.2), and History of Science (G.3).

Content Standards

1 Electricity

National Science Education Standards

B.3.a | Energy
  • is often associated with electricity
  • is transferred in many ways

B.3.e | In most chemical reactions, energy is transferred into or out of the system. Heat, light, motion, or electricity might be involved in this transfer.

Project 2061 Benchmarks

3.A.3 | Engineers and others who work in design and technology use scientific knowledge to solve practical problems.

2 Circuits and Electronics

National Science Education Standards

B.3.a | Energy. is property of substances that is often associated with electricity. Energy is transferred in many ways.

B.3.d | Circuits transfer electrical energy. Heat, light, sound, and chemical changes are produced.

Project 2061 Benchmarks

1.C.6 | Computers are important in science.

3.A.2 | Technology is important in the computation and communication of information.

8.D.2 | The ability to code information or electric currents in wires has made communication many times faster than is possible by mail or sound.

8.E.1 | Computers use digital codes containing only two symbols to perform all operations. Analog signals must be converted into digital codes before a computer can process them.

9.A.6 | Numbers can be represented by using only 1 and 0.

CHAPTER 1

Electricity

the BIG idea

Moving electric charges transfer energy.

What keeps this dragon glowing brightly?

Key Concepts

SECTION 1.1 Materials can become electrically charged. Learn how the movement of electrons builds static charges and how static charges are used in technology.

SECTION 1.2 Charges can move from one place to another. Learn what factors control the movement of charges.

SECTION 1.3 Electric current is a flow of charge. Learn how electric current is measured and how it can be produced.

Internet Preview

CLASSZONE.COM

Chapter 1 online resources: Content Review, two Simulations, two Resource Centers, Math Tutorial, Test Practice.

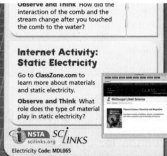

Observe and Think How did the interaction of the comb and the stream change after you touched the comb to the water?

Internet Activity: Static Electricity

Go to ClassZone.com to learn more about materials and static electricity.

Observe and Think What role does the type of material play in static electricity?

NSTA sciLINKS
scilinks.org

Electricity Code: MDL065

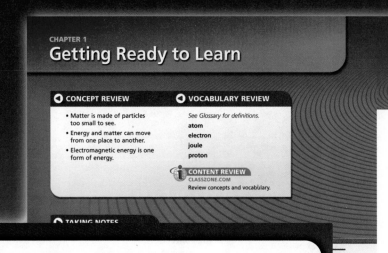

Getting Ready to Learn

◄ CONCEPT REVIEW

- Matter is made of particles too small to see.
- Energy and matter move from one place to another.
- Electromagnetic energy is one form of energy.

◄ VOCABULARY REVIEW

See Glossary for definitions.

atom
electron
joule
proton

CONTENT REVIEW
CLASSZONE.COM
Review concepts and vocabulary.

TAKING NOTES

Magnetism

Project 2061 Benchmarks

4.G.3 | Electric currents and magnets can exert a force on each other.
8.C.4 | Electrical energy can be produced from a variety of energy sources and can be transformed into almost any other form of energy. Electricity is used to distribute energy quickly and conveniently to distant locations.

Process and Skill Standards

	National Science Education Standards		Project 2061 Benchmarks
A.1	Identify questions that can be answered through investigation.	1.A.3	Some knowledge in science is very old and yet is still used today.
A.2	Design and conduct a scientific investigation.	1.C.1	Contributions to science and technology have been made by different people, in different cultures, at different times.
A.3	Use appropriate tools and techniques to gather and interpret data.	3.B.1	Design requires taking constraints into account.
A.4	Use evidence to describe, predict, explain, and model.	3.B.2	Technologies have effects other than those intended.
A.5	Use critical thinking to find relationships between results and interpretations.	8.B.1	The choice of materials for a job depends on their properties.
A.6	Consider alternative explanations and predictions.	9.A.3	How decimals should be written depends on how precise the measurements are.
A.7	Communicate procedures, results, and conclusions.	9.A.7	Computations can give more digits than make sense or are useful.
A.8	Use mathematics in scientific investigations.	9.B.3	Graphs can show the relationship between two variables.
E.1	Identify a problem to be reached.	9.C.4	Graphs show patterns and can be used to make predictions.
E.2	Design a solution or product.	11.C.4	Use equations to summarize observed changes.
E.3	Implement the proposed solution.	12.B.1	Find what percentage one number is of another.
E.4	Evaluate the solution or design.	12.B.8	Round a calculation to the correct number of significant figures.
E.5	Communicate the process of technological design.	12.C.1	Compare amounts proportionally.
F.4.c	Use systematic thinking to eliminate risks.	12.C.5	Inspect, disassemble, and reassemble simple devices and describe what the various parts are for.
F.4.d	Decisions are made based on estimated risks and benefits.	12.D.1	Use tables and graphs to organize information and identify relationships.
		12.D.2	Read, interpret, and describe tables and graphs.
		12.D.4	Understand information that includes different types of charts and graphs, including circle charts, bar graphs, line graphs, data tables, diagrams, and symbols.
		12.E.3	Be skeptical of arguments based on samples for which there was no control group.
		12.E.4	There may be more than one good way to interpret scientific findings.

Standards and Benchmarks xi

Effective Instructional Strategies

McDougal Littell Science incorporates strategies that research shows are effective in improving student achievement. These strategies include

- Notetaking and nonlinguistic representations (Marzano, Pickering, and Pollock)
- A focus on big ideas (Kameenui and Carnine)
- Background knowledge and active involvement (Project CRISS)

Robert J. Marzano, Debra J. Pickering, and Jane E. Pollock, *Classroom Instruction that Works; Research-Based Strategies for Increasing Student Achievement* (ASCD, 2001)

Edward J. Kameenui and Douglas Carnine, *Effective Teaching Strategies that Accommodate Diverse Learners* (Pearson, 2002)

Project CRISS (Creating Independence through Student Owned Strategies)

VOCA

electric
electric
static c
inducti

WHAT DO YOU THINK?
- How did the strips behave before step 2? How did they behave after step 2?
- How might you explain your observations?

Electric charge is a property of matter.

You are already familiar with electricity, static electricity, and magnetism. You know electricity as the source of power for many appliances, including lights, tools, and computers. Static electricity is what makes clothes stick together when they come out of a dryer and gives you a shock when you touch a metal doorknob on a dry, winter day. Magnetism can hold an invitation or report card on the door of your refrigerator.

COMBINATION NOTES
As you read this section, write down important ideas about electric charge and static charges. Make sketches to help you remember these concepts.

You may not know, however, that electricity, static electricity, and magnetism are all related. All three are the result of a single property of matter—electric charge.

Chapter 1: Electricity 9 **E**

Comprehensive Research, Review, and Field Testing

An ongoing program of research and review guided the development of *McDougal Littell Science*.

- Program plans based on extensive data from classroom visits, research surveys, teacher panels, and focus groups
- All pupil edition activities and labs classroom-tested by middle school teachers and students
- All chapters reviewed for clarity and scientific accuracy by the Content Reviewers listed on page T5
- Selected chapters field-tested in the classroom to assess student learning, ease of use, and student interest

Content Organized Around Big Ideas

Each chapter develops a big idea of science, helping students to place key concepts in context.

CHAPTER

1 Electricity

the **BIG** idea

Moving electric charges transfer energy.

Key Concepts

SECTION
1.1 Materials can become electrically charged.
Learn how the movement of electrons builds static charges and how static charges are used in technology.

SECTION
1.2 Charges can move from one place to another.
Learn what factors control the movement of charges.

SECTION
1.3 Electric current is a flow of charge.
Learn how electric current is measured and how it can be produced.

Internet Preview

CLASSZONE.COM
Chapter 1 online resources:
Content Review, two Simulations, two Resource Centers, Math Tutorial, Test Practice.

What keeps this dragon glowing brightly?

Chapter Opener

- Provides an advance organizer of the chapter Big Idea and Key Concepts
- Connects the Big Idea to the real world through an engaging photo and related question

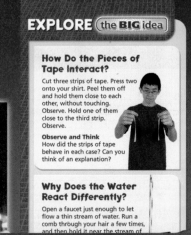

EXPLORE the **BIG** idea

How Do the Pieces of Tape Interact?
Cut three strips of tape. Press two onto your shirt. Peel them off and hold them close to each other, without touching. Observe. Hold one of them close to the third strip. Observe.

Observe and Think
How did the strips of tape behave in each case? Can you think of an explanation?

Why Does the Water React Differently?
Open a faucet just enough to let flow a thin stream of water. Run a comb through your hair a few times, and then hold it near the stream of

CHAPTER 1
Getting Ready to Learn

CONCEPT REVIEW
- Matter is made of particles too small to see.
- Energy and matter can move from one place to another.
- Electromagnetic energy is one form of energy.

VOCABULARY REVIEW
See Glossary for definitions.
atom
electron
joule
proton

CONTENT REVIEW
CLASSZONE.COM
Review concepts and vocabulary.

TAKING NOTES

COMBINATION NOTES
To take notes about a new concept, first make an informal outline of the information. Then make a sketch of the concept and label it so you can study it later.

VOCABULARY STRATEGY
Write each new vocabulary term in the center of a **four square** diagram. Write notes in the squares around each term. Include a definition, some characteristics, and some examples of the term. If possible, write some things that are not examples of the term.

SCIENCE NOTEBOOK

NOTES
How static charges are built
• Contact
• Induction
• Charge polarization

Definition
imbalance of charge in material

Character
results movemen electrons; affe by type of ma

STATIC CHARGE

Examples
clinging laundry, doorknob shock, lightning

Nonexam
electricity fror
electrical o

See the Note-Taking Handbook on pages R45–R51.

Visual Summary

- Summarizes Key Concepts using both text and visuals
- Reinforces the connection of Key Concepts to the Big Idea

Section Opener

- Highlights the Key Concept
- Connects new learning to prior knowledge
- Previews important vocabulary

The Big Idea Questions

- Help students connect their new learning back to the Big Idea
- Prompt students to synthesize and apply the Big Idea and Key Concepts

---

the BIG idea
Moving electric charges transfer energy.

 CONTENT REVIEW
CLASSZONE.COM

KEY CONCEPTS SUMMARY

1.1 Materials can become electrically charged.

Electric charge is a property of matter.

Electrons have a negative charge. Protons have a positive charge. Unlike charges attract. Like charges repel.

Static charges are caused by the movement of electrons, resulting in an imbalance of positive and negative charges.

VOCABULARY
electric charge p. 10
electric field p. 10
static charge p. 11
induction p. 13

1.2 Charges can move from one place to another.

Charge movement is affected by
- electric potential, measured in volts
- resistance, measured in ohms

A conductor has low resistance.
An insulator has high resistance.
A ground is the path of least resistance.

VOCABULARY
electric potential p. 19
volt p. 19
conductor p. 22
insulator p. 22
resistance p. 23
ohm p. 23
grounding p. 25

1.3 Electric current is a flow of charge.

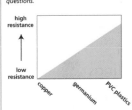

Electric current is
Ohm's law states
divided by resista
Electrochemical c
through chemical

VOCABULARY

Reviewing Vocabulary

Copy the chart below, and write each term
definition. Use the meanings of the underli
roots to help you.

Word	Root	Definition
EXAMPLE current	to run	continuous flow of charg
1. static charge	standing	
2. induction	into + to lead	
3. electric cell	chamber	
4. conductor	with + to lead	
5. insulator	island	
6. resistance	to stop	
7. electric potential	power	
8. grounding	surface of Earth	

Write a vocabulary term to match each clue

9. In honor of scientist Alessandro Volta (1745–1827)

10. In honor of the scientist who discovered th
relationship among voltage, resistance, and

past a giv

of the

Thinking Critically

Use the diagram of an electrochemical cell
below to answer the next three questions.

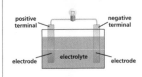

positive terminal   negative terminal

electrode   electrolyte   electrode

22. ANALYZE In which direction do electrons flow between the two terminals?

23. PREDICT What changes will occur in the cell as it discharges?

24. ANALYZE What determines whether the cell is rechargeable or not?

Use the graph below to answer the next three questions.

high resistance

low resistance

copper   germanium   PVC plastics

25. INFER Which material could you probably use as an insulator?

26. INFER Which material could be used in a lightning rod?

27. APPLY Materials that conduct electrons under some—but not all—conditions are known as semiconductors. Which material is probably a semiconductor?

Using Math in Science

Use the formula for Ohm's law to answer the next four questions.

$$I = \frac{V}{R}$$

28. An electrical pathway has a voltage of 240 volts and a current of 10 amperes. What is the resistance?

29. A 240-volt air conditioner has a resistance of 8 ohms. What is the current?

30. An electrical pathway has a current of 1.2 amperes and resistance of 40 ohms. What is the voltage?

31. An electrical pathway has a voltage of 400 volts and resistance of 2000 ohms. What is the current?

the BIG idea

32. INFER Look back at the photograph on pages 6 and 7. Based on what you have learned in this chapter, describe what you think is happening to keep the dragon lit.

33. COMPARE AND CONTRAST Draw two simple diagrams to compare and contrast static charges and electric current. Add labels and captions to make your comparison clear. Then write a paragraph summarizing the comparison.

UNIT PROJECTS

If you are doing a unit project, make a folder for your project. Include in your folder a list of the resources you will need, the date on which the project is due, and a schedule to keep track of your progress. Begin gathering data.

KEY CONCEPT

1.1 Materials can become electrically charged.

▶ **BEFORE, you learned**
- Atoms are made up of particles called protons, neutrons, and electrons
- Protons and electrons are electrically charged

▶ **NOW, you will learn**
- How charged particles behave
- How electric charges build up in materials
- How static electricity is used in technology

VOCABULARY
electric charge p. 10
electric field p. 10
static charge p. 11
induction p. 13

EXPLORE Static Electricity

How can materials interact electrically?

PROCEDURE

1. Hold the newspaper strips firmly together at one end and let the free ends hang down. Observe the strips.

2. Put the plastic bag over your other hand, like a mitten. Slide the plastic down the entire length of the strips and then let go. Repeat several times.

3. Notice how the strips of paper are hanging. Describe what you observe.

MATERIALS
- 2 strips of newspaper
- plastic bag

WHAT DO YOU THINK?
- How did the strips behave before step 2? How did they behave after step 2?
- How might you explain your observations?

Electric charge is a property of matter.

You are already familiar with electricity, static electricity, and magnetism. You know electricity as the source of power for many appliances, including lights, tools, and computers. Static electricity is what makes clothes stick together when they come out of a dryer and gives you a shock when you touch a metal doorknob on a dry, winter day. Magnetism can hold an invitation or report card on the door of your refrigerator.

You may not know, however, that electricity, static electricity, and magnetism are all related. All three are the result of a single property of matter—electric charge.

COMBINATION NOTES
As you read this section, write down important ideas about electric charge and static charges. Make sketches to help you remember these concepts.

Chapter 1: Electricity 9 E

Many Ways to Learn

Because students learn in so many ways, *McDougal Littell Science* gives them a variety of experiences with important concepts and skills. Text, visuals, activities, and technology all focus on Big Ideas and Key Concepts.

Integrated Technology

- Interaction with Key Concepts through Simulations and Visualizations
- Easy access to relevant Web resources through Resource Centers and SciLinks
- Opportunities for review through Content Review and Math Tutorials

Considerate Text

- Clear structure of meaningful headings
- Information clearly connected to main ideas
- Student-friendly writing style

Motors use electromagnets.

Because magnetism is a force, magnets can be used to move things. Electric motors convert the energy of an electric current into motion by taking advantage of the interaction between current and magnetism.

There are hundreds of devices that contain electric motors. Examples include power tools, electrical kitchen appliances, and the small fans in a computer. Almost anything with moving parts that uses current has an electric motor.

VISUALIZATION
CLASSZONE.COM

See a motor in motion.

Motors

Page 93 shows how a simple motor works. The photograph at the top of the page shows a motor that turns the blades of a fan. The illustration in the middle of the page shows the main parts of a simple motor. Although they may look different from each other, all motors have similar parts and work in a similar way. The main parts of an electrical motor include a voltage source, a shaft, an electromagnet, and at least one additional magnet. The shaft of the motor turns other parts of the device.

Recall that an electromagnet consists of a coil of wire with current flowing through it. Find the electromagnet in the illustration on page 93. The electromagnet is placed between the poles of another magnet.

When current from the voltage source flows through the coil, a magnetic field is produced around the electromagnet. The poles of the magnet interact with the poles of the electromagnet, causing the motor to turn.

1. The poles of the magnet push on the like poles of the electromagnet, causing the electromagnet to turn.

2. As the motor turns, the opposite poles pull on each other.

3. When the poles of the electromagnet line up with the opposite poles of the magnet, a part of the motor called the commutator reverses the polarity of the electromagnet. Now, the poles push on each other again and the motor continues to turn.

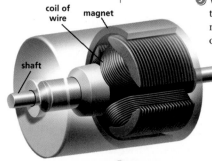

coil of wire   magnet

shaft

The illustration of the motor on page 93 is simplified so that you can see all of the parts. If you saw the inside of an actual motor, it might look like the illustration on the left. Notice that the wire is coiled many times. The electromagnet in a strong motor may coil hundreds of times. The more coils, the stronger the motor.

CHECK YOUR READING What causes the electromagnet in a motor to turn?

Visuals that Teach

- Information-rich visuals directly connected to the text
- Thoughtful pairing of diagrams and real-world photos
- Reading Visuals questions to support student learning

How a Motor Works

Although motors may look different from each other, they all have similar parts and work in a similar way.

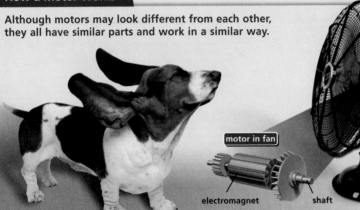

motor in fan

electromagnet   shaft

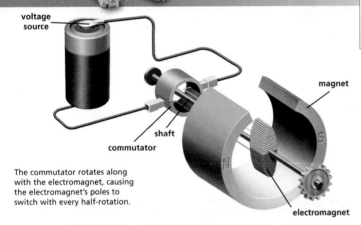

voltage source

magnet

shaft

commutator

electromagnet

The commutator rotates along with the electromagnet, causing the electromagnet's poles to switch with every half-rotation.

① Like poles of the magnets push on each other.

② As the motor turns, opposite poles attract.

③ The electromagnet's poles are switched, and like poles again repel.

READING VISUALS Would a motor work without an electromagnet? Why or why not?

Making an Electromagnet

Recall that a piece of iron in a strong magnetic field becomes a magnet itself. An **electromagnet** is a magnet made by placing a piece of iron or steel inside a coil of wire. As long as the coil carries a current, the metal acts as a magnet and increases the magnetic field of the coil. But when the current is turned off, the magnetic domains in the metal become random again and the magnetic field disappears.

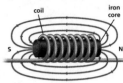

coil   iron core

S   N

By increasing the number of loops in the coil, you can increase the strength of the electromagnet. Electromagnets exert a much more powerful magnetic field than a coil of wire without a metal core. They can also be much stronger than the strongest permanent magnets made of metal alone. You can increase the field strength of an electromagnet by adding more coils or a stronger current. Some of the most powerful magnets in the world are huge electromagnets that are used in scientific instruments.

CHECK YOUR READING How can you increase the strength of an electromagnet?

INVESTIGATE Electromagnets

How can you make an electromagnet?

SKILL FOCUS
Observing

PROCEDURE

① Starting about 25 cm from one end of the wire, wrap the wire in tight coils around the nail. The coils should cover the nail from the head almost to the point.

② Tape the two batteries together as shown. Tape one end of the wire to a free battery terminal.

③ Touch the point of the nail to a paper clip and record your observations.

④ Connect the other end of the wire to the other battery terminal. Again touch the point of the nail to a paper clip. Disconnect the wire from the battery. Record your observations.

WHAT DO YOU THINK?

- What did you observe?
- Did you make an electromagnet? How do you know?

CHALLENGE Do you think the result would be different if you used an aluminum nail instead of an iron nail? Why?

MATERIALS
- insulated wire
- large iron nail
- 2 D cells
- electrical tape
- paper clip

TIME
20 minutes

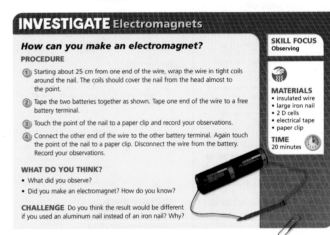

Hands-on Learning

- Activities that reinforce Key Concepts
- Skill Focus for important inquiry and process skills
- Multiple activities in every chapter, from quick Explores to full-period Chapter Investigations

Differentiated Instruction

A full spectrum of resources for differentiating instruction supports you in reaching the wide range of learners in your classroom.

1.1 INSTRUCT

History of Science

The first recorded observation of static charge is from ancient Greece. The Greeks noticed that fossilized tree sap—the material known as amber—attracted objects such as feathers after it was rubbed with fur or certain other materials. Many words with the root *electr-*, such as *electron*, *electricity*, and *electronic*, come from the Greek word *elecktron*, which means "amber."

Teach from Visuals

Point out that the charges in the diagram of electric charge are equal in size but opposite in sign. Ask:

• Which has more mass, an electron or a proton? *proton*

• Is the charge on the proton equal to, larger than, or smaller than the charge on the electron? *equal to*

Ongoing Assessment

Describe how charged particles behave.

Ask: If a balloon has a negative charge and a rod has a positive charge, will the balloon and the rod repel or attract each other? *attract*

READING VISUALS *Answer: Each particle's force lines bend toward the other particle.*

The smallest unit of a material that still has the characteristics of that material is an atom or a molecule. A molecule is two or more atoms bonded together. Most of an atom's mass is concentrated in the nucleus at the center of the atom. The nucleus contains particles called protons and neutrons. Much smaller particles called electrons move at high speeds outside the nucleus.

Protons and electrons have electric charges. **Electric charge** is a property that allows an object to exert a force on another object without touching it. Recall that a force is a push or a pull. The space around a particle through which an electric charge can exert this force is called an **electric field**. The strength of the field is greater near the particle and weaker farther away.

All protons have a positive charge (+), and all electrons have a negative charge (−). Normally, an atom has an equal number of protons and electrons, so their charges balance each other, and the overall charge on the atom is neutral.

Particles with the same type of charge—positive or negative—are said to have like charges, and particles with different charges have unlike charges. Particles with like charges repel each other, that is, they push each other away. Particles with unlike charges attract each other, or pull on each other.

VOCABULARY
Make a four square diagram for the term *electric charge* and the other vocabulary terms in this section.

Electric Charge

Charged particles exert forces on each other through their electric fields.

Charged Particles
Electric charge can be either negative or positive.

The balloon and the cat's fur have unlike charges, so they attract each other.

① Attraction
Particles with unlike charges attract—pull on each other.

② Repulsion
Particles with like charges repel—push each other away.

◯ = electron
◯ = proton
— = lines of force

READING VISUALS How do the force lines change when particles attract?

Static charges are caused by the movement of electrons.

You have read that protons and electrons have electric charges. Objects and materials can also have charges. A **static charge** is a buildup of electric charge in an object caused by the presence of many particles with the same charge. Ordinarily, the atoms that make up a material have a balance of protons and electrons. A material develops a static charge—or becomes charged—when it contains more of one type of charged particle than another.

If there are more protons than electrons in a material, the material has a positive charge. If there are more electrons than protons in a material, it has a negative charge. The amount of the charge depends on how many more electrons or protons there are. The total number of unbalanced positive or negative charges in an object is the net charge of the object. Net charge is measured in coulombs (KOO-lahmz). One coulomb is equivalent to more than 10^{19} electrons or protons.

Electrons can move easily from one atom to another. Protons cannot. For this reason, charges in materials usually result from the movement of electrons. The movement of electrons through a material is called conduction. If electrons move from one atom to another, the atom they move to develops a negative charge. The atom they move away from develops a positive charge. Atoms with either a positive or a negative charge are called ions.

A static charge can build up in an uncharged material when it touches or comes near a charged material. Static charges also build up when some types of uncharged materials come into contact with each other.

READING TIP
The word *static* comes from the Greek word *statos*, which means "standing."

REMINDER
10^{19} is the same as 1 followed by 19 zeros.

Charging by Contact

When two uncharged objects made of certain materials—such as rubber and glass—touch each other, electrons move from one material to the other. This process is called charging by contact. It can be demonstrated by a balloon and a glass rod, as shown below.

① At first, a balloon and a glass rod each have balanced, neutral charges.

② When they touch, electrons move from the rod to the balloon.

③ Afterwards, the balloon has a negative charge, and the rod has a positive charge.

DIFFERENTIATE INSTRUCTION

❓ More Reading Support

A What type of charge does a proton have? *positive*

B If two objects are both positive, will they attract or repel? *repel*

English Learners The similar words, *buildup* and *build up*, may confuse English learners. Explain that *buildup*, is a noun, while *build up* is a verb. *Buildup* is an informal way of describing something that has gathered over time. *Build up* is a phrasal verb that refers to the action of accumulating something over time.

DIFFERENTIATE INSTRUCTION

❓ More Reading Support

C What is a buildup of electric charge in an object? *static charge*

D What is charging by contact? *charging through direct touching*

Inclusion To help students with visual impairments, enlarge the diagrams of attraction and repulsion on p. 10, and glue string on the lines of force. Have students use their sense of touch to examine the force of attraction between unlike charges and the force of repulsion between like charges.

Teacher's Edition

• More Reading Support for below-level readers

• Strategies for below-level and advanced learners, English learners, and inclusion students

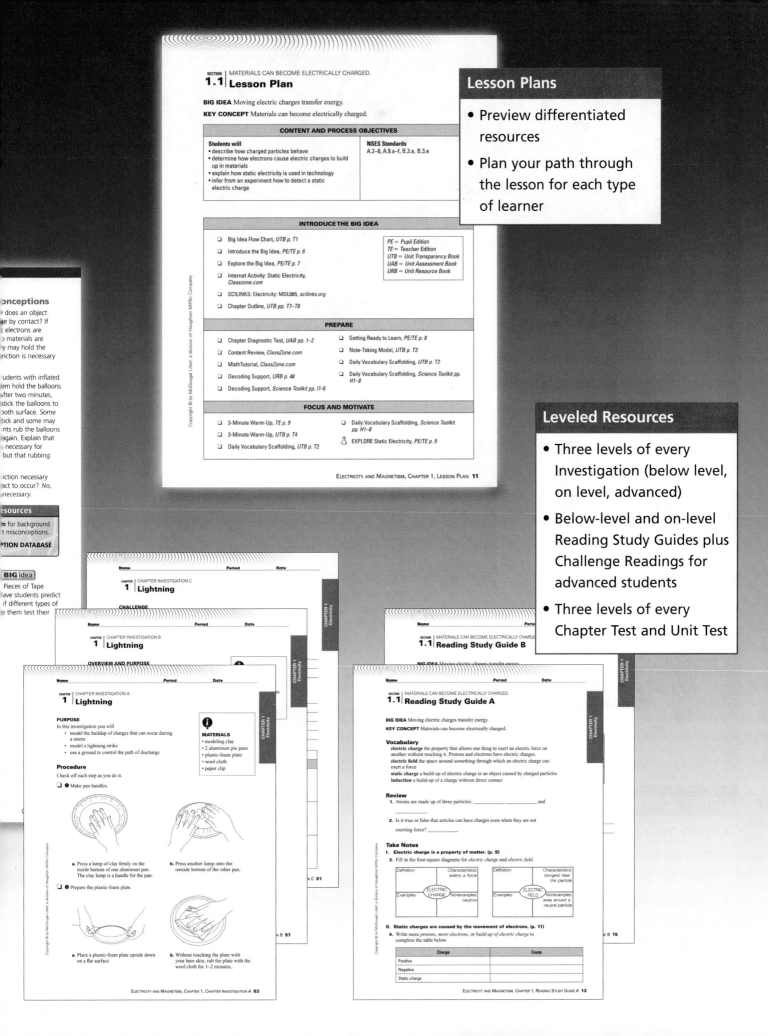

Lesson Plans

- Preview differentiated resources
- Plan your path through the lesson for each type of learner

Leveled Resources

- Three levels of every Investigation (below level, on level, advanced)
- Below-level and on-level Reading Study Guides plus Challenge Readings for advanced students
- Three levels of every Chapter Test and Unit Test

Effective Assessment

McDougal Littell Science incorporates a comprehensive set of resources for assessing student knowledge and performance before, during, and after instruction.

Diagnostic Tests

- Assessment of students' prior knowledge
- Readiness check for concepts and skills in the upcoming chapter

---

Ongoing Assessment

 Answer: *Particles in the air become charged and are attracted to an oppositely charged plate in an electrostatic filter.*

Reinforce (the **BIG idea**)
Have students relate the section to the Big Idea.

Reinforcing Key Concepts, p. 21

1.1 ASSESS & RETEACH

Assess
Section 1.1 Quiz, p. 3

Reteach
Have small groups of students outline the section by using heads as main topics. Tell students that vocabulary terms, visual titles and captions, and topic sentences can serve as clues to subtopics. When the groups complete their outlines, lead a discussion and bring the class to a consensus about which subtopics should be included in the outline. From their conclusions, create a master outline on the board.

Technology Resources
Have students visit ClassZone.com for reteaching of Key Concepts.

- CONTENT REVIEW
- CONTENT REVIEW CD-ROM

Static electricity is also used in making cars. When new cars are painted, the paint is given an electric charge and then sprayed onto the car in a fine mist. The tiny droplets of paint stick to the car more firmly than they would without the charge. This process results in a coat of paint that is very even and smooth.

Another example of the use of static electricity in technology is a device called an electrostatic air filter. This device cleans air inside buildings with the help of static charges. The filter gives a static charge to pollen, dust, germs, and other particles in the air. Then an oppositely charged plate inside the filter attracts these particles, pulling them out of the air. Larger versions of electrostatic filters are used to remove pollutants from industrial smokestacks.

CHECK YOUR READING How can static charges help clean air?

1.1 Review

KEY CONCEPTS
1. How do a positive and a negative particle interact?
2. Describe how the movement of electrons between two objects with balanced charges could cause the buildup of electric charge in both objects.
3. Describe one technological use of static electricity.

CRITICAL THINKING
4. **Infer** A sock and a shirt from the dryer stick together. What does this tell you about the charges on the sock and shirt?
5. **Analyze** You walk over a rug and get a shock from a door-knob. What do the materials of the rug and the shoes have to do with the type of charge your body had?

CHALLENGE
6. **Apply** Assume you start with a negatively charged rod and two balloons. Describe a series of steps you could take to create a positively charged balloon, pick up negatively charged powder with the balloon, and drop the powder from the balloon.

16 Unit: Electricity and Magnetism

ANSWERS
1. They attract each other.
2. The object that the electrons move to acquires a net negative charge and the object they move from acquires a positive charge.
3. photocopying, air filtering, or painting cars

4. They are oppositely charged.
5. If the shoe material had a greater attraction for electrons, your body acquired a negative charge. If the carpet material had a greater attraction for electrons, your body acquired a positive charge.

6. Place the two balloons side by side, touching. Touch one of the balloons with the rod, and then withdraw the rod. Separate the balloons. Use the balloon you touched with the rod to attract the powder. Touch the rod to the balloon again to release the powder.

E 16 Unit: Electricity and Magnetism

Ongoing Assessment

- Check Your Reading questions for student self-check of comprehension
- Consistent Teacher Edition prompts for assessing understanding of Key Concepts

---

Reviewing Vocabulary

Copy the chart below, and write each term's definition. Use the meanings of the underlined roots to help you.

Word	Root	Definition
EXAMPLE <u>current</u>	to run	continuous flow of charge
1. <u>static</u> charge	standing	
2. <u>induction</u>	into + to lead	
3. <u>electric cell</u>	chamber	
4. <u>conductor</u>	with + to lead	
5. <u>insulator</u>	island	
6. <u>resistance</u>	to stop	
7. <u>electric potential</u>	power	
8. <u>grounding</u>	surface of Earth	

Write a vocabulary term to match each clue.

9. In honor of scientist Alessandro Volta (1745–1827)
10. In honor of the scientist who discovered the relationship among voltage, resistance, and current
11. The amount of charge that flows past a given point in a unit of time.

Reviewing Key Concepts

Multiple Choice *Choose the letter of the best answer.*

12. An electric charge is a
   a. kind of liquid
   b. reversible chemical reaction
   c. type of matter
   d. force acting at a distance

13. A static charge is different from electric current in that a static charge
   a. never moves
   b. can either move or not move
   c. moves only when resistance is low enough
   d. moves only when voltage is high enough

14. Charging by induction means charging
   a. with battery power
   b. by direct contact
   c. at a distance
   d. using solar power

15. Electric potential refers to
   a. the amount of energy a charge has
   b. the number of electrons that make up a charge
   c. whether a charge is positive or negative
   d. whether or not a charge can move

16. A superconductor is a material that, when very cold, has no
   a. amperage
   b. resistance
   c. electric charge
   d. electric potential

17. Ohm's law says that when resistance goes up, current
   a. increases   c. stays the same
   b. decreases   d. matches voltage

18. Electrochemical cells include
   a. all materials that build up a charge
   b. primary cells and storage cells
   c. batteries and solar cells
   d. storage cells and lightning rods

Short Answer *Write a short answer to each question.*

19. What determines whether a charge you get when walking across a rug is positive or negative?
20. What is the difference between resistance and insulation?
21. What is one disadvantage of solar cells?

Chapter 1: Electricity **37**

Section and Chapter Reviews

- Focus on Key Concepts and critical thinking skills
- A full range of question types and levels of thinking

T14

Leveled Chapter and Unit Tests

- Three levels of test for every chapter and unit
- Same Big Ideas, Key Concepts, and essential skills assessed on all levels

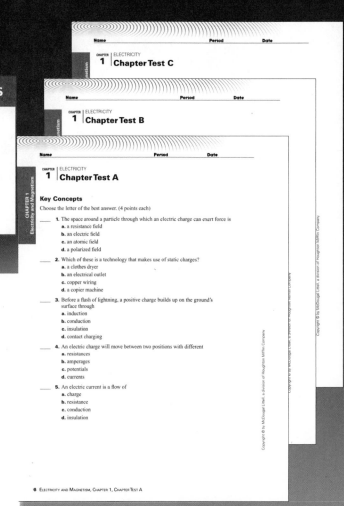

Name   Period   Date

CHAPTER 1 ELECTRICITY
1 Chapter Test C

Name   Period   Date

CHAPTER 1 ELECTRICITY
1 Chapter Test B

Name   Period   Date

CHAPTER 1 ELECTRICITY
1 Chapter Test A

Key Concepts

Choose the letter of the best answer. (4 points each)

___ 1. The space around a particle through which an electric charge can exert force is
  a. a resistance field
  b. an electric field
  c. an atomic field
  d. a polarized field

___ 2. Which of these is a technology that makes use of static charges?
  a. a clothes dryer
  b. an electrical outlet
  c. copper wiring
  d. a copier machine

___ 3. Before a flash of lightning, a positive charge builds up on the ground's surface through
  a. induction
  b. conduction
  c. insulation
  d. contact charging

___ 4. An electric charge will move between two positions with different
  a. resistances
  b. amperages
  c. potentials
  d. currents

___ 5. An electric current is a flow of
  a. charge
  b. resistance
  c. conduction
  d. insulation

6 ELECTRICITY AND MAGNETISM, CHAPTER 1, CHAPTER TEST A

Thinking Critically

Use the diagram of an electrochemical cell below to answer the next three questions.

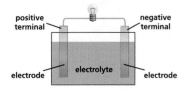

positive terminal — negative terminal

electrode — electrolyte — electrode

22. ANALYZE In which direction do electrons flow between the two terminals?

23. PREDICT What changes will occur in the cell as it discharges?

24. ANALYZE What determines whether the cell is rechargeable or not?

Use the graph below to answer the next three questions.

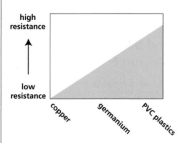

high resistance

low resistance

copper   germanium   PVC plastics

25. INFER Which material could you probably use as an insulator?

26. INFER Which material could be used in a lightning rod?

27. APPLY Materials that conduct electrons under some—but not all—conditions are known as semiconductors. Which material is probably a semiconductor?

Using Math in Science

Use the formula for Ohm's law to answer the next four questions.

$$I = \frac{V}{R}$$

28. An electrical pathway has a voltage of 240 volts and a current of 10 amperes. What is the resistance?

29. A 240-volt air conditioner has a resistance of 8 ohms. What is the current?

30. An electrical pathway has a current of 1.2 amperes and resistance of 40 ohms. What is the voltage?

31. An electrical pathway has a voltage of 400 volts and resistance of 2000 ohms. What is the current?

the BIG idea

32. INFER Look back at the photograph on pages 6 and 7. Based on what you have learned in this chapter, describe what you think is happening to keep the dragon lit.

33. COMPARE AND CONTRAST Draw two simple diagrams to compare and contrast static charges and electric current. Add labels and captions to make your comparison clear. Then write a paragraph summarizing the comparison.

UNIT PROJECTS

If you are doing a unit project, make a folder for your project. Include in your folder a list of the resources you will need, the date on which the project is due, and a schedule to keep track of your progress. Begin gathering data.

Rubrics

- Rubrics in Teacher Edition for all extended response questions
- Rubrics for all Unit Projects
- Alternative Assessment with rubric for each chapter
- A wide range of additional rubrics in the Science Toolkit

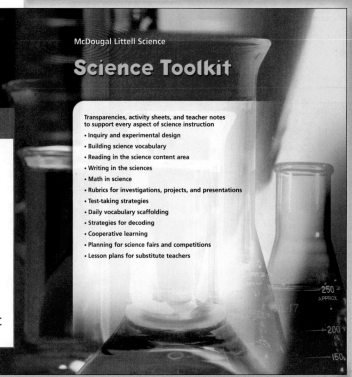

McDougal Littell Science

Science Toolkit

Transparencies, activity sheets, and teacher notes to support every aspect of science instruction

- Inquiry and experimental design
- Building science vocabulary
- Reading in the science content area
- Writing in the sciences
- Math in science
- Rubrics for investigations, projects, and presentations
- Test-taking strategies
- Daily vocabulary scaffolding
- Strategies for decoding
- Cooperative learning
- Planning for science fairs and competitions
- Lesson plans for substitute teachers

McDougal Littell Science Modular Series

McDougal Littell Science lets you choose the titles that match your curriculum. Each module in this flexible 15-book series takes an in-depth look at a specific area of life, earth, or physical science.

- Flexibility to match your curriculum
- Convenience of smaller books
- Complete Student Resource Handbooks in every module

Life Science Titles

A ▶ Cells and Heredity
1. The Cell
2. How Cells Function
3. Cell Division
4. Patterns of Heredity
5. DNA and Modern Genetics

B ▶ Life Over Time
1. The History of Life on Earth
2. Classification of Living Things
3. Population Dynamics

C ▶ Diversity of Living Things
1. Single-Celled Organisms and Viruses
2. Introduction to Multicellular Organisms
3. Plants
4. Invertebrate Animals
5. Vertebrate Animals

D ▶ Ecology
1. Ecosystems and Biomes
2. Interactions Within Ecosystems
3. Human Impact on Ecosystems

E ▶ Human Biology
1. Systems, Support, and Movement
2. Absorption, Digestion, and Exchange
3. Transport and Protection
4. Control and Reproduction
5. Growth, Development, and Health

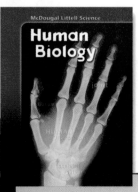

Earth Science Titles

A ▶ **Earth's Surface**
1. Views of Earth Today
2. Minerals
3. Rocks
4. Weathering and Soil Formation
5. Erosion and Deposition

B ▶ **The Changing Earth**
1. Plate Tectonics
2. Earthquakes
3. Mountains and Volcanoes
4. Views of Earth's Past
5. Natural Resources

C ▶ **Earth's Waters**
1. The Water Planet
2. Freshwater Resources
3. Ocean Systems
4. Ocean Environments

D ▶ **Earth's Atmosphere**
1. Earth's Changing Atmosphere
2. Weather Patterns
3. Weather Fronts and Storms
4. Climate and Climate Change

E ▶ **Space Science**
1. Exploring Space
2. Earth, Moon, and Sun
3. Our Solar System
4. Stars, Galaxies, and the Universe

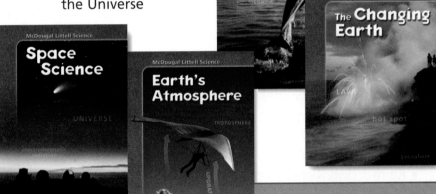

Physical Science Titles

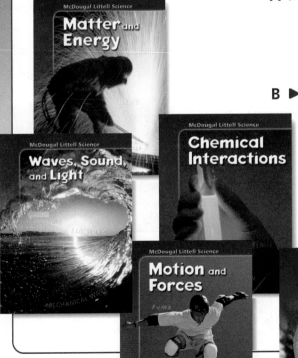

A ▶ **Matter and Energy**
1. Introduction to Matter
2. Properties of Matter
3. Energy
4. Temperature and Heat

B ▶ **Chemical Interactions**
1. Atomic Structure and the Periodic Table
2. Chemical Bonds and Compounds
3. Chemical Reactions
4. Solutions
5. Carbon in Life and Materials

C ▶ **Motion and Forces**
1. Motion
2. Forces
3. Gravity, Friction, and Pressure
4. Work and Energy
5. Machines

D ▶ **Waves, Sound, and Light**
1. Waves
2. Sound
3. Electromagnetic Waves
4. Light and Optics

E ▶ **Electricity and Magnetism**
1. Electricity
2. Circuits and Electronics
3. Magnetism

Teaching Resources

A wealth of print and technology resources help you adapt the program to your teaching style and to the specific needs of your students.

Book-Specific Print Resources

Unit Resource Book provides all of the teaching resources for the unit organized by chapter and section.

- Family Letters
- *Scientific American Frontiers* Video Guide
- Unit Projects
- Lesson Plans
- Reading Study Guides (Levels A and B)
- Spanish Reading Study Guides
- Challenge Readings
- Challenge and Extension Activities
- Reinforcing Key Concepts
- Vocabulary Practice
- Math Support and Practice
- Investigation Datasheets
- Chapter Investigations (Levels A, B, and C)
- Additional Investigations (Levels A, B, and C)
- Summarizing the Chapter

Unit Assessment Book contains complete resources for assessing student knowledge and performance.

- Chapter Diagnostic Tests
- Section Quizzes
- Chapter Tests (Levels A, B, and C)
- Alternative Assessments
- Unit Tests (Levels A, B, and C)

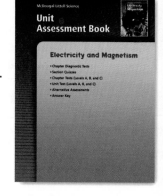

Unit Transparency Book includes instructional visuals for each chapter.

- Three-Minute Warm-Ups
- Note-Taking Models
- Daily Vocabulary Scaffolding
- Chapter Outlines
- Big Idea Flow Charts
- Chapter Teaching Visuals

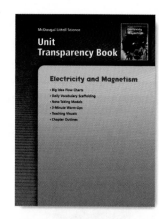

Unit Lab Manual

Unit Note-Taking/Reading Study Guide

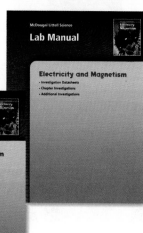

McDougal Littell Science

Unit Resource Book

Electricity and Magnetism

- Family Letters (English and Spanish)
- *Scientific American Frontiers* Video Guides
- Unit Projects (with Rubrics)
- Lesson Plans
- Reading Study Guides (Levels A and B and Spanish)
- Challenge Activities and Readings
- Reinforcing Key Concepts
- Vocabulary Practice and Decoding Support
- Math Support and Practice
- Investigation Datasheets
- Chapter Investigations (Levels A, B, and C)
- Additional Investigations (Levels A, B, and C)

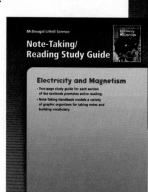

Program-Wide Print Resources

Process and Lab Skills

Problem Solving and Critical Thinking

Standardized Test Practice

Science Toolkit

City Science

Visual Glossary

Multi-Language Glossary

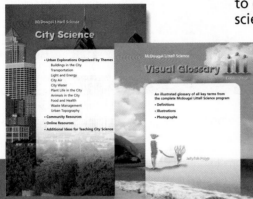

English Learners Package

Scientific American Frontiers Video Guide

How Stuff Works Express
This quarterly magazine offers opportunities to explore current science topics.

Technology Resources

Scientific American Frontiers **Video Program**
Each specially-tailored segment from this award-winning PBS series correlates to a unit; available on VHS and DVD

Audio CDs Complete chapter texts read in both English and Spanish

Lab Generator CD-ROM
A searchable database of all activities from the program plus additional labs for each unit; edit and print your own version of labs

Test Generator CD-ROM

eEdition CD-ROM

EasyPlanner CD-ROM

Content Review CD-ROM

Power Presentations CD-ROM

Online Resources

ClassZone.com

Content Review Online

eEdition Plus Online

EasyPlanner Plus Online

eTest Plus Online

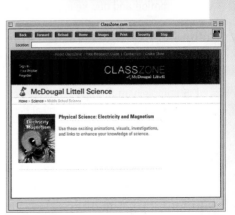

Correlation to National Science Education Standards

This chart provides an overview of how the five Physical Science modules of *McDougal Littell Science* address the National Science Education Standards.

A Matter and Energy
B Chemical Interactions
C Motion and Forces
D Waves, Sound, and Light
E Electricity and Magnetism

A. Science as Inquiry

	Book, Chapter, and Section
A.1– A.8 **Abilities necessary to do scientific inquiry** Identify questions for investigation; design and conduct investigations; use evidence; think critically and logically; analyze alternative explanations; communicate; use mathematics.	All books (pp. R2–R44), All Chapter Investigations, All Think Science features
A.9 **Understandings about scientific inquiry** Different kinds of investigations for different questions; investigations guided by current scientific knowledge; importance of mathematics and technology for data gathering and analysis; importance of evidence, logical argument, principles, models, and theories; role of legitimate skepticism; scientific investigations lead to new investigations.	All books (pp. xxii–xxv) A3.1, B2.2, C4.2, D3.2, E3.1

B. Physical Science

	Book, Chapter, and Section
B.1 **Properties and changes of properties in matter** Physical properties; substances, elements, and compounds; chemical reactions.	A1.1, A1.2, A1.3, A1.4, A2.1, A2.2, B1, B3.2, B4.1, B4.2, B4.3, B5.1, B5.3, C3.4
B.2 **Motions and forces** Position, speed, direction of motion; balanced and unbalanced forces.	C1.1, C1.2, C1.3, C2.1, C2.3, C3.1, C3.2, C3.3, C3.4, C4.1, E5.1
B.3 **Transfer of energy** Energy transfer; forms of energy; heat and light; electrical circuits; sun as source of Earth's energy.	A3.1, A3.2, A3.3, A4.1, A4.2, A4.3, B3.3, C4.2, D1.1, D3.3, D3.4, D4.1, D4.2, D4.3, E1.1, E1.2, E1.3, E2.1, E2.2

C. Life Science

	Book, Chapter, and Section
C.1 **Structure and function in living systems** Systems; structure and function; levels of organization; cells and cell activities; specialization; human body systems; disease.	B1.1 (Connecting Sciences), B5.2, C5.2 (Connecting Sciences)

D. Earth and Space Science

	Book, Chapter, and Section
D.1 **Earth's changing atmosphere**	B4.2 (Connecting Sciences)
D.3 **Earth in the solar system** Sun, planets, asteroids, comets; regular and predictable motion and day, year, phases of the moon, and eclipses; gravity and orbits; sun as source of energy for earth; cause of seasons.	A3.2, A3.3, C3.1, D3.3

E. Science and Technology

	Book, Chapter, and Section
E.1– E.5 **Abilities of technological design** Identify problems; design a solution or product; implement a proposed design; evaluate completed designs or products; communicate the process of technological design.	A2.3, A3.1, A4.3, B (p. 5), B4.2, C1.2, C2.1, C3.2, C5.3, D2.4, D3.1, D3.3, D4.4, E2.3
E.6 **Understandings about science and technology** Similarities and differences between scientific inquiry and technological design; contributions of people in different cultures; reciprocal nature of science and technology; nonexistence of perfectly designed solutions; constraints, benefits, and unintended consequences of technological designs.	All books (pp. xxvi–xxvii) All books (Frontiers in Science, Timelines in Science) A.1.2, A3.3, B3.4, B3.1, B3.3, C5.3, D4.4, E1.2, E1.3

F.	Science in Personal and Social Perspectives	Book, Chapter, and Section
F.1	**Personal health** Exercise; fitness; hazards and safety; tobacco, alcohol, and other drugs; nutrition; STDs; environmental health	B4.2, B5.2
F.2	**Populations, resources, and environments** Overpopulation and resource depletion; environmental degradation.	A3.1
F.3	**Natural hazards** Earthquakes, landslides, wildfires, volcanic eruptions, floods, storms; hazards from human activity; personal and societal challenges.	B3.2, D3.2, E1.2
F.4	**Risks and benefits** Risk analysis; natural, chemical, biological, social, and personal hazards; decisions based on risks and benefits.	B4.3
F.5	**Science and technology in society** Science's influence on knowledge and world view; societal challenges and scientific research; technological influences on society; contributions from people of different cultures and times; work of scientists and engineers; ethical codes; limitations of science and technology.	All books (Timelines in Science) A1.2, A3.2, A3.3, B3.4, B4.4, C5.3, D2, D3.2, D3.3, D4.4, E1.2, E1.3

G.	History and Nature of Science	Book, Chapter, and Section
G.1	**Science as a human endeavor** Diversity of people w.orking in science, technology, and related fields; abilities required by science	All books (pp. xxii–xxv; Frontiers in Science)
G.2	**Nature of science** Observations, experiments, and models; tentative nature of scientific ideas; differences in interpretation of evidence; evaluation of results of investigations, experiments, observations, theoretical models, and explanations; importance of questioning, response to criticism, and communication.	B1.2, B2.1, B2.3, E3.2
G.3	**History of science** Historical examples of inquiry and relationships between science and society; scientists and engineers as valued contributors to culture; challenges of breaking through accepted ideas.	All books (Frontiers in Science; Timelines in Science) B1.2, B3.2, C2.1, D2.4

Correlations to Benchmarks

This chart provides an overview of how the five Physical Science modules of *McDougal Littell Science* address the National Science Education Standards.

A Matter and Energy
B Chemical Interactions
C Motion and Forces
D Waves, Sound, and Light
E Electricity and Magnetism

1. The Nature of Science	Book, Chapter, and Section
	The Nature of Science (pp. xxii–xxv); E2.3; Think Science Features: A3.1, B2.2, C2.1, C4.2, D3.2, E3.1; Scientific Thinking Handbook (pp. R2–R9); Lab Handbook (pp. R10–R35)

3. The Nature of Technology	Book, Chapter, and Section
	The Nature of Technology (pp. xxvi–xxvii); A3.3, B4.4, D4.4, E1, E2.3, E3.2, E3.3, E3.4; Timelines in Science Features

4. The Physical Setting	Book, Chapter, and Section
4.B THE EARTH	A3.1, A4.3, C3.1
4.D STRUCTURE OF MATTER	
4.D.1 All matter is made of atoms; atoms of any element are alike but different from atoms of other elements; different arrangements of atoms into groups compose all substances.	A1.2, A1.3, B1.1, B2.1, B2.2
4.D.2 Equal volumes of different substances usually have different weights.	A2.1, A2.3
4.D.3 Atoms and molecules are perpetually in motion; increased temperature means greater average energy of motion; states of matter: solids, liquids, gases.	A1.2, A4.1
4.D.4 Temperature and acidity of a solution influence reaction rates. Many substances dissolve in water, which may facilitate reactions between them.	B3.1, B4.2, B4.3
4.D.5 Greek philosopheres' scientific ideas about elements; most elements tend to combine with others, so few elements are found in their pure form.	B1.1
4.D.6 Groups of elements have similar properties; oxidation; some elements, like carbon and hydrogen, don't fit into any category and are essential elements of living matter.	B1.2, B1.3, B3.1, B5.1, B5.2
4.D.7 Conservation of matter: the total weight of a closed system remains the same because the total number of atoms stays the same regardless of how they interact with one another.	B3.2
4.E ENERGY TRANSFORMATIONS	
4.E.1 Energy cannot be created or destroyed, but only changed from one form into another.	A3.2, C4.2
4.E.2 Most of what goes on in the universe involves energy transformations.	A3, A4.2, A4.3
4.E.3 Heat can be transferred through materials by the collisions of atoms or across space by radiation; convection currents transfer heat in fluid materials.	A4.2, A4.3
4.E.4 Energy appears in many different forms, including heat energy, chemical energy, mechanical energy, and gravitational energy.	A3.1, A4.2, A4.3, C4.2
4.F MOTION	
4.F.1 Light from the Sun is made up of many different colors of light; objects that give off or reflect light have a different mix of colors.	D3.3, D3.4
4.F.2 Something can be "seen" when light waves emitted or reflected by it enter the eye.	D4.1, D4.3

4.F.3 An unbalanced force acting on an object changes its speed or direction of motion, or both. If the force acts toward a single center, the object's path may curve into an orbit around the center.	C2.1, C2.2, C3.1
4.F.4 Vibrations in materials set up wavelike disturbances (such as sound) that spread away from the source; waves move at different speeds in different materials.	D1, D2.1, D2.2, D3.1, D3.4
4.F.5 Human eyes respond to only a narrow range of wavelengths of electromagnetic radiation—visible light. Differences of wavelengths within that range are perceived as differences in color.	D3.2, D3.4, D4.3
4.G FORCES OF NATURE	
4.G.1 Objects exerts gravitational forces on one another, but these forces depend on the mass and distance of objects, and may be too small to detect.	C3.1
4.G.2 The Sun's gravitational pull holds Earth and other planets in their orbits; planets' gravitational pull keeps their moons in orbit around them.	C3.1
4.G.3 Electric currents and magnets can exert a force on each other.	E3.1, E3.2, E3.3

5. The Living Environment	Book, Chapter, and Section
5.E Flow of Matter and Energy	B5.2

8. The Designed World	A2.1, A2.3, A3.3, B3.4, B4.4, B5.3, C5.3, E2.3, E3.2

9. The Mathematical World	All Math in Science Features, E2.3

10. Historical Perspectives	B1, B2, B3.2, C1.1, C2, D4.4

12. Habits of Mind	Book, Chapter, and Section
12.A VALUES AND ATTITUDES	Think Science Features: A3.1, B2.2, C2.1, C4.2, D3.2, E3.1
12.B Computation and Estimation	All Math in Science Features, Lab Handbook (pp. R10–R35)
12.C Manipulation and Observation	All Investigates and Chapter Investigations
12.D Communication Skills	All Chapter Investigations, Lab Handbook (pp. R10–R35)
12.E Critical-Response Skills	Think Science Features: A3.1, B2.2, C2.1, C4.2, D3.2, E3.1; Scientific Thinking Handbook (pp. R2–R9)

Planning the Unit

The Pacing Guide provides suggested pacing for all chapters in the unit as well as the two unit features shown below.

Frontiers in Science

- Features cutting-edge research as an engaging point of entry into the unit
- Connects to an accompanying *Scientific American Frontiers* video and viewing guide
- Introduces three options for unit projects.

Timelines in Science

- Traces the history of key scientific discoveries
- Highlights interactions between science and technology.

Electricity and Magnetism Pacing Guide

The following pacing guide shows how the chapters in *Electricity and Magnetism* can be adapted to fit your specific course needs.

	TRADITIONAL SCHEDULE (DAYS)	BLOCK SCHEDULE (DAYS)
Frontiers in Science: Electronics in Music	1	0.5
Chapter 1 Electricity		
1.1 Materials can become electrically charged.	2	1
1.2 Charges can move from one place to another.	2	1
1.3 Electrical current is a flow of charge.	3	1.5
Chapter Investigation	1	0.5
Chapter 2 Circuits and Electronics		
2.1 Charge needs a continuous path to flow.	2	1
2.2 Circuits make electric current useful.	2	1
2.3 Electronic technology is based on circuits.	3	1.5
Chapter Investigation	1	0.5
Timelines in Science: The Story of Electronics	1	0.5
Chapter 3 Magnetism		
3.1 Magnetism is a force that acts at a distance.	2	1
3.2 Current can produce magnetism.	2	1
3.3 Magnetism can produce current.	2	1
3.4 Generators supply electrical energy.	3	1.5
Chapter Investigation	1	0.5
Total Days for Module	**28**	**14**

Planning the Chapter

Complete planning support precedes each chapter.

Previewing Content

- Section-by-section science background notes
- Common Misconceptions notes

CHAPTER

1 Electricity

Physical Science
UNIFYING PRINCIPLES

PRINCIPLE 1	PRINCIPLE 2	PRINCIPLE 3	PRINCIPLE 4
Matter is made of particles too small to see.	Matter changes form and moves from place to place.	Energy cha... one form t... but it cann... created or ...	

Unit: Electricity and Magnetism
BIG IDEAS

CHAPTER 1 Electricity	CHAPTER 2 Circuits and Electronics	CHAPTER 3 Magnetism
Moving electric charges transfer energy.	Circuits control the flow of electric charge.	Current can produce ma... magnetism can produce...

CHAPTER 1
KEY CONCEPTS

SECTION 1.1	SECTION 1.2
Materials can become electrically charged.	**Charges can move from one place to another.**
1. Electric charge is a property of matter.	1. Static charges have potential energy
2. Static charges are caused by the movement of electrons.	2. Materials affect charge movement.
3. Technology uses static electricity.	

 The Big Idea Flow Chart is available on p. T1 in the **UNIT TRANSPARENCY BOOK**

Previewing Content

SECTION

1.1 Materials can become electrically charged. pp. 9–17

1. Electric charge is a property of matter.
Objects can have a positive, negative, or neutral charge. A proton has a positive charge. An electron has a negative charge. Electric charge is measured in a unit called coulombs. Two objects with **electric charge** exert an electric force on each other. Electric force can be exerted even if the objects are sepa...

SECTION

1.2

1. Sta
A ch
tial
one
is n
have

Previewing Content

SECTION

1.3 Electric current is a flow of charge. pp. 28–35

1. Electric charge can flow continuously.
If a constant potential difference, or voltage, is maintained between two points and there is a path along which electrons can move between those two points, **electric current** results.

- Electric current, the continuous flow of charge between two points at different potentials, is measured in **amperes.** One ampere is equal to a flow rate of one coulomb per second.
- You can measure current using an ammeter, voltage using a voltmeter, and resistance using an ohmmeter. A multimeter can measure all three values. **Ohm's law** shows the relationships among current (*I*), voltage (*V*), and resistance (*R*).

$$I = \frac{V}{R}$$

2. Electric cells supply electric current.
An **electric cell** maintains a constant voltage between its two terminals by using the physical and chemical properties of different materials.

- An electrochemical cell contains two electrodes suspended in an electrolyte, which undergoes chemical reactions with the electrodes. Batteries contain two or more electrochemical cells.
- Primary cells are either wet cells or dry cells in which chemical reactions continue until at least one reactant is used up. Most household batteries are primary dry cells, like the one shown below.

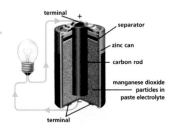

- Some batteries are storage cells in which the chemical reactions can be reversed. Such batteries are rechargeable.
- Solar cells contain materials that absorb energy from the Sun or other sources of light and then release electrons to create an electric current.

Common Misconceptions

SPEED OF ELECTRON MOVEMENT IN CIRCUITS Students may think that electrons move through a circuit instantly or at the speed of light. Electrons actually move very slowly through a circuit. The energy changes in the electric field, however, do move at the speed of light.

T E This misconception is addressed in Teach Difficult Concepts on p. 29.

MISCONCEPTION DATABASE
CLASSZONE.COM Background on student misconceptions

Previewing Chapter Resources

KEY TO ICONS			
🖳 CLASSZONE.COM	💿 CD/CD-ROM	TE Teacher Edition	
🛈 INTERNET	PE Pupil Edition	R UNIT RESOURCE BOOK	

Chapter 1 Electricity

INTEGRATED TECHNOLOGY		**READING AND REINFORCEMENT**	**ASSESSMENT**
🖳 **CLASSZONE.COM** • eEdition Plus • EasyPlanner • Misconception Database • Content Review • Test Practice • Simulations • Resource Centers • Internet Activity: Static Electricity • Math Tutorial 🖳 **SCILINKS.ORG** SC/LINKS	💿 **CD-ROM** • eEdition • EasyPlanner • Power Presentations • Content Review • Lab Generator • Test Generator 🎧 **AUDIO CDS** • Audio Readings • Audio Readings in Spanish	• Four Square, B22–23 • Combination Notes, C36 • Daily Vocabulary Scaffolding, H1–8 R **UNIT RESOURCE BOOK** • Vocabulary Practice, pp. 46–47 • Decoding Support, p. 48 • Summarizing the Chapter, pp. 71–72	PE • Chapter Re • Standardize A **UNIT ASSESS** • Diagnostic • Chapter Te • Alternative SP A Spanish Char
	How Do the Pieces of Tape Interact? Why Does the Water React Differently? • Internet Activity: Static Electricity R **UNIT RESOURCE BOOK** • Family Letter, p. vii • Spanish Family Letter, p. viii • Unit Projects, pp. 5–10	💿 **Audio Readings CD** Listen to Pupil Edition. 💿 **Audio Readings in Spanish CD** Listen to Pupil Edition in Spanish.	💿 **Test Genera** Generate cus 🔬 **Lab Generat** Rubrics for La
	🔬 **Lab Generator CD-ROM** Generate customized labs.		

SECTION 1.1 Materials can become electrically charged

UNIT TRANSPARENCY BOOK • Big Idea Flow Chart, p. T1 • Daily Vocabulary Scaffolding, p. T2 • Note-Taking Model, p. T3	PE • EXPLORE Static Electricity, p. 9 • INVESTIGATE Making a Static Detector, p. 14 • Connecting Sciences, p. 17	R **UNIT RESOURCE BOOK** • Reading Study Guide, A & B, pp. 13–16 • Spanish Reading Study Guide, pp. 17–18 • Challenge and Extension, p. 19 • Reinforcing Key Concepts, p. 21	TE Ongoing Asse PE Section 1.1 R A **UNIT ASSESS** Section 1.1 Q
	K tic Detector, p. 20		

🔬 **Lab Generator CD-ROM** — Edit these Pupil Edition labs and generate alternative labs.

rge, p. 18 tors and Insulators, ION, Lightning,	R **UNIT RESOURCE BOOK** • Reading Study Guide, A & B, pp. 24–27 • Spanish Reading Study Guide, pp. 28–29 • Challenge and Extension, p. 30 • Reinforcing Key Concepts, p. 32 • Challenge Reading, pp. 44–45	TE Ongoing Asse PE Section 1.2 R A **UNIT ASSESS** Section 1.2 Q
s and Insulators, ION, Lightning,		
28 Cells, p. 31 K lls, p. 42 , 51 , 52 TION, Making a C, pp. 62–70	R **UNIT RESOURCE BOOK** • Reading Study Guide, A & B, pp. 35–38 • Spanish Reading Study Guide, pp. 39–40 • Challenge and Extension, p. 41 • Reinforcing Key Concepts, p. 43	TE Ongoing Asse PE Section 1.3 Re A **UNIT ASSESS** Section 1.3 Q

---

ove from one pla
8–27

ial energy.
other charged object has
to the ground when it is l
or away from another cha
placed near a charged ob

Previewing Labs

EXPLORE the **BIG** idea

How Do the Pieces of Tape Interact? p. 7
Students produce static electricity in strips of tape and note its effects.
TIME 10 minutes
MATERIALS 3 strips of transparent tape

Why Does the Water React Differently? p. 7
Students observe the effect of a charged and discharged comb on a stream of water coming from a faucet.
TIME 10 minutes
MATERIALS water faucet, comb

Internet Activity: Static Electricity, p. 7
Students observe how different types of materials affect charging by contact.
TIME 20 minutes
MATERIALS computer with Internet access

SECTION 1.1

EXPLORE Static Electricity, p. 9
Students explore how paper and plastic interact electrically.
TIME 10 minutes
MATERIALS 2 strips of newspaper, plastic bag

INVESTIGATE Making a Static Detector, p. 14
Students construct a static detector and infer the presence of static electric charge.
TIME 20 minutes
MATERIALS metal paper clip, clear plastic cup with hole, small piece of modeling clay, ball of aluminum foil, 2 strips of aluminum foil, balloon

SECTION 1.2

EXPLORE Static Discharge, p. 18
Students use a fluorescent bulb to observe static discharge.
TIME 10 minutes
MATERIALS inflated balloon, wool cloth, fluorescent light bulb

INVESTIGATE Conductors and Insulators, p. 22
Students interpret data to determine what materials conduct electricity.
TIME 20 minutes
MATERIALS D cell battery, 3 pieces of low-voltage wire (20 cm each), 25 cm duct tape, flashlight bulb, bulb holder, objects made from different materials

CHAPTER INVESTIGATION
Lightning, pp. 26–27
Students model the buildup of charges that can occur during a storm and a lightning strike, and use a ground to control the path of discharge.
TIME 40 minutes
MATERIALS small piece of modeling clay, 2 aluminum pie pans, Styrofoam plate, wool cloth, metal paper clip

SECTION 1.3

EXPLORE Current, p. 28
Students relate resistance to flow of charge through different lengths of graphite.
TIME 20 minutes
MATERIALS pencil lead, posterboard, 25 cm electrical tape, 3 lengths of low-voltage wire (20 cm each), D cell battery, flashlight bulb, bulb holder

INVESTIGATE Electric Cells, p. 31
Students infer how electric current can be produced by a lemon and different types of metal.
TIME 20 minutes
MATERIALS metal paper clip, penny, large lemon, multimeter, additional fruits or vegetables, additional metal objects

R Additional **INVESTIGATION,** Making a Coin Battery, A, B, & C, pp. 62–70; Teacher Instructions, pp. 206–207

Planning the Lesson

Point-of-use support for each lesson provides a wealth of teaching options.

1. Prepare

- Concept and vocabulary review
- Note-taking and vocabulary strategies

2. Focus

- Set Learning Goals
- 3-Minute Warm-up

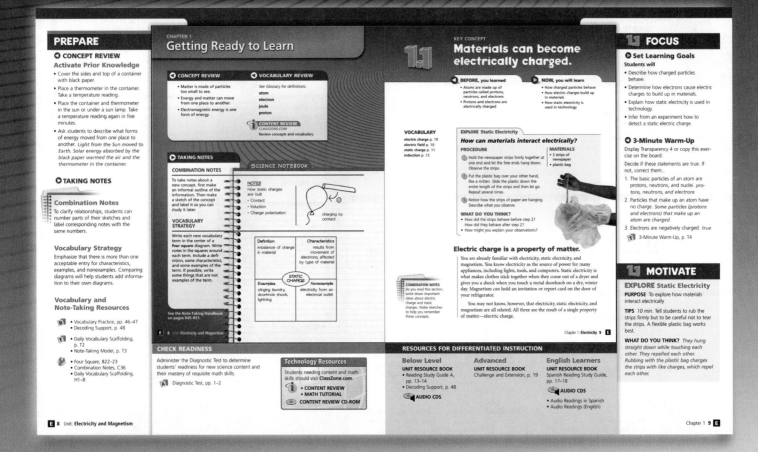

3. Motivate

- Engaging entry into the section
- Explore activity or Think About question

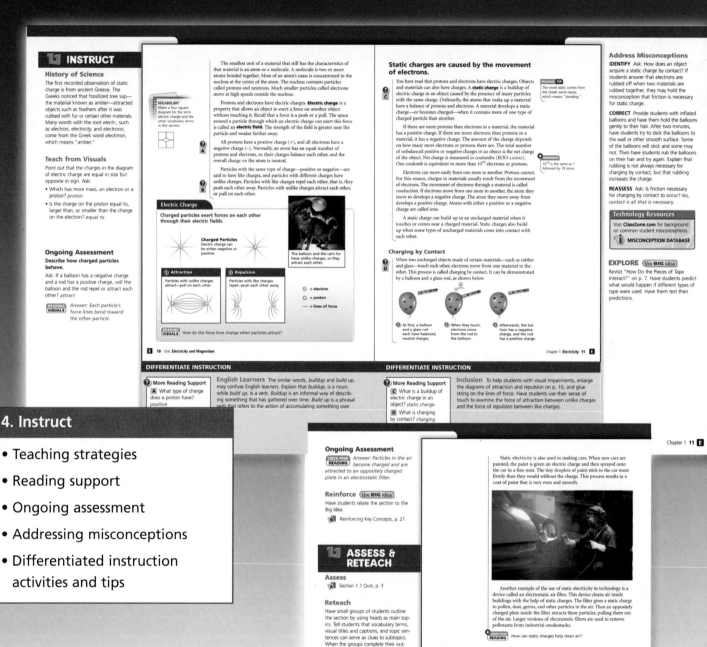

4. Instruct

- Teaching strategies
- Reading support
- Ongoing assessment
- Addressing misconceptions
- Differentiated instruction activities and tips

5. Assess & Reteach

- Answers to Section Review
- Reteaching activity
- Resources for review and assessment

Lab Materials List

The following charts list the consumables, nonconsumables, and equipment needed for all activities. Quantities are per group of four students. Lab aprons, goggles, water, books, paper, pens, pencils, and calculators are assumed to be available for all activities.

Materials kits are available. For more information, please call McDougal Littell at 1-800-323-5435.

Consumables

Description	Quantity per Group	Explore page	Investigate page	Chapter Investigation page
aluminum foil	60 cm	43	14, 85	66
bag, plastic grocery	1	9		
balloon	2	18	14	
cardboard, 20 cm x 20 cm	1			66
clay, modeling	1 stick		14	26
craft stick, jumbo	2			66
cup, clear plastic	1		14	
cup, Styrofoam	1			100
fruit or vegetable, various	1–3		31	
lemon	1		31	
marker, colored	1			100
mechanical pencil lead	10 cm	28		
newspaper, 3 cm x 30 cm	2	9		
paper, graph 8.5" x 11"	2		59, 105	
paper, white 8.5" x 11"	1		59	
paper clip	10		14, 31, 90	26, 66, 100
pencil, colored	4–6		105	
pie plate, aluminum	2			26
plate, Styrofoam	1			26
posterboard, 10 cm x 20 cm	1	28		
rubber band, large	1			66
sandpaper	1		98	
steel wool	1 strand		48	
tape, duct	1 roll		22	
tape, electrical	1 roll	28, 43, 88	90, 98	66
tape, masking	1 roll		48	100
toothpick	1			66
wire, copper, insulated	1 meter		90	
wire, copper, uninsulated	100 cm	88		66
wire, low voltage	150 cm	28	22	
wire, magnet	4 meters		98	100

Nonconsumables

Description	Quantity per Group	Explore *page*	Investigate *page*	Chapter Investigation *page*
battery holder, D cell	2		54	
battery, 6 volt	1		48	
battery, AA	1	88, 95		
battery, D cell	2	28, 43	22, 54, 90, 98	66
bowl, small plastic	1		85	
cloth, wool, 8" x 8"	1	18		26
clothespin, wood, lever type	1	79		66
coin, penny	1		31	
compass	1	88	98	
dowel rod, 1/4" diameter	15 cm	79		
jar, baby food	1		48	
light bulb holder	1	28, 95	22, 54	66
light bulb, flashlight	1	28, 43, 95	22, 54	66
light bulb, small fluorescent tube	1	18		
magnet, bar	1		85	
magnet, high strength	1			100
magnet, low strength	1			100
magnet, medium strength	1			100
magnet, ring	3	79		
motor, small ultra-sensitive	1	95		
multimeter	1		31	
nail, iron, 3"	1		90	
ruler, metric	1	79	98	100
scissors	1			66
sewing needle	1		85	
stereo system or boom box	1			100
wire cutters	1			66
wire lead with alligator clips	4		48, 54	66, 100

Unit Resource Book Datasheets

Description		Explore *page*	Investigate *page*	Chapter Investigation *page*
Morse Code Chart				66

McDougal Littell Science

Electricity and Magnetism

magnet

magnetic field

ATTRACT

i

PHYSICAL SCIENCE

A ▶ Matter and Energy
B ▶ Chemical Interactions
C ▶ Motion and Forces
D ▶ Waves, Sound, and Light
E ▶ Electricity and Magnetism

LIFE SCIENCE

A ▶ Cells and Heredity
B ▶ Life Over Time
C ▶ Diversity of Living Things
D ▶ Ecology
E ▶ Human Biology

EARTH SCIENCE

A ▶ Earth's Surface
B ▶ The Changing Earth
C ▶ Earth's Waters
D ▶ Earth's Atmosphere
E ▶ Space Science

ISBN: 0-618-33440-8   1 2 3 4 5 6 7 8 VJM 08 07 06 05 04

Internet Web Site: http://www.mcdougallittell.com

Science Consultants

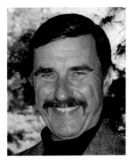

Chief Science Consultant

James Trefil, Ph.D. is the Clarence J. Robinson Professor of Physics at George Mason University. He is the author or co-author of more than 25 books, including *Science Matters* and *The Nature of Science*. Dr. Trefil is a member of the American Association for the Advancement of Science's Committee on the Public Understanding of Science and Technology. He is also a fellow of the World Economic Forum and a frequent contributor to *Smithsonian* magazine.

Rita Ann Calvo, Ph.D. is Senior Lecturer in Molecular Biology and Genetics at Cornell University, where for 12 years she also directed the Cornell Institute for Biology Teachers. Dr. Calvo is the 1999 recipient of the College and University Teaching Award from the National Association of Biology Teachers.

Kenneth Cutler, M.S. is the Education Coordinator for the Julius L. Chambers Biomedical Biotechnology Research Institute at North Carolina Central University. A former middle school and high school science teacher, he received a 1999 Presidential Award for Excellence in Science Teaching.

Instructional Design Consultants

Douglas Carnine, Ph.D. is Professor of Education and Director of the National Center for Improving the Tools of Educators at the University of Oregon. He is the author of seven books and over 100 other scholarly publications, primarily in the areas of instructional design and effective instructional strategies and tools for diverse learners. Dr. Carnine also serves as a member of the National Institute for Literacy Advisory Board.

Linda Carnine, Ph.D. consults with school districts on curriculum development and effective instruction for students struggling academically. A former teacher and school administrator, Dr. Carnine also co-authored a popular remedial reading program.

Donald Steely, Ph.D. serves as principal investigator at the Oregon Center for Applied Science (ORCAS) on federal grants for science and language arts programs. His background also includes teaching and authoring of print and multimedia programs in science, mathematics, history, and spelling.

Sam Miller, Ph.D. is a middle school science teacher and the Teacher Development Liaison for the Eugene, Oregon, Public Schools. He is the author of curricula for teaching science, mathematics, computer skills, and language arts.

Vicky Vachon, Ph.D. consults with school districts throughout the United States and Canada on improving overall academic achievement with a focus on literacy. She is also co-author of a widely used program for remedial readers.

Content Reviewers

John Beaver, Ph.D.
Ecology
Professor, Director of Science Education Center
College of Education and Human Services
Western Illinois University
Macomb, IL

Donald J. DeCoste, Ph.D.
Matter and Energy, Chemical Interactions
Chemistry Instructor
University of Illinois
Urbana-Champaign, IL

Dorothy Ann Fallows, Ph.D., MSc
Diversity of Living Things, Microbiology
Partners in Health
Boston, MA

Michael Foote, Ph.D.
The Changing Earth, Life Over Time
Associate Professor
Department of the Geophysical Sciences
The University of Chicago
Chicago, IL

Lucy Fortson, Ph.D.
Space Science
Director of Astronomy
Adler Planetarium and Astronomy Museum
Chicago, IL

Elizabeth Godrick, Ph.D.
Human Biology
Professor, CAS Biology
Boston University
Boston, MA

Isabelle Sacramento Grilo, M.S.
The Changing Earth
Lecturer, Department of the Geological Sciences
Montana State University
Bozeman, MT

David Harbster, MSc
Diversity of Living Things
Professor of Biology
Paradise Valley Community College
Phoenix, AZ

Richard D. Norris, Ph.D.
Earth's Waters
Professor of Paleobiology
Scripps Institution of Oceanography
University of California, San Diego
La Jolla, CA

Donald B. Peck, M.S.
*Motion and Forces; Waves, Sound, and Light;
 Electricity and Magnetism*
Director of the Center for Science Education (retired)
Fairleigh Dickinson University
Madison, NJ

Javier Penalosa, Ph.D.
Diversity of Living Things, Plants
Associate Professor, Biology Department
Buffalo State College
Buffalo, NY

Raymond T. Pierrehumbert, Ph.D.
Earth's Atmosphere
Professor in Geophysical Sciences (Atmospheric Science)
The University of Chicago
Chicago, IL

Brian J. Skinner, Ph.D.
Earth's Surface
Eugene Higgins Professor of Geology and Geophysics
Yale University
New Haven, CT

Nancy E. Spaulding, M.S.
Earth's Surface, The Changing Earth, Earth's Waters
Earth Science Teacher (retired)
Elmira Free Academy
Elmira, NY

Steven S. Zumdahl, Ph.D.
Matter and Energy, Chemical Interactions
Professor Emeritus of Chemistry
University of Illinois
Urbana-Champaign, IL

Susan L. Zumdahl, M.S.
Matter and Energy, Chemical Interactions
Chemistry Education Specialist
University of Illinois
Urbana-Champaign, IL

Safety Consultant

Juliana Texley, Ph.D.
Former K–12 Science Teacher and School Superintendent
Boca Raton, FL

English Language Advisor

Judy Lewis, M.A.
Director, State and Federal Programs for reading proficiency
and high risk populations
Rancho Cordova, CA

Teacher Panel Members

Carol Arbour
Tallmadge Middle School,
Tallmadge, OH

Patty Belcher
Goodrich Middle School,
Akron, OH

Gwen Broestl
Luis Munoz Marin Middle School,
Cleveland, OH

Al Brofman
Tehipite Middle School,
Fresno, CA

John Cockrell
Clinton Middle School,
Columbus, OH

Jenifer Cox
Sylvan Middle School,
Citrus Heights, CA

Linda Culpepper
Martin Middle School,
Charlotte, NC

Kathleen Ann DeMatteo
Margate Middle School,
Margate, FL

Melvin Figueroa
New River Middle School,
Ft. Lauderdale, FL

Doretha Grier
Kannapolis Middle School,
Kannapolis, NC

Robert Hood
Alexander Hamilton Middle School,
Cleveland, OH

Scott Hudson
Coverdale Elementary School,
Cincinnati, OH

Loretta Langdon
Princeton Middle School,
Princeton, NC

Carlyn Little
Glades Middle School,
Miami, FL

Ann Marie Lynn
Amelia Earhart Middle School,
Riverside, CA

James Minogue
Lowe's Grove Middle School,
Durham, NC

Joann Myers
Buchanan Middle School,
Tampa, FL

Barbara Newell
Charles Evans Hughes Middle School,
Long Beach, CA

Anita Parker
Kannapolis Middle School,
Kannapolis, NC

Greg Pirolo
Golden Valley Middle School,
San Bernardino, CA

Laura Pottmyer
Apex Middle School,
Apex, NC

Lynn Prichard
Booker T. Washington Middle Magnet
School, Tampa, FL

Jacque Quick
Walter Williams High School,
Burlington, NC

Robert Glenn Reynolds
Hillman Middle School,
Youngstown, OH

Theresa Short
Abbott Middle School,
Fayetteville, NC

Rita Slivka
Alexander Hamilton Middle School,
Cleveland, OH

Marie Sofsak
B F Stanton Middle School,
Alliance, OH

Nancy Stubbs
Sweetwater Union Unified School District,
Chula Vista, CA

Sharon Stull
Quail Hollow Middle School,
Charlotte, NC

Donna Taylor
Okeeheelee Middle School,
West Palm Beach, FL

Sandi Thompson
Harding Middle School,
Lakewood, OH

Lori Walker
Audubon Middle School & Magnet Center,
Los Angeles, CA

Teacher Lab Evaluators

Jill Brimm-Byrne
Albany Park Academy,
Chicago, IL

Gwen Broestl
Luis Munoz Marin Middle School,
Cleveland, OH

Al Brofman
Tehipite Middle School,
Fresno, CA

Michael A. Burstein
The Rashi School,
Newton, MA

Trudi Coutts
Madison Middle School,
Naperville, IL

Jenifer Cox
Sylvan Middle School,
Citrus Heights, CA

Larry Cwik
Madison Middle School,
Naperville, IL

Jennifer Donatelli
Kennedy Junior High School,
Lisle, IL

Paige Fullhart
Highland Middle School,
Libertyville, IL

Sue Hood
Glen Crest Middle School,
Glen Ellyn, IL

Ann Min
Beardsley Middle School,
Crystal Lake, IL

Aileen Mueller
Kennedy Junior High School,
Lisle, IL

Nancy Nega
Churchville Middle School,
Elmhurst, IL

Oscar Newman
Sumner Math and Science Academy,
Chicago, IL

Marina Penalver
Moore Middle School,
Portland, ME

Lynn Prichard
Booker T. Washington Middle Magnet
School, Tampa, FL

Jacque Quick
Walter Williams High School,
Burlington, NC

Seth Robey
Gwendolyn Brooks Middle School,
Oak Park, IL

Kevin Steele
Grissom Middle School,
Tinley Park, IL

Electricity and Magnetism

How can circuits control the flow of charge? page 40

What force is acting on this compass needle? page 76

Features

Visual Highlights

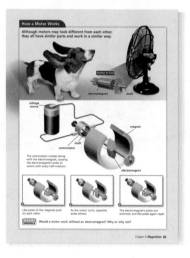

Internet Resources @ ClassZone.com

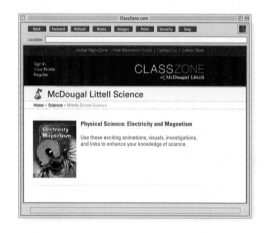

INVESTIGATIONS AND ACTIVITIES

Standards and Benchmarks

Each chapter in **Electricity and Magnetism** covers some of the learning goals that are described in the *National Science Education Standards* (NSES) and the Project 2061 *Benchmarks for Science Literacy*. Selected content and skill standards are shown below in shortened form. The following National Science Education Standards are covered on pages xii–xxvii, in Frontiers in Science, and in Timelines in Science, as well as in chapter features and laboratory investigations: Understandings About Scientific Inquiry (A.9), Understandings About Science and Technology (E.6), Science and Technology in Society (F.5), Science as a Human Endeavor (G.1), Nature of Science (G.2), and History of Science (G.3).

Content Standards

1 Electricity

National Science Education Standards

B.3.a	Energy • is often associated with electricity • is transferred in many ways
B.3.e	In most chemical reactions, energy is transferred into or out of the system. Heat, light, motion, or electricity might be involved in this transfer.

Project 2061 Benchmarks

3.A.3	Engineers and others who work in design and technology use scientific knowledge to solve practical problems.

2 Circuits and Electronics

National Science Education Standards

B.3.a	Energy. is property of substances that is often associated with electricity. Energy is transferred in many ways.
B.3.d	Circuits transfer electrical energy. Heat, light, sound, and chemical changes are produced.

Project 2061 Benchmarks

1.C.6	Computers are important in science.
3.A.2	Technology is important in the computation and communication of information.
8.D.2	The ability to code information or electric currents in wires has made communication many times faster than is possible by mail or sound.
8.E.1	Computers use digital codes containing only two symbols to perform all operations. Analog signals must be converted into digital codes before a computer can process them.
9.A.6	Numbers can be represented by using only 1 and 0.

Magnetism

Project 2061 Benchmarks

4.G.3	Electric currents and magnets can exert a force on each other.
8.C.4	Electrical energy can be produced from a variety of energy sources and can be transformed into almost any other form of energy. Electricity is used to distribute energy quickly and conveniently to distant locations.

Process and Skill Standards

National Science Education Standards

A.1	Identify questions that can be answered through investigation.
A.2	Design and conduct a scientific investigation.
A.3	Use appropriate tools and techniques to gather and interpret data.
A.4	Use evidence to describe, predict, explain, and model.
A.5	Use critical thinking to find relationships between results and interpretations.
A.6	Consider alternative explanations and predictions.
A.7	Communicate procedures, results, and conclusions.
A.8	Use mathematics in scientific investigations.
E.1	Identify a problem to be reached.
E.2	Design a solution or product.
E.3	Implement the proposed solution.
E.4	Evaluate the solution or design.
E.5	Communicate the process of technological design.
F.4.c	Use systematic thinking to eliminate risks.
F.4.d	Decisions are made based on estimated risks and benefits.

Project 2061 Benchmarks

1.A.3	Some knowledge in science is very old and yet is still used today.
1.C.1	Contributions to science and technology have been made by different people, in different cultures, at different times.
3.B.1	Design requires taking constraints into account.
3.B.2	Technologies have effects other than those intended.
8.B.1	The choice of materials for a job depends on their properties.
9.A.3	How decimals should be written depends on how precise the measurements are.
9.A.7	Computations can give more digits than make sense or are useful.
9.B.3	Graphs can show the relationship between two variables.
9.C.4	Graphs show patterns and can be used to make predictions.
11.C.4	Use equations to summarize observed changes.
12.B.1	Find what percentage one number is of another.
12.B.8	Round a calculation to the correct number of significant figures.
12.C.1	Compare amounts proportionally.
12.C.5	Inspect, disassemble, and reassemble simple devices and describe what the various parts are for.
12.D.1	Use tables and graphs to organize information and identify relationships.
12.D.2	Read, interpret, and describe tables and graphs.
12.D.4	Understand information that includes different types of charts and graphs, including circle charts, bar graphs, line graphs, data tables, diagrams, and symbols.
12.E.3	Be skeptical of arguments based on samples for which there was no control group.
12.E.4	There may be more than one good way to interpret scientific findings.

Introducing Physical Science

Scientists are curious. Since ancient times, they have been asking and answering questions about the world around them. Scientists are also very suspicious of the answers they get. They carefully collect evidence and test their answers many times before accepting an idea as correct.

In this book you will see how scientific knowledge keeps growing and changing as scientists ask new questions and rethink what was known before. The following sections will help get you started.

What Is Physical Science?

In the simplest terms, physical science is the study of what things are made of and how they change. It combines the studies of both physics and chemistry. Physics is the science of matter, energy, and forces. It includes the study of topics such as motion, light, and electricity and magnetism. Chemistry is the study of the structure and properties of matter, and it especially focuses on how substances change into different substances.

The text and pictures in this book will help you learn key concepts and important facts about physical science. A variety of activities will help you investigate these concepts. As you learn, it helps to have a big picture of physical science as a framework for this new information. The four unifying principles listed below will give you this big picture. Read the next few pages to get an overview of each of these principles and a sense of why they are so important.

- **Matter is made of particles too small to see.**

- **Matter changes form and moves from place to place.**

- **Energy changes from one form to another, but it cannot be created or destroyed.**

- **Physical forces affect the movement of all matter on Earth and throughout the universe.**

the **BIG** idea

Each chapter begins with a big idea. Keep in mind that each big idea relates to one or more of the unifying principles.

UNIFYING PRINCIPLE

Matter is made of particles too small to see.

This simple statement is the basis for explaining an amazing variety of things about the world. For example, it explains why substances can exist as solids, liquids, and gases, and why wood burns but iron does not. Like the tiles that make up this mosaic picture, the particles that make up all substances combine to make patterns and structures that can be seen. Unlike these tiles, the individual particles themselves are far too small to see.

What It Means

To understand this principle better, let's take a closer look at the two key words: *matter* and *particles*.

Matter

Objects you can see and touch are all around you. The materials that these objects are made of are called **matter.** All living things—even you—are also matter. Even though you can't see it, the air around you is matter too. Scientists often say that matter is anything that has mass and takes up space. **Mass** is a measure of the amount of matter in an object. We use the word **volume** to refer to the amount of space an object or a substance takes up.

Particles

The tiny particles that make up all matter are called **atoms.** Just how tiny are atoms? They are far too small to see, even through a powerful microscope. In fact, an atom is more than a million times smaller than the period at the end of this sentence.

There are more than 100 basic kinds of matter called **elements.** For example, iron, gold, and oxygen are three common elements. Each element has its own unique kind of atom. The atoms of any element are all alike but different from the atoms of any other element.

Many familiar materials are made of particles called molecules. In a **molecule,** two or more atoms stick together to form a larger particle. For example, a water molecule is made of two atoms of hydrogen and one atom of oxygen.

Why It's Important

Understanding atoms and molecules makes it possible to explain and predict the behavior of matter. Among other things, this knowledge allows scientists to

- explain why different materials have different characteristics
- predict how a material will change when heated or cooled
- figure out how to combine atoms and molecules to make new and useful materials

Matter changes form and moves from place to place.

You see matter change form every day. You see the ice in your glass of juice disappear without a trace. You see a black metal gate slowly develop a flaky, orange coating. Matter is constantly changing and moving.

What It Means

Remember that matter is made of tiny particles called atoms. Atoms are constantly moving and combining with one another. All changes in matter are the result of atoms moving and combining in different ways.

Matter Changes and Moves

You can look at water to see how matter changes and moves. A block of ice is hard like a rock. Leave the ice out in sunlight, however, and it changes into a puddle of water. That puddle of water can eventually change into water vapor and disappear into the air. The water vapor in the air can become raindrops, which may fall on rocks, causing them to weather and wear away. The water that flows in rivers and streams picks up tiny bits of rock and carries them from one shore to another. Understanding how the world works requires an understanding of how matter changes and moves.

Matter Is Conserved

No matter was lost in any of the changes described above. The ice turned to water because its molecules began to move more quickly as they got warmer. The bits of rock carried away by the flowing river were not gone forever. They simply ended up farther down the river. The puddles of rainwater didn't really disappear; their molecules slowly mixed with molecules in the air.

Under ordinary conditions, when matter changes form, no matter is created or destroyed. The water created by melting ice has the same mass as the ice did. If you could measure the water vapor that mixes with the air, you would find it had the same mass as the water in the puddle did.

Why It's Important

Understanding how mass is conserved when matter changes form has helped scientists to

- describe changes they see in the world
- predict what will happen when two substances are mixed
- explain where matter goes when it seems to disappear

UNIFYING PRINCIPLE

Energy changes from one form to another, but it cannot be created or destroyed.

When you use energy to warm your food or to turn on a flashlight, you may think that you "use up" the energy. Even though the camp-stove fuel is gone and the flashlight battery no longer functions, the energy they provided has not disappeared. It has been changed into a form you can no longer use. Understanding how energy changes forms is the basis for understanding how heat, light, and motion are produced.

What It Means

Changes that you see around you depend on energy. **Energy,** in fact, means the ability to cause change. The electrical energy from an outlet changes into light and heat in a light bulb. Plants change the light energy from the Sun into chemical energy, which animals use to power their muscles.

Energy Changes Forms

Using energy means changing energy. You probably have seen electric energy changing into light, heat, sound, and mechanical energy in household appliances. Fuels like wood, coal, and oil contain chemical energy that produces heat when burned. Electric power plants make electrical energy from a variety of energy sources, including falling water, nuclear energy, and fossil fuels.

Energy Is Conserved

Energy can be converted into forms that can be used for specific purposes. During the conversion, some of the original energy is converted into unwanted forms. For instance, when a power plant converts the energy of falling water into electrical energy, some of the energy is lost to friction and sound.

Similarly, when electrical energy is used to run an appliance, some of the energy is converted into forms that are not useful. Only a small percentage of the energy used in a light bulb, for instance, produces light; most of the energy becomes heat. Nonetheless, the total amount of energy remains the same through all these conversions.

The fact that energy does not disappear is a law of physical science. The **law of conservation of energy** states that energy cannot be created or destroyed. It can only change form.

Why It's Important

Understanding that energy changes form but does not disappear has helped scientists to

- predict how energy will change form
- manage energy conversions in useful ways
- build and improve machines

Physical forces affect the movement of all matter on Earth and throughout the universe.

What makes the world go around? The answer is simple: forces. Forces allow you to walk across the room, and forces keep the stars together in galaxies. Consider the forces acting on the rafts below. The rushing water is pushing the rafts forward. The force from the people paddling helps to steer the rafts.

What It Means

A **force** is a push or a pull. Every time you push or pull an object, you're applying a force to that object, whether or not the object moves. There are several forces—several pushes and pulls—acting on you right now. All these forces are necessary for you to do the things you do, even sitting and reading.

- You are already familiar with the force of gravity. **Gravity** is the force of attraction between two objects. Right now gravity is at work pulling you to Earth and Earth to you. The Moon stays in orbit around Earth because gravity holds it close.

- A contact force occurs when one object pushes or pulls another object by touching it. If you kick a soccer ball, for instance, you apply a contact force to the ball. You apply a contact force to a shopping cart that you push down a grocery aisle or a sled that you pull up a hill.

- **Friction** is the force that resists motion between two surfaces pressed together. If you've ever tried to walk on an icy sidewalk, you know how important friction can be. If you lightly rub your finger across a smooth page in a book and then across a piece of sandpaper, you can feel how the different surfaces produce different frictional forces. Which is easier to do?

- There are other forces at work in the world too. For example, a compass needle responds to the magnetic force exerted by Earth's magnetic field, and objects made of certain metals are attracted by magnets. In addition to magnetic forces, there are electrical forces operating between particles and between objects. For example, you can demonstrate electrical forces by rubbing an inflated balloon on your hair. The balloon will then stick to your head or to a wall without additional means of support.

Why It's Important

Although some of these forces are more obvious than others, physical forces at work in the world are necessary for you to do the things you do. Understanding forces allows scientists to

- predict how objects will move
- design machines that perform complex tasks
- predict where planets and stars will be in the sky from one night to the next

The Nature of Science

You may think of science as a body of knowledge or a collection of facts. More important, however, science is an active process that involves certain ways of looking at the world.

Scientific Habits of Mind

Scientists are curious. They are always asking questions. Scientists have asked questions such as, "What is the smallest form of matter?" and "How do the smallest particles behave?" These and other important questions are being investigated by scientists around the world.

Scientists are observant. They are always looking closely at the world around them. Scientists once thought the smallest parts of atoms were protons, neutrons, and electrons. Later, protons and neutrons were found to be made of even smaller particles called quarks.

Scientists are creative. They draw on what they know to form possible explanations for a pattern, an event, or an interesting phenomenon that they have observed. Then scientists create a plan for testing their ideas.

Scientists are skeptical. Scientists don't accept an explanation or answer unless it is based on evidence and logical reasoning. They continually question their own conclusions and the conclusions suggested by other scientists. Scientists trust only evidence that is confirmed by other people or methods.

Scientists cannot always make observations with their own eyes. They have developed technology, such as this particle detector, to help them gather information about the smallest particles of matter.

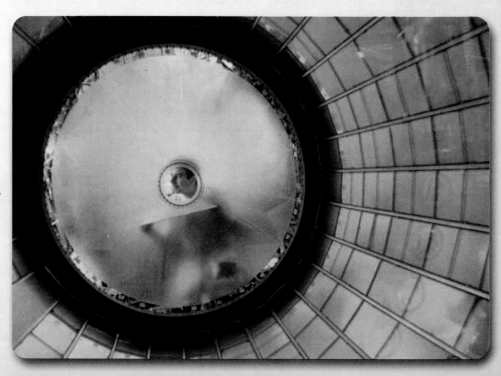

Scientists ask questions about the physical world and seek answers through carefully controlled procedures. Here a researcher works with supercooled magnets.

Science Processes at Work

You can think of science as a continuous cycle of asking and seeking answers to questions about the world. Although there are many processes that scientists use, scientists typically do each of the following:

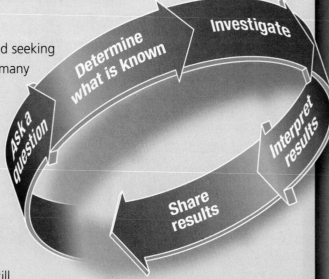

- Ask a question
- Determine what is known
- Investigate
- Interpret results
- Share results

Ask a Question

It may surprise you that asking questions is an important skill. A scientific process may start when a scientist asks a question. Perhaps scientists observe an event or a process that they don't understand, or perhaps answering one question leads to another.

Determine What Is Known

When beginning an inquiry, scientists find out what is already known about a question. They study results from other scientific investigations, read journals, and talk with other scientists. A scientist working on subatomic particles is most likely a member of a large team using sophisticated equipment. Before beginning original research, the team analyzes results from previous studies.

Investigate

Investigating is the process of collecting evidence. Two important ways of investigating are observing and experimenting.

Observing is the act of noting and recording an event, a characteristic, or anything else detected with an instrument or with the senses. A researcher may study the properties of a substance by handling it, finding its mass, warming or cooling it, stretching it, and so on. For information about the behavior of subatomic particles, however, a researcher may rely on technology such as scanning tunneling microscopes, which produce images of structures that cannot be seen with the eye.

An **experiment** is an organized procedure to study something under controlled conditions. In order to study the effect of wing shape on the motion of a glider, for instance, a researcher would need to conduct controlled studies in which gliders made of the same materials and with the same masses differed only in the shape of their wings.

Scanning tunneling microscopes create images that allow scientists to observe molecular structure.

Physical chemists have found a way to observe chemical reactions at the atomic level. Using lasers, they can watch bonds breaking and new bonds forming.

Forming hypotheses and making predictions are two of the skills involved in scientific investigations. A **hypothesis** is a tentative explanation for an observation, a phenomenon, or a scientific problem that can be tested by further investigation. For example, in the mid-1800s astronomers noticed that the planet Uranus departed slightly from its expected orbit. One astronomer hypothesized that the irregularities in the planet's orbit were due to the gravitational effect of another planet—one that had not yet been detected. A **prediction** is an expectation of what will be observed or what will happen. A prediction can be used to test a hypothesis. The astronomers predicted that they would discover a new planet in the position calculated, and their prediction was confirmed with the discovery of the planet Neptune.

Interpret Results

As scientists investigate, they analyze their evidence, or data, and begin to draw conclusions. **Analyzing data** involves looking at the evidence gathered through observations or experiments and trying to identify any patterns that might exist in the data. Scientists often need to make additional observations or perform more experiments before they are sure of their conclusions. Many times scientists make new predictions or revise their hypotheses.

Often scientists use computers to help them analyze data. Computers reveal patterns that might otherwise be missed.

Scientists use computers to create models of objects or processes they are studying. This model shows carbon atoms forming a sphere.

Share Results

An important part of scientific investigation is sharing results of experiments. Scientists read and publish in journals and attend conferences to communicate with other scientists around the world. Sharing data and procedures gives them a way to test one another's results. They also share results with the public through newspapers, television, and other media.

The Nature of Technology

When you think of technology, you may think of cars, computers, and cell phones, as well as refrigerators, radios, and bicycles. Technology is not only the machines and devices that make modern lives easier, however. It is also a process in which new methods and devices are created. Technology makes use of scientific knowledge to design solutions to real-world problems.

Science and Technology

Science and technology go hand in hand. Each depends upon the other. Even designing a device as simple as a toaster requires knowledge of how heat flows and which materials are the best conductors of heat. Just as technology based on scientific knowledge makes our lives easier, some technology is used to advance scientific inquiry itself. For example, researchers use a number of specialized instruments to help them collect data. Microscopes, telescopes, spectrographs, and computers are just a few of the tools that help scientists learn more about the world. The more information these tools provide, the more devices can be developed to aid scientific research and to improve modern lives.

The Process of Technological Design

The process of technology involves many choices. For example, how does an automobile engineer design a better car? Is a better car faster? safer? cheaper? Before designing any new machine, the engineer must decide exactly what he or she wants the machine to do as well as what may be given up for the machine to do it. A faster car may get people to their destinations more quickly, but it may cost more and be less safe. As you study the technological process, think about all the choices that were made to build the technologies you use.

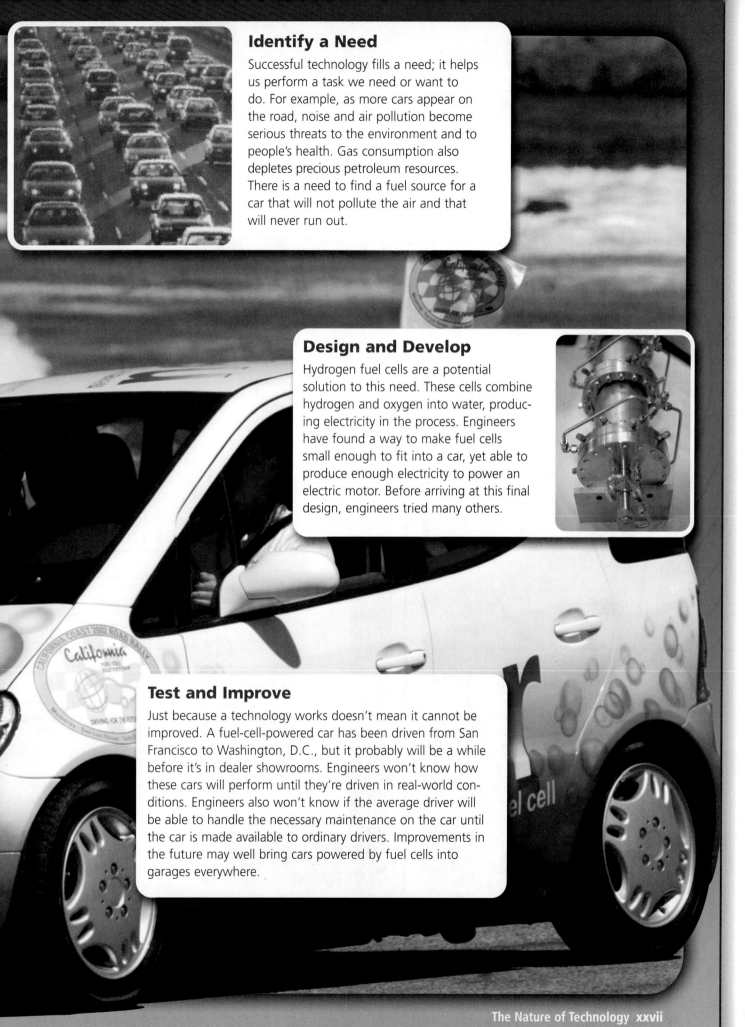

Identify a Need

Successful technology fills a need; it helps us perform a task we need or want to do. For example, as more cars appear on the road, noise and air pollution become serious threats to the environment and to people's health. Gas consumption also depletes precious petroleum resources. There is a need to find a fuel source for a car that will not pollute the air and that will never run out.

Design and Develop

Hydrogen fuel cells are a potential solution to this need. These cells combine hydrogen and oxygen into water, producing electricity in the process. Engineers have found a way to make fuel cells small enough to fit into a car, yet able to produce enough electricity to power an electric motor. Before arriving at this final design, engineers tried many others.

Test and Improve

Just because a technology works doesn't mean it cannot be improved. A fuel-cell-powered car has been driven from San Francisco to Washington, D.C., but it probably will be a while before it's in dealer showrooms. Engineers won't know how these cars will perform until they're driven in real-world conditions. Engineers also won't know if the average driver will be able to handle the necessary maintenance on the car until the car is made available to ordinary drivers. Improvements in the future may well bring cars powered by fuel cells into garages everywhere.

Using McDougal Littell Science

Reading Text and Visuals

This book is organized to help you learn. Use these boxed pointers as a path to help you learn and remember the **Big Ideas** and **Key Concepts**.

Take notes.

Use the strategies on the **Getting Ready to Learn** page.

Read the Big Idea.

As you read **Key Concepts** for the chapter, relate them to **the Big Idea**.

CHAPTER 2
Circui   Electr

the BIG idea

Circuits control the flow of electric charge.

Key Concepts

SECTION 2.1
Charge needs a continuous path to flow.
Learn how circuits are used to control the flow of charge.

SECTION 2.2
Circuits make electric current useful.
Learn about series circuits and parallel circuits.

SECTION 2.3
Electronic technology is based on circuits.
Learn about computers and other electronic devices.

Internet Preview
CLASSZONE.COM
Chapter 2 online resources:
Content Review, Simulation,
Visualization, two Resource
Centers, Math Tutorial,
Test Practice

E 40 Unit: Electricity and Magnetism

CHAPTER 2
Getting Ready to Learn

◄ **CONCEPT REVIEW**

- Energy can change from one form to another.
- Energy can move from one place to another.
- Current is the flow of charge through a conductor.

◄ **VOCABULARY REVIEW**

electric current p. 28
electric potential p. 18
conductor p. 22
resistance p. 23
ampere p. 29

CONTENT REVIEW
CLASSZONE.COM
Review concepts and vocabulary.

► **TAKING NOTES**

OUTLINE

As you read, copy the headings on your paper in the form of an outline. Then add notes in your own words that summarize what you read.

VOCABULARY STRATEGY

Write each new vocabulary term in the center of a **frame game** diagram. Decide what information to frame it with. Use examples, descriptions, parts, sentences that use the term in context, or pictures. You can change the frame to fit each term.

See the Note-Taking Handbook on pages R45–R51.

SCIENCE NOTEBOOK

I. ELECTRIC CHARGE FLOWS IN A LOOP.
  A. THE PARTS OF A CIRCUIT
    1. voltage source
    2. connection
    3. electrical device
    4. switch

	Electrical device	
Part of a circuit	**RESISTOR**	Light bulb an examp
	Slows the flow of charge	

E 42 Unit: Electricity and Magnetism

Read each heading.

See how it fits in the outline of the chapter.

KEY CONCEPT

2.1 Charge needs a continuous path to flow.

Remember what you know.

Think about concepts you learned earlier and preview what you'll learn now.

◀ **BEFORE,** you learned

- Current is the flow of charge
- Voltage is a measure of electric potential
- Materials affect the movement of charge

▶ **NOW,** you will learn

- About the parts of a circuit
- How a circuit functions
- How safety devices stop current

VOCABULARY

...uit p. 43
...istor p. 44
...rt circuit p. 46

EXPLORE Circuits

How can you light the bulb?

PROCEDURE

① Tape one end of a strip of foil to the negative terminal, or the flat end, of the battery. Tape the other end of the foil to the tip at the base of the light bulb, as shown.

② Tape the second strip of foil to the positive terminal, or the raised end, of the battery.

③ Find a way to make the bulb light.

WHAT DO YOU THINK?

- How did you make the bulb light?
- Can you find other arrangements that make the bulb light?

MATERIALS

- 2 strips of aluminum foil
- electrical tape
- D cell (battery)
- light bulb

Try the activities.

They will introduce you to science concepts.

VOCABULARY

...se a frame game dia-
...am to record the term
...rcuit in your notebook.

Electric charge flows in a loop.

In the last chapter, you read that current is electric charge that flows from one place to another. Charge does not flow continuously through a material unless the material forms a closed path, or loop. A **circuit** is a closed path through which a continuous charge can flow. The path is provided by a low-resistance material, or conductor, usually wire. Circuits are designed to do specific jobs, such as light a bulb.

Circuits can be found all around you and serve many different purposes. In this chapter, you will read about simple circuits, such as the ones in flashlights, and more complex circuits, such as the ones that run toys, cameras, computers, and more.

CHECK YOUR READING How are circuits related to current?

Learn the vocabulary.

Take notes on each term.

Answer the questions.

Check Your Reading questions will help you remember what you read.

Chapter 2: Circuits and Electronics **43** **E**

Reading Text and Visuals

Open and Closed Circuits

Current in a circuit is similar to water running through a hose. The flow of charge differs from the flow of water in an important way, however. The water does not require a closed path to flow. If you cut the hose, the water continues to flow. If you cut a wire, the charge stops flowing.

Batteries have connections at both ends so that charge can follow a closed path to and from the battery. The cords that you see on appliances might look like single cords but actually contain at least two wires. The wires connect the device to a power plant and back to make a closed path.

Switches work by opening and closing the circuit. A switch that is on closes the circuit and allows charge to flow through the electrical devices. A switch that is off opens the circuit and stops the current.

 REMINDER
Current requires a closed loop.

CHECK YOUR READING How are switches used to control the flow of charge through a circuit?

Standard symbols are used to represent the parts of a circuit. Some common symbols are shown in the circuit diagrams below. The diagrams represent the circuit shown on page 44 with the switch in both open and closed positions. Electricians and architects use diagrams such as these to plan the wiring of a building.

Circuit Diagrams

Symbols are used to represent the parts of a circuit. The circuit diagrams below show the circuit from page 44 in both an open and closed position.

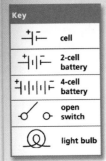

Key	
	cell
	2-cell battery
	4-cell battery
	open switch
	light bulb

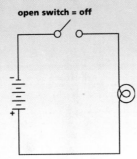 open switch = off

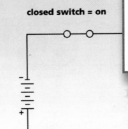 closed switch = on

READING VISUALS Would charge flow through the circuit diagrammed on the left? Why or why not?

Read one paragraph at a time.

Look for a topic sentence that explains the main idea of the paragraph. Figure out how the details relate to that idea. One paragraph might have several important ideas; you may have to reread to understand.

Answer the questions.

Check Your Reading questions will help you remember what you read.

Study the visuals.

- Read the title.
- Read all labels and captions.
- Figure out what the picture is showing. Notice colors, arrows, and lines.

Doing Labs

To understand science, you have to see it in action. Doing labs helps you understand how things really work.

① Read the entire lab first.

② Form a hypothesis.

③ Follow the procedure.

④ Record the data.

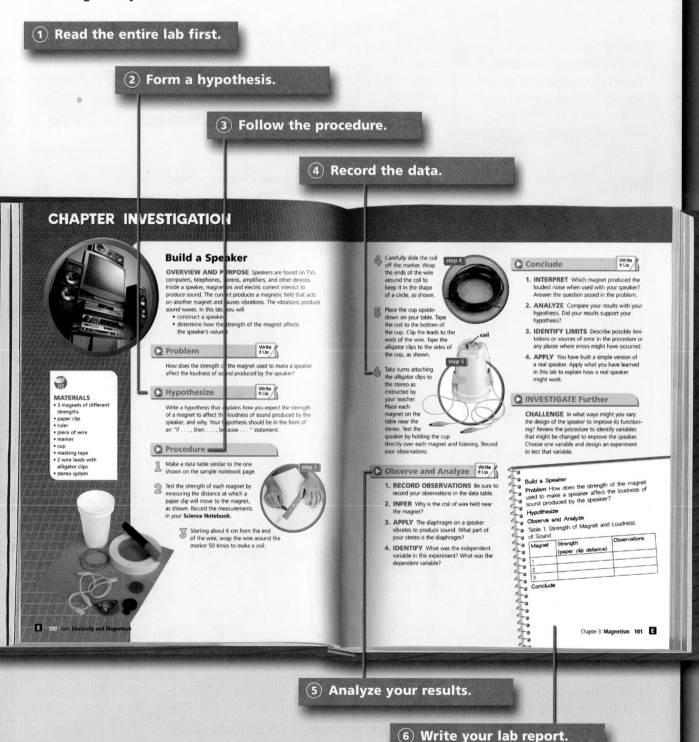

CHAPTER INVESTIGATION

Build a Speaker

OVERVIEW AND PURPOSE Speakers are found on TVs, computers, telephones, stereos, amplifiers, and other devices. Inside a speaker, magnetism and electric current interact to produce sound. The current produces a magnetic field that acts on another magnet and causes vibrations. The vibrations produce sound waves. In this lab, you will
- construct a speaker
- determine how the strength of the magnet affects the speaker's volume

▶ Problem _Write It Up_

How does the strength of the magnet used to make a speaker affect the loudness of sound produced by the speaker?

▶ Hypothesize _Write It Up_

Write a hypothesis that explains how you expect the strength of a magnet to affect the loudness of sound produced by the speaker, and why. Your hypothesis should be in the form of an "if . . . , then . . . , because . . . " statement.

▶ Procedure

MATERIALS
- 3 magnets of different strengths
- paper clip
- ruler
- piece of wire
- marker
- cup
- masking tape
- 2 wire leads with alligator clips
- stereo system

1. Make a data table similar to the one shown on the sample notebook page.

2. Test the strength of each magnet by measuring the distance at which a paper clip will move to the magnet, as shown. Record the measurements in your Science Notebook.

step 2

3. Starting about 6 cm from the end of the wire, wrap the wire around the marker 50 times to make a coil.

4. Carefully slide the coil off the marker. Wrap the ends of the wire around the coil to keep it in the shape of a circle, as shown.

step 4

5. Place the cup upside-down on your table. Tape the coil to the bottom of the cup. Clip the leads to the ends of the wire. Tape the alligator clips to the sides of the cup, as shown.

coil

step 5

6. Take turns attaching the alligator clips to the stereo as instructed by your teacher. Place each magnet on the table near the stereo. Test the speaker by holding the cup directly over each magnet and listening. Record your observations.

▶ Observe and Analyze _Write It Up_

1. **RECORD OBSERVATIONS** Be sure to record your observations in the data table.

2. **INFER** Why is the coil of wire held near the magnet?

3. **APPLY** The diaphragm on a speaker vibrates to produce sound. What part of your stereo is the diaphragm?

4. **IDENTIFY** What was the independent variable in this experiment? What was the dependent variable?

▶ Conclude _Write It Up_

1. **INTERPRET** Which magnet produced the loudest noise when used with your speaker? Answer the question posed in the problem.

2. **ANALYZE** Compare your results with your hypothesis. Did your results support your hypothesis?

3. **IDENTIFY LIMITS** Describe possible limitations or sources of error in the procedure or any places where errors might have occurred.

4. **APPLY** You have built a simple version of a real speaker. Apply what you have learned in this lab to explain how a real speaker might work.

▶ INVESTIGATE Further

CHALLENGE In what ways might you vary the design of the speaker to improve its functioning? Review the procedure to identify variables that might be changed to improve the speaker. Choose one variable and design an experiment to test that variable.

Build a Speaker
Problem How does the strength of the magnet used to make a speaker affect the loudness of sound produced by the speaker?
Hypothesize
Observe and Analyze
Table 1. Strength of Magnet and Loudness of Sound

Magnet	Strength (paper clip distance)	Observations
1		
2		
3		

Conclude

E 100 Unit: Electricity and Magnetism

Chapter 3: Magnetism 101 E

⑤ Analyze your results.

⑥ Write your lab report.

Using Technology

The Internet is a great source of up-to-date science. The ClassZone Website and SciLinks have exciting sites for you to explore. Video clips and simulations can make science come alive.

Look for red banners.

Go to **classzone.com** to see simulations, visualizations, and content review.

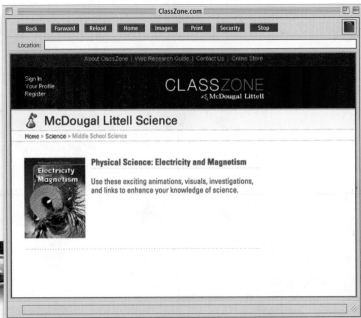

Watch the video.

See science at work in the **Scientific American Frontiers video.**

Look up SciLinks.

Go to **scilinks.org** to explore the topic.

Atmospheric Pressure and Winds **CODE: MDL010**

Electricity and Magnetism
Contents Overview

Unit Features

FRONTIERS in Science

SCIENTIFIC AMERICAN FRONTIERS

"Toy Symphony," a segment of the *Scientific American Frontiers* series that aired on PBS stations, documents new technologies being used in music composition and performance. At MIT's Media Lab, music professor Tod Machover has developed a "hyperviolin," which can record subtle aspects of violin playing such as bow tilt, speed, and pressure, and use these data to create new sounds. As violinist Joshua Bell plays Machover's *Toy Symphony*, a computer program picks up the sound and doubles the melody an octave lower. Another musical invention demonstrated in the video is the Beatbug, an electronic device that loops and replays simple rhythms made when the player strikes the outer shell of the instrument. In another part of the lab, students create music with the Hyperscore computer program, which uses shapes, colors, and textures instead of traditional musical notation.

National Science Education Standards

A.9.a–d Understandings About Scientific Inquiry

E.6.a–f Understandings About Science and Technology

F.5.a–e Science and Technology in Society

G.1.a–b Science as a Human Endeavor

G.2.a Nature of Science

FRONTIERS in Science

Electronics in Music

How are electronics changing the way we make and listen to music?

SCIENTIFIC AMERICAN FRONTIERS

View the video segment "Toy Symphony" to learn about some creative new ways in which music and electronics can be combined.

E 2 Unit: Electricity and Magnetism

ADDITIONAL RESOURCES

Technology Resources

 Scientific American Frontiers Video: *Toy Symphony:* 11-minute video segment that introduces the unit.

 ClassZone.com
CAREER LINK, recording engineer, audio engineer

Guide student viewing and comprehension of the video:

 Video Teaching Guide, pp. 1–2; Video Viewing Guide, p. 3; Video Wrap-Up, p. 4

Scientific American Frontiers Video Guide, pp. 55–58

Unit projects procedures and rubrics:

 Unit Projects, pp. 5–10

The quality of amplified sound waves can be controlled using electronics. Controls on this soundboard are adjusted in preparation for an outdoor concert.

Catching a Sound Wave

Everyone knows that music and electronics go together. If you want to hear music, you turn on a radio or TV, choose a CD or DVD to play, or listen to a computer file downloaded in MP3. All of these formats use electronics to record, play, and amplify music. Some of the most recent developments in music also use electronics to produce the music in the first place. For example, the orchestral music playing in the background of the last blockbuster movie you saw may not have been played by an orchestra at all. It may have been produced electronically on a computer.

Music consists of sound, and sound is a wave. Inside your TV or stereo equipment, electronic circuits represent sound waves as analog signals or digital signals. In analog recordings a peak in the original sound wave corresponds to a peak in the current. Radio and TV broadcasts are usually analog signals. The sound wave is converted to electromagnetic waves sent out through the air. Your radio or TV set receives these waves and converts them back to a sound wave.

A

In digital sound recordings the system samples the incoming sound wave at frequent intervals of time, such as 44,100 times per second. The system measures the height of each wave and assigns it a number. The numbers form a digital signal. This information can then be stored and transmitted. The playback electronics, such as CD players and DVD players, convert the signal back to a sound wave for you to hear.

B

DIFFERENTIATE INSTRUCTION

(?) More Reading Support

A How do electronic circuits represent sound waves? *as analog signals or digital signals*

B What electronic devices use digital signals? *CD and DVD players*

English Learners English learners may be unfamiliar with some of the terms used in reference to music and electronics. Have them look up the terms *download*, *amplify*, *interval*, and *convert*. Help them understand how the definitions they find relate to the context of music and electronics.

◉ Set Learning Goals
Students will

- Learn the difference between analog and digital recordings
- Compare the types of digital devices used to store sound
- Design and produce a product that involves electricity or electronics

Remind students that frontiers are "new" places to explore, even within existing technologies. Point out that in the video segment "Toy Symphony," the scientists at MIT have examined new ways for people to interact with music. Using their knowledge of electronics, they invented electronic instruments for creating music. Tell students that frontiers may include looking at new possibilities for existing technology.

INSTRUCT

Teach from Visuals

Have students look at the photographs of the soundboard and the live concert. Point out to students that the controls on the soundboard are what adjust the sound and the quality of the music being played. Ask students what other electric devices they see in the photographs that might affect the sound they hear at a concert. *Sample answers: amplifiers, microphones, electric guitars*

Technology

Point out that the sound we hear as music consists of waves and that these waves can be represented as analog signals or digital signals. Ask: What is the difference between analog recordings and digital recordings? *In an analog recording, the peak in the original sound wave matches the peak in the electric current. In a digital recording, the incoming sound wave is recorded at intervals of time and the height of each wave is assigned a number.*

Technology

Focus students' attention on "Making Music" on pages 4–5. Discuss how electronic technology provides ways to change or create music. Ask:

• What device did recording engineers use in the past to record sound? What can they use now? *large electronic consoles; Musical Instrument Digital Interface (MIDI)*

• How is MIDI technology different from CD, DVD, or MP3 players? *MIDI represents the instructions for another device to play the music, while digital devices such as CD players represent the actual sound waves.*

Design Technology

Ask: In developing the technology to record and store sound, what questions might engineers have asked? *Sample answer: How much computer space is needed to store a song in a digital format?*

Digital Devices

In a compact disc (CD), the numbers representing the sound wave are coded into a series of microscopic pits in a long spiral track burned into the plastic of the CD. A laser beam scans the track and reads the pits, converting the data back into numbers. This information is then converted into sound waves by an electronic circuit in the CD player. CDs can store up to 74 minutes of music because the pits are only a few millionths of a meter in size. Digital video discs (DVDs) often have several layers, each with a separate data

MP3 players store digital files that are compressed in size.

track, and use even smaller tracks and pits than CDs use. As a result, a DVD can store seven times as much information as a CD.

The amount of computer space needed to represent a song in normal digital format is too large to store very many songs on a single device. However, the development of a compression program called MP3 decreases the size of a typical song to one-tenth its original size. This enables you to buy and download a song from the Internet in minutes instead of hours and store files on your computer or MP3 player without taking up too much space.

Making Music

These advances in recording and playing music enables you to listen to music, whatever your taste in music happens to be. Electronic technology also allows you to change the music or even generate your own music, as shown in the video. Recording engineers used to work with large electronic consoles with hundreds of switches in order to blend different singers

View the "Toy Symphony" segment of your *Scientific American Frontiers* video to learn how electronic devices allow people to interact with music in new ways.

IN THIS SCENE FROM THE VIDEO ▶ Kids play with beat bugs at MIT's Media Lab.

PLAYING WITH SOUNDS At the Massachusetts Institute of Technology (MIT) Media Lab, Tod Machover and his colleagues have invented several new musical instruments that are based on electronics. One such invention is the hyperviolin,

demonstrated by concert violinist Joshua Bell. As Joshua plays the violin, a computer registers the movements of the bow and produces new and different sounds from the movements. Other musical electronic devices in the lab are designed to allow someone with little or no experience with an instrument to play and compose music.

With beat bugs—small interactive devices—kids can play music and collaborate with others. They can also compose and edit their own music. Using new computer software, a ten-year-old boy was able to compose an entire symphony played by the German Symphony Orchestra.

DIFFERENTIATE INSTRUCTION

 **More Reading Support**

C About how many minutes of music can a CD store? *74 minutes*

D What does the compression program MP3 do to a song file? *decreases its size*

Below Level Group students in pairs and ask each pair to make a chart comparing the characteristics of digital devices. Have students write down the name of each digital device, such as a CD, and then organize them in order of storage capacity, least to greatest. Then have students describe each device's characteristics on the chart.

Home recording studios are possible because of new electronic technology.

and background instruments or to add special effects such as echoes or distortion. Now this can all be done on a laptop computer, using the Musical Instrument Digital Interface (MIDI).

MIDI technology is different from that used in digital devices such as CD, DVD, and MP3 players. Whereas digital formats represent the sound waves themselves, MIDI formats represent the instructions for another device or instrument to play the music. With MIDI, you can connect an electronic keyboard directly to a computer and compose and edit your own music, layer in the sounds of different instruments, and dub in special effects. Once you understand how to use electronics to produce the sound waves you want, you can become your own favorite band.

? UNANSWERED Questions

Every year, scientists develop new technologies affecting the way we produce and listen to music. As advances in music technology are made, new question arise.

• Are there electronic sounds that no one has heard before?

• How will the development of music technology affect who is producing music?

• What type of devices will people be using to listen to music in 50 years?

UNIT PROJECTS

As you study this unit, work alone or with a group on one of these projects.

Multimedia Presentation

Put together an informative presentation that explains how electric guitars work.

• Gather information about electric guitars. Learn how they use both electricity and magnetism.

• Give a presentation that uses mixed media, such as a computer slide show, model, poster, or tape recording.

Build a Radio

Build a working radio from simple materials.

• Find instructions for building a simple crystal radio using books or the Internet.

• Collect the materials and assemble the radio. Modify the design of the radio to improve it.

• Demonstrate the radio to the class and explain how it works.

Design an Invention

Design an electronic invention.

• Select a purpose for your invention, such as a toy, a fan, or a burglar alarm. Write a paragraph that explains the purpose of your invention.

• Draw a sketch of your design and modify it if necessary.

• Make a pamphlet to advertise your invention. If possible, build the invention and include photographs of it the pamphlet.

 **CAREER CENTER**
CLASSZONE.COM

Learn more about careers in music and computer science.

? UNANSWERED Questions

Have students read the questions and think of some of their own. Remind them that scientists usually end up with more questions—that inquiry is the driving force of science.

• With the class, generate on the board a list of new questions.

• Students can add to the list after they watch the Scientific American Frontiers Video.

• Students can use the list as a springboard for choosing their Unit Projects.

UNIT PROJECTS

Encourage students to pick the project that most appeals to them. Point out that each is long-term and will take several weeks to complete. You might group or pair students to work on projects and in some cases guide student choice. Some of the projects have student choice built into them.

Each project has two worksheet pages, including a rubric. Use the pages to guide students through criteria, process, and schedule.

 Unit Projects, pp. 5–10

REVISIT concepts introduced in this article:

Chapter 1
• Materials can become electrically charged, pp. 9–16
• Charges can move, pp. 18–25
• Electric current, pp. 28–34

Chapter 2
• Charge needs a continuous path, pp. 43–49
• Circuits, pp. 51–55
• Electronic technology, pp. 57–65

Chapter 3
• Magnetism is a force, pp. 79–86
• Current can produce magnetism, pp. 88–94
• Magnetism can produce current, pp. 95–99
• Generators supply electrical energy, pp. 102–106

DIFFERENTIATE INSTRUCTION

? More Reading Support

E What does MIDI allow musicians and recording engineers to do to music? *Sample answers: compose and edit music; layer in different sounds from various instruments*

Differentiate Unit Projects Projects are appropriate for varying abilities. Allow students to choose the ones that interest them most. Encourage them to vary the products they produce throughout the year.

Below Level Encourage students to try "Design an Invention."

Advanced Challenge students to complete "Build a Radio."

Electricity

Physical Science
UNIFYING PRINCIPLES

PRINCIPLE 1

Matter is made of particles too small to see.

PRINCIPLE 2

Matter changes form and moves from place to place.

PRINCIPLE 3

Energy changes from one form to another, but it cannot be created or destroyed.

PRINCIPLE 4

Physical forces affect the movement of all matter on Earth and throughout the universe.

Unit: Electricity and Magnetism
BIG IDEAS

CHAPTER 1
Electricity

Moving electric charges transfer energy.

CHAPTER 2
Circuits and Electronics

Circuits control the flow of electric charge.

CHAPTER 3
Magnetism

Current can produce magnetism, and magnetism can produce current.

CHAPTER 1
KEY CONCEPTS

SECTION 1.1

Materials can become electrically charged.

1. Electric charge is a property of matter.

2. Static charges are caused by the movement of electrons.

3. Technology uses static electricity.

SECTION 1.2

Charges can move from one place to another.

1. Static charges have potential energy.

2. Materials affect charge movement.

SECTION 1.3

Electric current is a flow of charge.

1. Electric charge can flow continuously.

2. Electric cells supply electric current.

The Big Idea Flow Chart is available on p. T1 in the **UNIT TRANSPARENCY BOOK.**

Previewing Content

1.1 Materials can become electrically charged. pp. 9–17

1. Electric charge is a property of matter.

Objects can have a positive, negative, or neutral charge. A proton has a positive charge. An electron has a negative charge. Electric charge is measured in a unit called coulombs. Two objects with **electric charge** exert an electric force on each other. Electric force can be exerted even if the objects are separated by empty space. The amount of force is greater if the charges are greater and less if the distance between the objects is greater. Like charges, such as two positive charges, will repel each other, while unlike charges will attract each other. (Neutrally charged objects do not exert an electric force on each other.) The space around a charge where an electric force can be exerted if another charged particle is present is called an **electric field.**

2. Static charges are caused by the movement of electrons.

Objects that have equal numbers of electrons and protons have a neutral, or zero, charge. An object with more protons than electrons has a positive charge, and an object with more electrons than protons has a negative charge. The greater the imbalance, the greater the charge an object has. In most cases, it is the movement of electrons that causes an object to gain a **static charge.**

- Charging by contact occurs when electrons move because materials are touching. In this case, one material attracts electrons more strongly than the other.
- Charging by **induction** occurs when a charged object produces a temporary movement of electrons in another object. This movement creates a temporary charge imbalance within the object.

3. Technology uses static electricity.

Differences in charge attract toner to letters or images on paper in a photocopier. In electrostatic air filters, charged plates attract oppositely charged particles from the air. Static charges are used to make paint stick better to new cars.

1.2 Charges can move from one place to another. pp. 18–27

1. Static charges have potential energy.

A charged object held near another charged object has potential energy. Just as a rock falls to the ground when it is let go, one charge will move toward or away from another charge if it is not fixed in place. A proton placed near a charged object will have a certain amount of potential energy due to its position. The **electric potential** of a charged object is the potential energy per unit charge that another charge would have if it were in the object's electric field. Electric potential is measured in **volts.** The difference in the electric potential at two different points is called the voltage between those two points. If there is a path for the charges to follow, they will move from higher potential energy to a lower potential energy.

2. Materials affect charge movement.

Electrical **resistance** is a measure of how easily charge moves through a material. Resistance is measured in **ohms.** Resistance depends on the size, the shape, and the type of material.
- Charge moves easily through a **conductor.**
- Charge does not move easily through an **insulator.**
- Superconductors have almost no resistance at extremely low temperatures.

Grounding provides a charge with a safe path through a low-resistance material instead of one with a high resistance. In the picture below, the cable has less resistance than the building.

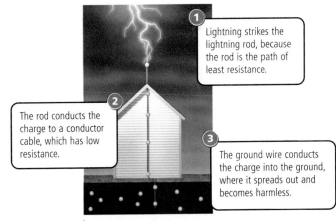

1. Lightning strikes the lightning rod, because the rod is the path of least resistance.

2. The rod conducts the charge to a conductor cable, which has low resistance.

3. The ground wire conducts the charge into the ground, where it spreads out and becomes harmless.

Common Misconceptions

CHARGING BY CONTACT Students may think that charging by contact can occur only when two objects are rubbed together. In fact, electrons can move from one object to another without rubbing, although rubbing does increase the contact, and therefore the amount of charge transferred.

TE This misconception is addressed on p. 11.

MISCONCEPTION DATABASE
CLASSZONE.COM Background on student misconceptions

MOVING STATIC CHARGES Students may think that static charges don't move. In reality, static charges move when a path is available and there is a difference in electric potential.

TE This misconception is addressed on p. 20.

Previewing Content

 1.3 ## Electric current is a flow of charge.
pp. 28–35

1. Electric charge can flow continuously.

If a constant potential difference, or voltage, is maintained between two points and there is a path along which electrons can move between those two points, **electric current** results.

- Electric current, the continuous flow of charge between two points at different potentials, is measured in **amperes.** One ampere is equal to a flow rate of one coulomb per second.
- You can measure current using an ammeter, voltage using a voltmeter, and resistance using an ohmmeter. A multimeter can measure all three values. **Ohm's law** shows the relationships among current (*I*), voltage (*V*), and resistance (*R*).

$$I = \frac{V}{R}$$

2. Electric cells supply electric current.

An **electric cell** maintains a constant voltage between its two terminals by using the physical and chemical properties of different materials.

- An electrochemical cell contains two electrodes suspended in an electrolyte, which undergoes chemical reactions with the electrodes. Batteries contain two or more electrochemical cells.
- Primary cells are either wet cells or dry cells in which chemical reactions continue until at least one reactant is used up. Most household batteries are primary dry cells, like the one shown below.

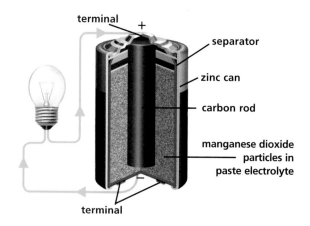

terminal
+
separator
zinc can
carbon rod
manganese dioxide particles in paste electrolyte
terminal

- Some batteries are storage cells in which the chemical reactions can be reversed. Such batteries are rechargeable.
- Solar cells contain materials that absorb energy from the Sun or other sources of light and then release electrons to create an electric current.

Common Misconceptions

SPEED OF ELECTRON MOVEMENT IN CIRCUITS Students may think that electrons move through a circuit instantly or at the speed of light. Electrons actually move very slowly through a circuit. The energy changes in the electric field, however, do move at the speed of light.

 This misconception is addressed in Teach Difficult Concepts on p. 29.

 MISCONCEPTION DATABASE
CLASSZONE.COM Background on student misconceptions

Previewing Labs

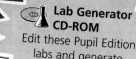

Lab Generator
CD-ROM
Edit these Pupil Edition labs and generate alternative labs.

EXPLORE the BIG idea

How Do the Pieces of Tape Interact? p. 7
Students produce static electricity in strips of tape and note its effects.

TIME 10 minutes
MATERIALS 3 strips of transparent tape

Why Does the Water React Differently? p. 7
Students observe the effect of a charged and discharged comb on a stream of water coming from a faucet.

TIME 10 minutes
MATERIALS water faucet, comb

Internet Activity: Static Electricity, p. 7
Students observe how different types of materials affect charging by contact.

TIME 20 minutes
MATERIALS computer with Internet access

 SECTION **1.1**

EXPLORE Static Electricity, p. 9
Students explore how paper and plastic interact electrically.

TIME 10 minutes
MATERIALS 2 strips of newspaper, plastic bag

INVESTIGATE Making a Static Detector, p. 14
Students construct a static detector and infer the presence of static electric charge.

TIME 20 minutes
MATERIALS metal paper clip, clear plastic cup with hole, small piece of modeling clay, ball of aluminum foil, 2 strips of aluminum foil, balloon

---

 SECTION **1.2**

EXPLORE Static Discharge, p. 18
Students use a fluorescent bulb to observe static discharge.

TIME 10 minutes
MATERIALS inflated balloon, wool cloth, fluorescent light bulb

INVESTIGATE Conductors and Insulators, p. 22
Students interpret data to determine what materials conduct electricity.

TIME 20 minutes
MATERIALS D cell battery, 3 pieces of low-voltage wire (20 cm each), 25 cm duct tape, flashlight bulb, bulb holder, objects made from different materials

CHAPTER INVESTIGATION
Lightning, pp. 26-27
Students model the buildup of charges that can occur during a storm and a lightning strike, and use a ground to control the path of discharge.

TIME 40 minutes
MATERIALS small piece of modeling clay, 2 aluminum pie pans, Styrofoam plate, wool cloth, metal paper clip

---

 SECTION **1.3**

EXPLORE Current, p. 28
Students relate resistance to flow of charge through different lengths of graphite.

TIME 20 minutes
MATERIALS pencil lead, posterboard, 25 cm electrical tape, 3 lengths of low-voltage wire (20 cm each), D cell battery, flashlight bulb, bulb holder

INVESTIGATE Electric Cells, p. 31
Students infer how electric current can be produced by a lemon and different types of metal.

TIME 20 minutes
MATERIALS metal paper clip, penny, large lemon, multimeter, additional fruits or vegetables, additional metal objects

---

R **Additional INVESTIGATION,** Making a Coin Battery, A, B, & C, pp. 62–70; Teacher Instructions, pp. 206–207

Previewing Chapter Resources

	INTEGRATED TECHNOLOGY	LABS AND ACTIVITIES

Chapter 1 Electricity

 CLASSZONE.COM
- eEdition Plus
- EasyPlanner
- Misconception Database
- Content Review
- Test Practice
- Simulations
- Resource Centers
- Internet Activity: Static Electricity
- Math Tutorial

 SCILINKS.ORG
 SCI LINKS

 CD-ROMS
- eEdition
- EasyPlanner
- Power Presentations
- Content Review
- Lab Generator
- Test Generator

AUDIO CDS
- Audio Readings
- Audio Readings in Spanish

P E EXPLORE the Big Idea, p. 7
- How Do the Pieces of Tape Interact?
- Why Does the Water React Differently?
- Internet Activity: Static Electricity

R **UNIT RESOURCE BOOK**
- Family Letter, p. vii
- Spanish Family Letter, p. viii
- Unit Projects, pp. 5–10

 Lab Generator CD-ROM
Generate customized labs.

SECTION
 1.1 Materials can become electrically charged.
pp. 9–17

Time: 2 periods (1 block)
 R Lesson Plan, pp. 11–12

 T **UNIT TRANSPARENCY BOOK**
- Big Idea Flow Chart, p. T1
- Daily Vocabulary Scaffolding, p. T2
- Note-Taking Model, p. T3
- 3-Minute Warm-Up, p. T4

P E
- EXPLORE Static Electricity, p. 9
- INVESTIGATE Making a Static Detector, p. 14
- Connecting Sciences, p. 17

R **UNIT RESOURCE BOOK**
Datasheet, Making a Static Detector, p. 20

SECTION
 1.2 Charges can move from one place to another.
pp. 18–27

Time: 3 periods (1.5 blocks)
 R Lesson Plan, pp. 22–23

 RESOURCE CENTER, Lightning and Lightning Safety

 T **UNIT TRANSPARENCY BOOK**
- Daily Vocabulary Scaffolding, p. T2
- 3-Minute Warm-Up, p. T4
- "How Lightning Forms" Visual, p. T6

P E
- EXPLORE Static Discharge, p. 18
- INVESTIGATE Conductors and Insulators, p. 22
- CHAPTER INVESTIGATION, Lightning, pp. 26–27

R **UNIT RESOURCE BOOK**
- Datasheet, Conductors and Insulators, p. 31
- CHAPTER INVESTIGATION, Lightning, A, B, & C, pp. 53–61

SECTION
 1.3 Electric current is a flow of charge.
pp. 28–35

Time: 3 periods (1.5 blocks)
 R Lesson Plan, pp. 33–34

 • **SIMULATION,** Ohm's Law
• **RESOURCE CENTER,** Electrochemical Cells
• **MATH TUTORIAL**

 T **UNIT TRANSPARENCY BOOK**
- Big Idea Flow Chart, p. T1
- Daily Vocabulary Scaffolding, p. T2
- 3-Minute Warm-Up, p. T5
- Chapter Outline, pp. T7–T8

P E
- EXPLORE Current, p. 28
- INVESTIGATE Electric Cells, p. 31
- Math in Science, p. 35

R **UNIT RESOURCE BOOK**
- Datasheet, Electric Cells, p. 42
- Math Support, pp. 49, 51
- Math Practice, pp. 50, 52
- Additional INVESTIGATION, Making a Coin Battery, A, B, & C, pp. 62–70

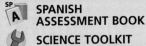

READING AND REINFORCEMENT

ASSESSMENT

STANDARDS

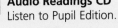
- Four Square, B22–23
- Combination Notes, C36
- Daily Vocabulary Scaffolding, H1–8

 UNIT RESOURCE BOOK
- Vocabulary Practice, pp. 46–47
- Decoding Support, p. 48
- Summarizing the Chapter, pp. 71–72

 Audio Readings CD
Listen to Pupil Edition.

 Audio Readings in Spanish CD
Listen to Pupil Edition in Spanish.

- Chapter Review, pp. 37–38
- Standardized Test Practice, p. 39

 UNIT ASSESSMENT BOOK
- Diagnostic Test, pp. 1–2
- Chapter Test, A, B, & C, pp. 6–17
- Alternative Assessment, pp. 18–19

 Spanish Chapter Test, pp. 301–304

 Test Generator CD-ROM
Generate customized tests.

 Lab Generator CD-ROM
Rubrics for Labs

National Standards
A.2–8, A.9.a–f, B.3.a, B.3.e, E.6.c–f, F.5.c

See p. 7 for the standards.

 UNIT RESOURCE BOOK
- Reading Study Guide, A & B, pp. 13–16
- Spanish Reading Study Guide, pp. 17–18
- Challenge and Extension, p. 19
- Reinforcing Key Concepts, p. 21

 Ongoing Assessment, pp. 10, 12, 15–16

 Section 1.1 Review, p. 16

 UNIT ASSESSMENT BOOK
Section 1.1 Quiz, p. 3

National Standards
A.2–8, A.9.a–f, B.3.a, B.3.e

 UNIT RESOURCE BOOK
- Reading Study Guide, A & B, pp. 24–27
- Spanish Reading Study Guide, pp. 28–29
- Challenge and Extension, p. 30
- Reinforcing Key Concepts, p. 32
- Challenge Reading, pp. 44–45

 Ongoing Assessment, pp. 18–25

 Section 1.2 Review, p. 25

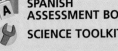 **UNIT ASSESSMENT BOOK**
Section 1.2 Quiz, p. 4

National Standards
A.2–8, A.9.a–f, B.3.a, B.3.e, E.6.c–f, F.5.c

 UNIT RESOURCE BOOK
- Reading Study Guide, A & B, pp. 35–38
- Spanish Reading Study Guide, pp. 39–40
- Challenge and Extension, p. 41
- Reinforcing Key Concepts, p. 43

 Ongoing Assessment, pp. 29–30, 32–34

 Section 1.3 Review, p. 34

UNIT ASSESSMENT BOOK
Section 1.3 Quiz, p. 5

National Standards
A.2–8, A.9.a–f, B.3.a, B.3.e, E.6.c–f, F.5.c

Previewing Resources for Differentiated Instruction

CHAPTER INVESTIGATION

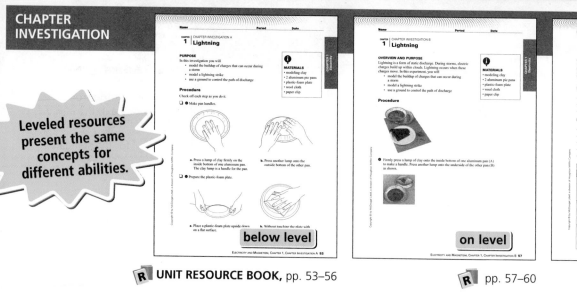

Leveled resources present the same concepts for different abilities.

below level

on level

advanced

R **UNIT RESOURCE BOOK,** pp. 53–56

R pp. 57–60

R pp. 57–61

READING STUDY GUIDE

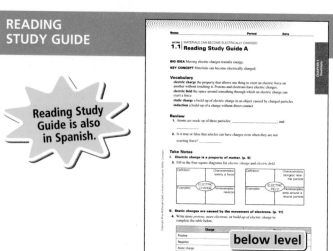
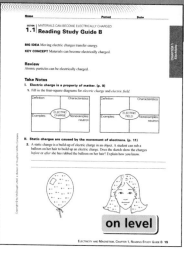
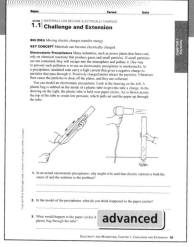

Reading Study Guide is also in Spanish.

below level

on level

advanced

R **UNIT RESOURCE BOOK,** pp. 13–14

R pp. 15–16

R p. 19

CHAPTER TEST

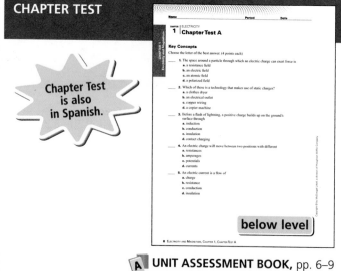

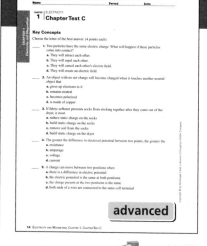

Chapter Test is also in Spanish.

below level

on level

advanced

A **UNIT ASSESSMENT BOOK,** pp. 6–9

A pp. 10–13

A pp. 14–17

TECHNOLOGY

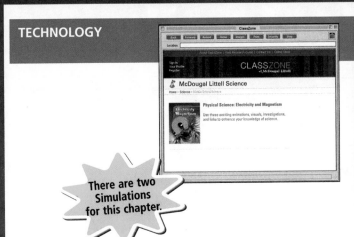

There are two Simulations for this chapter.

CLASSZONE.COM   CD/CD-ROMS   CLASSZONE.COM

VISUAL CONTENT

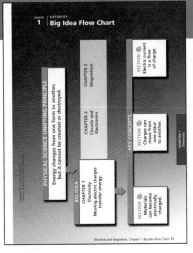

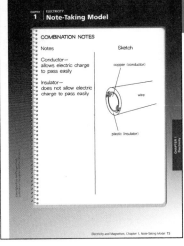

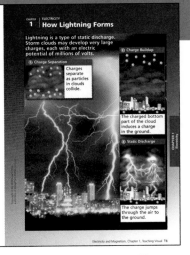

 **UNIT TRANSPARENCY BOOK,** p. T1

 p. T3

 p. T6

MORE SUPPORT

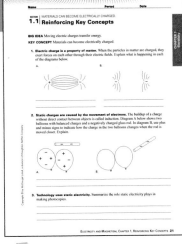

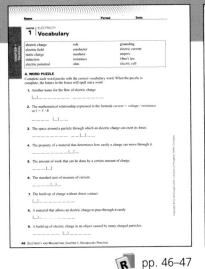

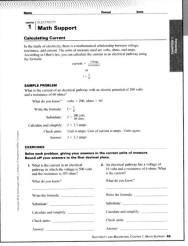

Reinforcing Key Concepts for each section

 **UNIT RESOURCE BOOK,** p. 21

pp. 46–47

p. 49

INTRODUCE

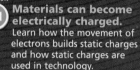

the **BIG** idea

Have students look at the photograph of the glowing dragon. Discuss how the question in the box links to the Big Idea:

- What might stop the dragon from glowing?

- What might make the dragon glow more brightly?

National Science Education Standards

Content

B.3.a Energy is a property of many substances and is associated with heat, light, electricity, mechanical motion, sound, nuclei, and the nature of a chemical. Energy is transferred in many ways.

B.3.e In most chemical and nuclear reactions, energy is transferred into or out of a system. Heat, light, mechanical motion, or electricity might all be involved in such transfers.

Process

A.2–8 Design and conduct an investigation; use tools to gather and interpret data; use evidence to describe, predict, explain, model; think critically to make relationships between evidence and explanation; recognize different explanations and predictions; communicate scientific procedures and explanations; use mathematics.

A.9.a–f Understand scientific inquiry by using different investigations, methods, mathematics, technology, explanations based on logic, evidence, and skepticism.

E.6.c–f Understandings about science and technology

F.5.c Technology influences society through its products and processes.

CHAPTER

1 Electricity

the **BIG** idea

Moving electric charges transfer energy.

Key Concepts

SECTION
1.1 Materials can become electrically charged.
Learn how the movement of electrons builds static charges and how static charges are used in technology.

SECTION
1.2 Charges can move from one place to another.
Learn what factors control the movement of charges.

SECTION
1.3 Electric current is a flow of charge.
Learn how electric current is measured and how it can be produced.

Internet Preview

CLASSZONE.COM

Chapter 1 online resources: Content Review, two Simulations, two Resource Centers, Math Tutorial, Test Practice.

E 6 Unit: Electricity and Magnetism

What keeps this dragon glowing brightly?

INTERNET PREVIEW

CLASSZONE.COM For student use with the following pages:

Review and Practice
- Content Review, pp. 8, 36
- Math Tutorial: Equations, p. 35
- Test Practice, p. 39

Activities and Resources
- Internet Activity: Static Electricity, p. 7
- Resource Centers: Lightning, p. 20; Electrochemical Cells, p. 32
- Simulation: Ohm's Law, p. 29

NSTA
scilinks.org

SCi LINKS

Electricity **Code: MDL065**

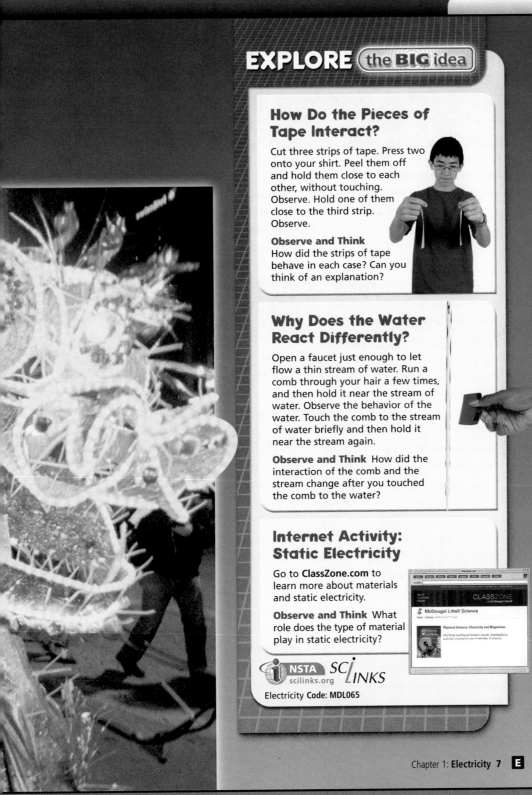

How Do the Pieces of Tape Interact?

Cut three strips of tape. Press two onto your shirt. Peel them off and hold them close to each other, without touching. Observe. Hold one of them close to the third strip. Observe.

Observe and Think
How did the strips of tape behave in each case? Can you think of an explanation?

Why Does the Water React Differently?

Open a faucet just enough to let flow a thin stream of water. Run a comb through your hair a few times, and then hold it near the stream of water. Observe the behavior of the water. Touch the comb to the stream of water briefly and then hold it near the stream again.

Observe and Think How did the interaction of the comb and the stream change after you touched the comb to the water?

Internet Activity: Static Electricity

Go to **ClassZone.com** to learn more about materials and static electricity.

Observe and Think What role does the type of material play in static electricity?

NSTA sclinks.org
SCiLINKS

Electricity Code: MDL065

TEACHING WITH TECHNOLOGY

CBL and Probeware Use probeware to measure current in "Investigate Electric Cells" on p. 31.

Multimeter Once students are familiar with multimeters, have them check voltage and resistance for the setup in "Explore Current" on p. 28.

EXPLORE (the BIG idea)

These inquiry-based activities are appropriate for use at home or as a supplement to classroom instruction.

How Do the Pieces of Tape Interact?

PURPOSE To introduce students to the production and effects of a static charge through contact. Students note repulsion and attraction in charged tape.

TIP *10 min.* Different types of tape might produce different results. Check effectiveness of the tape before class.

Answer: The strips from the shirt repel each other but attract the third piece. The tape becomes charged from contact with the shirt. Like charges repel; unlike charges attract.

REVISIT after p. 11.

Why Does the Water React Differently?

PURPOSE To introduce students to charge induction and grounding. Students observe the effects of a charged and discharged comb on a stream of water.

TIP *10 min.* If possible, perform this activity on a cool, dry day to maximize the amount of charge on the comb.

Answer: The comb attracts the stream of water before it touches it. After touching the water, the comb no longer attracts the water.

REVISIT after p. 13.

Internet Activity: Static Electricity

PURPOSE To have students relate types of materials to charging by contact.

TIP *20 min.* After students try a couple of examples, have them predict the results before trying additional materials.

Answer: It determines whether objects charge spontaneously on contact.

REVISIT after p. 12.

◐ CONCEPT REVIEW

Activate Prior Knowledge

- Cover the sides and top of a container with black paper.

- Place a thermometer in the container. Take a temperature reading.

- Place the container and thermometer in the sun or under a sun lamp. Take a temperature reading again in five minutes.

- Ask students to describe what forms of energy moved from one place to another. *Light from the Sun moved to Earth. Solar energy absorbed by the black paper warmed the air and the thermometer in the container.*

▶ TAKING NOTES

Combination Notes

To clarify relationships, students can number parts of their sketches and label corresponding notes with the same numbers.

Vocabulary Strategy

Emphasize that there is more than one acceptable entry for characteristics, examples, and nonexamples. Comparing diagrams will help students add information to their own diagrams.

Vocabulary and Note-Taking Resources

- Vocabulary Practice, pp. 46–47
- Decoding Support, p. 48

- Daily Vocabulary Scaffolding, p. T2
- Note-Taking Model, p. T3

- Four Square, B22–23
- Combination Notes, C36
- Daily Vocabulary Scaffolding, H1–8

◐ CONCEPT REVIEW

- Matter is made of particles too small to see.
- Energy and matter can move from one place to another.
- Electromagnetic energy is one form of energy.

◐ VOCABULARY REVIEW

See Glossary for definitions.

atom

electron

joule

proton

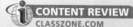

CONTENT REVIEW
CLASSZONE.COM
Review concepts and vocabulary.

▶ TAKING NOTES

COMBINATION NOTES

To take notes about a new concept, first make an informal outline of the information. Then make a sketch of the concept and label it so you can study it later.

VOCABULARY STRATEGY

Write each new vocabulary term in the center of a **four square** diagram. Write notes in the squares around each term. Include a definition, some characteristics, and some examples of the term. If possible, write some things that are not examples of the term.

See the Note-Taking Handbook on pages R45–R51.

SCIENCE NOTEBOOK

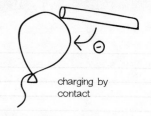

NOTES
How static charges are built
- Contact
- Induction
- Charge polarization

charging by contact

Definition	Characteristics
imbalance of charge in material	results from movement of electrons; affected by type of material
Examples	Nonexample
clinging laundry, doorknob shock, lightning	electricity from an electrical outlet

STATIC CHARGE

CHECK READINESS

Administer the Diagnostic Test to determine students' readiness for new science content and their mastery of requisite math skills.

 Diagnostic Test, pp. 1–2

Technology Resources

Students needing content and math skills should visit **ClassZone.com.**

- **CONTENT REVIEW**
- **MATH TUTORIAL**

- **CONTENT REVIEW CD-ROM**

1.1 Materials can become electrically charged.

◄ **BEFORE,** you learned

- Atoms are made up of particles called protons, neutrons, and electrons
- Protons and electrons are electrically charged

▶ **NOW,** you will learn

- How charged particles behave
- How electric charges build up in materials
- How static electricity is used in technology

VOCABULARY

electric charge p. 10
electric field p. 10
static charge p. 11
induction p. 13

EXPLORE Static Electricity

How can materials interact electrically?

PROCEDURE

① Hold the newspaper strips firmly together at one end and let the free ends hang down. Observe the strips.

② Put the plastic bag over your other hand, like a mitten. Slide the plastic down the entire length of the strips and then let go. Repeat several times.

③ Notice how the strips of paper are hanging. Describe what you observe.

WHAT DO YOU THINK?

- How did the strips behave before step 2? How did they behave after step 2?
- How might you explain your observations?

MATERIALS

- 2 strips of newspaper
- plastic bag

Electric charge is a property of matter.

You are already familiar with electricity, static electricity, and magnetism. You know electricity as the source of power for many appliances, including lights, tools, and computers. Static electricity is what makes clothes stick together when they come out of a dryer and gives you a shock when you touch a metal doorknob on a dry, winter day. Magnetism can hold an invitation or report card on the door of your refrigerator.

You may not know, however, that electricity, static electricity, and magnetism are all related. All three are the result of a single property of matter—electric charge.

COMBINATION NOTES
As you read this section, write down important ideas about electric charge and static charges. Make sketches to help you remember these concepts.

Chapter 1: Electricity **9** **E**

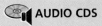
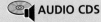

1.1 FOCUS

◉ Set Learning Goals

Students will

- Describe how charged particles behave.
- Determine how electrons cause electric charges to build up in materials.
- Explain how static electricity is used in technology.
- Infer from an experiment how to detect a static electric charge.

◉ 3-Minute Warm-Up

Display Transparency 4 or copy this exercise on the board:

Decide if these statements are true. If not, correct them.

1. The basic particles of an atom are protons, neutrons, and nuclei. *protons, neutrons, and electrons*

2. Particles that make up an atom have no charge. *Some particles (protons and electrons) that make up an atom are charged.*

3. Electrons are negatively charged. *true*

🖳 3-Minute Warm-Up, p. T4

1.1 MOTIVATE

EXPLORE Static Electricity

PURPOSE To explore how materials interact electrically

TIPS *10 min.* Tell students to rub the strips firmly but to be careful not to tear the strips. A flexible plastic bag works best.

WHAT DO YOU THINK? *They hung straight down while touching each other. They repelled each other. Rubbing with the plastic bag charges the strips with like charges, which repel each other.*

History of Science

The first recorded observation of static charge is from ancient Greece. The Greeks noticed that fossilized tree sap—the material known as amber—attracted objects such as feathers after it was rubbed with fur or certain other materials. Many words with the root *electr-*, such as *electron, electricity,* and *electronic,* come from the Greek word *elecktron,* which means "amber."

Teach from Visuals

Point out that the charges in the diagram of electric charge are equal in size but opposite in sign. Ask:

- Which has more mass, an electron or a proton? *proton*

- Is the charge on the proton equal to, larger than, or smaller than the charge on the electron? *equal to*

Ongoing Assessment

Describe how charged particles behave.

Ask: If a balloon has a negative charge and a rod has a positive charge, will the balloon and the rod repel or attract each other? *attract*

 READING VISUALS *Answer: Each particle's force lines bend toward the other particle.*

VOCABULARY
Make a four square diagram for the term *electric charge* and the other vocabulary terms in this section.

 A

 B

The smallest unit of a material that still has the characteristics of that material is an atom or a molecule. A molecule is two or more atoms bonded together. Most of an atom's mass is concentrated in the nucleus at the center of the atom. The nucleus contains particles called protons and neutrons. Much smaller particles called electrons move at high speeds outside the nucleus.

Protons and electrons have electric charges. **Electric charge** is a property that allows an object to exert a force on another object without touching it. Recall that a force is a push or a pull. The space around a particle through which an electric charge can exert this force is called an **electric field.** The strength of the field is greater near the particle and weaker farther away.

All protons have a positive charge (+), and all electrons have a negative charge (−). Normally, an atom has an equal number of protons and electrons, so their charges balance each other, and the overall charge on the atom is neutral.

Particles with the same type of charge—positive or negative—are said to have like charges, and particles with different charges have unlike charges. Particles with like charges repel each other, that is, they push each other away. Particles with unlike charges attract each other, or pull on each other.

Electric Charge

Charged particles exert forces on each other through their electric fields.

Charged Particles
Electric charge can be either negative or positive.

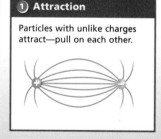

① **Attraction**
Particles with unlike charges attract—pull on each other.

② **Repulsion**
Particles with like charges repel—push each other away.

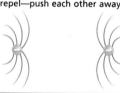
The balloon and the cat's fur have unlike charges, so they attract each other.

⊖ = electron

⊕ = proton

— = lines of force

 READING VISUALS How do the force lines change when particles attract?

DIFFERENTIATE INSTRUCTION

 More Reading Support

A What type of charge does a proton have? *positive*

B If two objects are both positive, will they attract or repel? *repel*

English Learners The similar words, *buildup* and *build up,* may confuse English learners. Explain that *buildup,* is a noun, while *build up,* is a verb. *Buildup* is an informal way of describing something that has gathered over time. *Build up* is a phrasal verb that refers to the action of accumulating something over time.

Static charges are caused by the movement of electrons.

You have read that protons and electrons have electric charges. Objects and materials can also have charges. A **static charge** is a buildup of electric charge in an object caused by the presence of many particles with the same charge. Ordinarily, the atoms that make up a material have a balance of protons and electrons. A material develops a static charge—or becomes charged—when it contains more of one type of charged particle than another.

If there are more protons than electrons in a material, the material has a positive charge. If there are more electrons than protons in a material, it has a negative charge. The amount of the charge depends on how many more electrons or protons there are. The total number of unbalanced positive or negative charges in an object is the net charge of the object. Net charge is measured in coulombs (KOO-LAHMZ). One coulomb is equivalent to more than 10^{19} electrons or protons.

Electrons can move easily from one atom to another. Protons cannot. For this reason, charges in materials usually result from the movement of electrons. The movement of electrons through a material is called conduction. If electrons move from one atom to another, the atom they move to develops a negative charge. The atom they move away from develops a positive charge. Atoms with either a positive or a negative charge are called ions.

A static charge can build up in an uncharged material when it touches or comes near a charged material. Static charges also build up when some types of uncharged materials come into contact with each other.

READING TiP

The word *static* comes from the Greek word *statos,* which means "standing."

REMINDER

10^{19} is the same as 1 followed by 19 zeros.

Charging by Contact

When two uncharged objects made of certain materials—such as rubber and glass—touch each other, electrons move from one material to the other. This process is called charging by contact. It can be demonstrated by a balloon and a glass rod, as shown below.

① At first, a balloon and a glass rod each have balanced, neutral charges.

② When they touch, electrons move from the rod to the balloon.

③ Afterwards, the balloon has a negative charge, and the rod has a positive charge.

Address Misconceptions

IDENTIFY Ask: How does an object acquire a static charge by contact? If students answer that electrons are rubbed off when two materials are rubbed together, they may hold the misconception that friction is necessary for static charge.

CORRECT Provide students with inflated balloons and have them hold the balloons gently to their hair. After two minutes, have students try to stick the balloons to the wall or other smooth surface. Some of the balloons will stick and some may not. Then have students rub the balloons on their hair and try again. Explain that rubbing is not always necessary for charging by contact, but that rubbing increases the charge.

REASSESS Ask: Is friction necessary for charging by contact to occur? *No, contact is all that is necessary.*

Technology Resources

Visit **ClassZone.com** for background on common student misconceptions.

MISCONCEPTION DATABASE

EXPLORE the BIG idea

Revisit "How Do the Pieces of Tape Interact?" on p. 7. Have students predict what would happen if different types of tape were used. Have them test their predictions.

DIFFERENTIATE INSTRUCTION

More Reading Support

C What is a buildup of electric charge in an object? *static charge*

D What is charging by contact? *charging through direct touching*

Inclusion To help students with visual impairments, enlarge the diagrams of attraction and repulsion on p. 10, and glue string on the lines of force. Have students use their sense of touch to examine the force of attraction between unlike charges and the force of repulsion between like charges.

Teach from Visuals

Have students examine the list of materials in the charging-by-contact chart. To help them better interpret the visual of the diagrams of wool and rubber contacts, ask:

- What type of charge would you have if you walked barefoot across a wool carpet? *positive*
- Across a rubber mat? *positive*

Develop Critical Thinking

SYNTHESIZE Tell students that a silk cloth acquires a negative charge when it rubs a neutral glass rod. If the same neutral glass rod rubs against a rubber rod, the rubber rod becomes negatively charged. If a rubber rod is rubbed with silk, the rubber rod becomes negatively charged.

- Ask: Which of the three materials—glass rod, rubber rod, or silk—has the greatest attraction for electrons? *rubber rod*
- Ask: Which has the least attraction? *glass*

Have students check their answers against the "Charging by Contact" chart on this page.

EXPLORE (the BIG idea)

Revisit "Internet Activity: Static Electricity" on p. 7. Have students explain their results.

Ongoing Assessment

Determine how electrons cause electric charges to build up in materials.

Ask: Electrons and protons are both charged. Why is charge buildup typically caused by the movement of electrons rather than protons? *Electrons can move from atom to atom but protons cannot.*

CHECK YOUR READING *Answer: The hair acquires a negative charge from the generator. Because the hairs contain like charges, they repel each other.*

As the sphere takes on a negative charge, electrons spread out over this student's skin and hair. Because her hairs all have the same charge, they repel one another.

metal globe

connection to globe

conveyor belt

source of electrons

A Van de Graaff generator is a device that builds up a strong static charge through contact. This device is shown at left. At the bottom of the device, a rubber conveyer belt rubs against a metal brush and picks up electrons. At the top, the belt rubs against metal connected to the sphere, transferring electrons to the sphere. As more and more electrons accumulate on the sphere, the sphere takes on a strong negative charge. In the photograph, the student touches the sphere as it is being charged. Some of the electrons spread across her arm to her head. The strands of her hair, which then all have a negative charge, repel one another.

CHECK YOUR READING How can a Van de Graaff generator make a person's hair stand on end?

How Materials Affect Static Charging

Charging by contact occurs when one material's electrons are attracted to another material more than they are attracted to their own. Scientists have determined from experience which materials are likely to give up or to accept electrons. For example, glass gives up electrons to wool. Wool accepts electrons from glass, but gives up electrons to rubber. The list at left indicates how some materials interact. Each material tends to give up electrons to anything below it on the list and to accept electrons from anything above it. The farther away two materials are from each other on the list, the stronger the interaction.

When you walk across a carpet, your body can become either positively or negatively charged. The type of charge depends on what materials the carpet and your shoes are made of. If you walk in shoes with rubber soles across a wool carpet, you will probably become negatively charged, because wool gives up electrons to rubber. But if you walk in wool slippers across a rubber mat, you will probably become positively charged.

E

Charging by Contact
skin
glass
hair
nylon
wool
fur
silk
paper
rubber
polyester

Materials higher on the list tend to give up electrons to materials lower on the list.

rubber

wool

Rubber soles on a wool carpet give a person a negative charge.

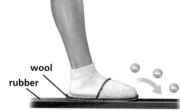

wool

rubber

Wool slippers on a rubber mat give a person a positive charge.

DIFFERENTIATE INSTRUCTION

 More Reading Support

E Fur becomes negatively charged when it touches skin. Which of these materials—fur or skin—has a stronger attraction for electrons? *fur*

Advanced Have interested students investigate the branch of science known as triboelectricity and report on it to the class. Encourage students to find lists known as triboelectric series on the Internet. These series rank different materials as to how easily they gain or lose electrons when they touch other materials. Have them plan and present a demonstration of the effects when different materials on the lists are placed in contact with each other.

R Challenge and Extension, p. 19

Charging by Induction

Charging can occur even when materials are not touching if one of the materials already has a charge. Remember that charged particles push and pull each other through their electric fields without touching. The pushing and pulling can cause a charge to build in another material. The first charge is said to induce the second charge. The buildup of a charge without direct contact is called **induction.**

Induction can produce a temporary static charge. Consider what happens when a glass rod with a negative charge is brought near a balloon, as shown below. The unbalanced electrons in the rod repel the electrons in the material of the balloon. Many electrons move to the side of the balloon that is farthest away from the rod. The side of the balloon that has more electrons becomes negatively charged. The side of the balloon with fewer electrons becomes positively charged. When the rod moves away, the electrons spread out evenly once again.

READING TiP

Induce and *induction* both contain the Latin root *ducere,* which means "to lead."

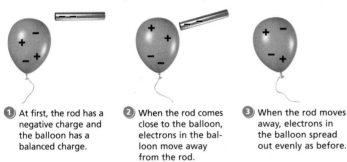

① At first, the rod has a negative charge and the balloon has a balanced charge.

② When the rod comes close to the balloon, electrons in the balloon move away from the rod.

③ When the rod moves away, electrons in the balloon spread out evenly as before.

If the electrons cannot return to their original distribution, however, induction can leave an object with a stable static charge. For example, if a negatively charged rod approaches two balloons that are touching each other, electrons will move to the balloon farther from the rod. If the balloons are then separated, preventing the electrons from moving again, the balloon with more electrons will have a negative charge and the one with fewer electrons will have a positive charge. When the rod is taken away, the balloons keep their new charges.

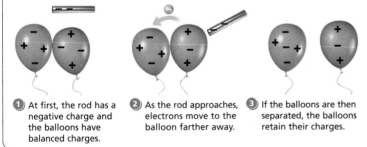

① At first, the rod has a negative charge and the balloons have balanced charges.

② As the rod approaches, electrons move to the balloon farther away.

③ If the balloons are then separated, the balloons retain their charges.

Integrate the Sciences

Television sets and computers can make allergies worse. The static charge on the screens of televisions and computer monitors attracts dust from the air. Accompanying the dust are dust mites and their feces, which are common allergens and can bring on asthma attacks and other bronchial problems.

Real World Example

When clothes rub together in a clothes dryer, they may become charged and stick together when they come out of the dryer. Fabric softeners solve this problem. Fabric softeners contain active ingredients made from long-chain molecules that have a positive charge. These positively charged molecules are attracted to negatively charged areas of fabric, with the result that smaller charge imbalances develop than there would be without the softener. Once the fabrics are out of the dryer, the softener attracts water molecules from the air, which further reduces the accumulation of charge.

Teach Difficult Concepts

Students may have difficulty distinguishing between the concepts represented by the sets of diagrams on the page in the textbook. Take the students step-by-step through the diagrams by using actual materials. You can give a rubber rod a negative charge by rubbing it with wool, nylon, or silk.

EXPLORE (the **BIG** idea)

Revisit "Why Does the Water React Differently?" on p. 7. Have students explain their results.

DIFFERENTIATE INSTRUCTION

(?) More Reading Support

F What is the buildup of charge without direct contact? *induction*

G Are induced charges temporary or stable? *They can be either.*

Inclusion Cut out paper models of balloons from construction paper. Use pennies to represent negative charges and buttons of the same size to represent positive charges. Model the diagrams on the page.

INVESTIGATE Making a Static Detector

PURPOSE To infer the presence of a static electric charge

TIPS *20 min.*

- Have students use a flow chart or sequence-diagram map to describe the procedure.
- Nothing at all should touch the ball of foil once the apparatus is assembled.

WHAT DO YOU THINK? *The strips moved apart. The balloon induced a charge on the ball. This charge passed through the paper clip into the strips. The strips then had a like charge, so they repelled each other.*

CHALLENGE *The strips would still repel because they would still have like charges.*

 Datasheet, Making a Static Detector, p. 20

Metacognitive Strategy

Ask: What generalizations about static charging do your lab results reinforce? Have students support their answers by citing the text. *Sample answers: Like charges repel each other—last paragraph, p. 10. Hair tends to give up electrons to rubber (the balloon)— "Charging by Contact" table, p. 12. Charged particles push and pull each other through their electric fields without touching, and the pushing and pulling can cause a charge to build in another material (the aluminum foil)— first paragraph, p. 13.*

Charge Polarization

Induction can build a charge by changing the position of electrons, even when electrons do not move between atoms. Have you ever charged a balloon by rubbing it on your head, and then stuck the balloon to a wall? When you bring the balloon close to the wall, the balloon's negative charge pushes against the electrons in the wall. If the electrons cannot easily move away from their atoms, the negative charges within the atoms may shift to the side away from the balloon. When this happens, the atoms are said to be polarized. The surface of the wall becomes positively charged, and the negatively charged balloon sticks to it.

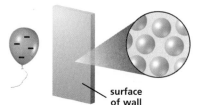 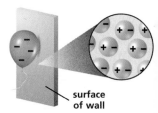

surface of wall   surface of wall

① Before the charged balloon comes near the wall, the atoms in the surface of the wall are not polarized.

② As the balloon nears the wall, atoms in the surface of the wall become polarized and attract the balloon.

INVESTIGATE Making a Static Detector

How can you detect a static electric charge?

PROCEDURE

① Straighten one end of the paper clip and insert it through the hole in the cup. Use clay to hold the paper clip in place. Stick the ball of foil onto the straight end. Hang both foil strips from the hook end.

② Give the balloon a static charge by rubbing it over your hair. Slowly bring the balloon near the ball of foil without letting them touch. Observe what happens to the foil strips inside the cup.

WHAT DO YOU THINK?

- What happened to the strips hanging inside the cup when the charged balloon came near the ball of foil?
- How can you explain what you observed?

CHALLENGE Suppose the balloon had the opposite charge of the one you gave it. What would happen to the strips if you brought the balloon near the ball of foil? Explain your answer.

SKILL FOCUS
Inferring

MATERIALS
- metal paper clip
- clear plastic cup with hole
- modeling clay
- ball of foil
- 2 strips of foil
- inflated balloon

TIME
20 minutes

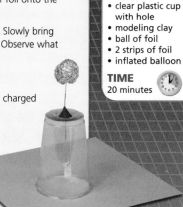

DIFFERENTIATE INSTRUCTION

? More Reading Support

H What happens when an atom is polarized? *Electrons shift to the side away from a negatively charged object*

Alternative Assessment Have students design an experiment that uses different materials in place of the foil ball (e.g., cork, clay, paper clip, copper wire) and report on which materials yield the same results. Have them organize their results in a chart or a table.

Technology uses static electricity.

Static charges can be useful in technology. An example is the photocopy machine. Photocopiers run on electricity that comes to them through wires from the power plant. But static charges play an important role in how they work.

How a Photocopier Works

A photocopier uses static charges to make copies.

Input An original document goes into the copier. A bright light shines on the page.

Inside the Copier The letters or images are transferred from the original to the copy, as shown in the box at right.

Output Heat fixes the toner to the paper, creating a permanent copy of the original.

mirror

original

lamp

toner cartridge

drum 1

drum 2

heating element

paper

Inside the Copier

light

1. A mirror reflects light from white areas of the original onto drum 1, which is positively charged. These lighted areas of the drum become negatively charged.

toner

2. Negatively charged toner (powdered ink) is attracted to the positive areas of drum 1 in the pattern of the original.

positively charged paper

3. Drum 1 rolls against a fresh, positively charged piece of paper on drum 2. The toner on drum 1 sticks to the paper.

READING VISUALS Why does the copy have the same pattern of light and dark areas as the original?

Teach from Visuals

To help students to better interpret the visual of how a photocopier works, ask:
- What is the purpose of the bright light in a copier? *The bright light provides the light that is reflected or absorbed by the original.*
- Why does a copier contain mirrors? *Mirrors reflect the light from the white areas of the original onto a metal drum.*

Ongoing Assessment

Explain how static electricity is used in technology.

Ask: How can static electricity that occurs naturally interfere with photocopying? *Static electricity might cause toner to be attracted to areas of the drum that are not in the pattern of what is being copied.*

READING VISUALS *Answer: The pattern on the original is reproduced on the drum with a positive static charge. This positive charge attracts the negatively charged toner.*

DIFFERENTIATE INSTRUCTION

 **More Reading Support**

I How does a photocopier tell the difference between the light and dark areas of the original? *by the difference in the type of charge*

Advanced Have students design a model of the way that a photocopier roll picks up toner that it then transfers to a piece of paper. They could use pepper as toner, cardboard tubes as the rollers, and other materials of their choosing. Have students present their model as a demonstration to the class.

Static electricity is also used in making cars. When new cars are painted, the paint is given an electric charge and then sprayed onto the car in a fine mist. The tiny droplets of paint stick to the car more firmly than they would without the charge. This process results in a coat of paint that is very even and smooth.

Another example of the use of static electricity in technology is a device called an electrostatic air filter. This device cleans air inside buildings with the help of static charges. The filter gives a static charge to pollen, dust, germs, and other particles in the air. Then an oppositely charged plate inside the filter attracts these particles, pulling them out of the air. Larger versions of electrostatic filters are used to remove pollutants from industrial smokestacks.

 CHECK YOUR READING How can static charges help clean air?

1.1 Review

KEY CONCEPTS

1. How do a positive and a negative particle interact?
2. Describe how the movement of electrons between two objects with balanced charges could cause the buildup of electric charge in both objects.
3. Describe one technological use of static electricity.

CRITICAL THINKING

4. **Infer** A sock and a shirt from the dryer stick together. What does this tell you about the charges on the sock and shirt?
5. **Analyze** You walk over a rug and get a shock from a doorknob. What do the materials of the rug and the shoes have to do with the type of charge your body had?

CHALLENGE

6. **Apply** Assume you start with a negatively charged rod and two balloons. Describe a series of steps you could take to create a positively charged balloon, pick up negatively charged powder with the balloon, and drop the powder from the balloon.

Electric Eels

An electric eel is a slow-moving fish with no teeth and poor eyesight. It lives in the murky waters of muddy rivers in South America. Instead of the senses that most animals use—vision, hearing, smell, and touch—an electric eel uses electricity to find its next meal. Since the fish that it eats often can swim much faster than the eel, it also uses electricity to catch its prey.

Electric Sense

An electric eel actually has three pairs of electric organs in its body. Two of them build electric charge for stunning prey and for self-defense. The third electric organ builds a smaller charge that helps in finding prey. The charge produces an electric field around the eel. Special sense organs on its body detect small changes in the electric field caused by nearby fish and other animals.

Shocking Organs

The electric eel builds an electric charge with a series of thousands of cells called electrocytes. Every cell in the series has a positive end and a negative end. Each electrocyte builds only a small charge. However, when all of the cells combine their charge, they can produce about five times as much electricity as a standard electrical outlet in a house. The charge is strong enough to paralyze or kill a human. Typically, though, the charge is used to stun or kill small fish, which the eel then swallows whole. Electric charge can also be used to scare away predators.

EXPLORE

1. **INFER** Electric eels live for 10 to 20 years, developing a stronger shock as they grow older. What could account for this increase in electric charge?

2. **CHALLENGE** Sharks and other animals use electricity also. Use the library or Internet to find out how.

An electric eel (Electrophorus electricus) can deliver a jolt five times as powerful as an electrical outlet.

Chapter 1: **Electricity 17** **E**

Set Learning Goal

To show how electricity is produced and used by some living organisms

Present the Science

The organ in an electric eel that produces low-voltage pulses is the Sach's organ and is near the back of the eel. The Main and Hunters' organs are in back of the head. They can produce over 500 volts. The eel's head is the positive terminal, and its tail is the negative terminal. This structure allows for voltage flow. Each electrocyte generates 0.15 volt. Thus, the more electrocytes the eel has, the greater the voltage produced.

Discussion Question

Ask: An electric eel has no teeth and poor eyesight and moves slowly. Why is it important that the eel produce both weak and strong voltage? *Because the eel has poor eyesight, it needs weak voltage to find food. Because it cannot move quickly or grab other fish without teeth, it must have another way to stun or kill its prey.*

DIFFERENTIATION TIP Have students sketch a cutaway drawing of an eel that shows its voltage-producing organs and uses arrows showing that electricity flows from the tail to the head.

Close

Ask: Why is it important for scientists to study fish that produce and use electricity? *They can learn how to use weak electric fields for locating objects and communicating.*

EXPLORE

1. **INFER** *Bigger eels have more electrocytes and therefore can produce a larger charge.*

2. **CHALLENGE** *Many of these animals produce weak electric fields and use these fields to find food and to communicate.*

1.2 FOCUS

⏵ Set Learning Goals

Students will

- Describe how charges move.
- Explain how charges store energy.
- Observe how differences in materials affect the movement of charges.
- Determine which materials conduct electricity in an experiment.

◀ 3-Minute Warm-Up

Display Transparency 4 or copy this exercise on the board:

Suppose you've blown up about 50 balloons for party decorations and want to put them on the walls. You've run out of tape. Explain how you could place the balloons without tape, and why your idea will work. *Rub a balloon on your hair or shirt, then put it on a wall. The balloon will have negative charge from the hair, will induce a positive charge in the wall, and will be attracted to the positive charge. This attraction will be strong enough to hold the balloon there against gravity for a short time.*

 3-Minute Warm-Up, p. T4

1.2 MOTIVATE

EXPLORE Static Discharge

PURPOSE To observe electrical energy

TIP *10 min.* Lighting the bulb can be best observed in a darkened room or against a dark background.

WHAT DO YOU THINK? *The bulb lit up briefly. A static charge moved from the balloon to the bulb and lit the bulb.*

Ongoing Assessment

 Answer: the force of attraction or repulsion between charged particles

---

1.2 Charges can move from one place to another.

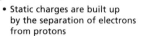

◀ **BEFORE, you learned**

- Static charges are built up by the separation of electrons from protons
- Materials affect how static charges are built
- Energy is the ability to cause change

▶ **NOW, you will learn**

- How charges move
- How charges store energy
- How differences in materials affect the movement of charges

VOCABULARY

electric potential p. 19
volt p. 19
conductor p. 22
insulator p. 22
resistance p. 23
ohm p. 23
grounding p. 25

EXPLORE Static Discharge

How can you observe electrical energy?

PROCEDURE

① Rub the balloon against the wool cloth several times to give the balloon a static charge.

② Slowly bring the balloon toward the middle part of the fluorescent bulb until a spark jumps between them.

WHAT DO YOU THINK?

- What happened in the fluorescent bulb when the spark jumped?
- How might you explain this observation?

MATERIALS

- inflated balloon
- wool cloth
- fluorescent light bulb

Static charges have potential energy.

You have read how a static charge is built up in an object such as a balloon. Once it is built up, the charge can stay where it is indefinitely. However, the charge can also move to a new location. The movement of a static charge out of an object is known as static discharge. When a charge moves, it transfers energy that can be used to do work.

What causes a charge to move is the same thing that builds up a charge in the first place—that is, the force of attraction or repulsion between charged particles. For example, suppose an object with a negative charge touches an object with a positive charge. The attraction of the unbalanced electrons in the first object to the unbalanced protons in the second object can cause the electrons to move to the second object.

⏺ **REMINDER**
Energy can be either kinetic (energy of motion) or potential (stored energy). Energy is measured in joules.

 What can cause a charge to move?

RESOURCES FOR DIFFERENTIATED INSTRUCTION

Below Level

UNIT RESOURCE BOOK
- Reading Study Guide A, pp. 24–25
- Decoding Support, p. 48

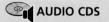 **AUDIO CDS**

Advanced

UNIT RESOURCE BOOK
- Challenge and Extension, p. 30
- Challenge Reading, pp. 44–45

English Learners

UNIT RESOURCE BOOK
Spanish Reading Study Guide, pp. 28–29

AUDIO CDS

- Audio Readings in Spanish
- Audio Readings (English)

Electric Potential Energy

Potential energy is stored energy an object may have because of its position. Water in a tower has gravitational potential energy because it is high above the ground. The kinetic energy—energy of motion—used to lift the water to the top of the tower is stored as potential energy. If you open a pipe below the tower, the water moves downward and its potential energy is converted back into kinetic energy.

Similarly, electric potential energy is the energy a charged particle has due to its position in an electric field. Because like charges repel, for example, it takes energy to push a charged particle closer to another particle with a like charge. That energy is stored as the electric potential energy of the first particle. When the particle is free to move again, it quickly moves away, and its electric potential energy is converted back into kinetic energy.

When water moves downward out of a tower and some of its potential energy is converted into kinetic energy, its potential energy decreases. Similarly, when a charged particle moves away from a particle with a like charge, its electric potential energy decreases. The water and the particle both move from a state of higher potential energy to one of lower potential energy.

Electric Potential

To push a charged particle closer to another particle with the same charge takes a certain amount of energy. To push two particles into the same position near that particle takes twice as much energy, and the two particles together have twice as much electric potential energy as the single particle. Although the amount of potential energy is higher, the amount of energy per unit charge at that position stays the same. **Electric potential** is the amount of electric potential energy per unit charge at a certain position in an electric field.

Electric potential is measured in units called volts, and voltage is another term for electric potential. A potential of one **volt** is equal to one joule of energy per coulomb of charge.

Just as water will not flow between two towers of the same height, a charge will not move between two positions with the same electric potential. For a charge to move, there must be a difference in potential between the two positions.

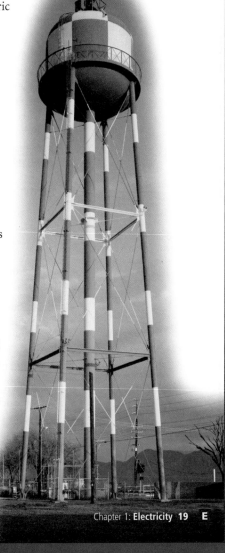

Like water in a tower, a static charge has potential energy. Just as gravity moves water down the supply pipe attached under the tank, the electric potential energy of a charge moves the charge along an electrical pathway.

IDENTIFY Ask: Do static charges move from one object to another, or stay in one place? If students answer that static charges stay in one place, they may hold the misconception that static charges cannot move.

CORRECT Have groups of students list examples of moving static charges. Compile a class list of the examples. Examples might include the static cling they sometimes feel from clothes.

REASSESS Ask: What has happened when you feel a shock on a doorknob after walking across a carpet? *A static charge has passed from your hand to the doorknob.*

Technology Resources

Visit **ClassZone.com** for background on common student misconceptions.

 MISCONCEPTION DATABASE

Integrate the Sciences

Sprites are phenomena that are associated with lightning storms and have been the subject of great interest in the scientific community over the past few years. Sprites are electrical discharges that occur between the clouds of thunderstorms and the lower ionosphere. Sprites differ from lightning in that they are dimmer and they seem to generate upward rather than downward. These high-altitude phenomena typically last just a few milliseconds.

Ongoing Assessment

Describe how charges move.

Ask: How does electric potential determine how charges move? *For a charge to move, there must be a difference in potential.*

CHECK YOUR READING *Answer: The charge must have a path to follow, and there must be a large enough difference in electric potential to move the charge through the path.*

Charge Movement

When water moves from a higher to a lower position, some of its potential energy is used to move it. Along the way, some of its potential energy can be used to do other work, such as turning a water wheel. Similarly, when a charge moves, some of its electric potential energy is used in moving the charge and some of it can be used to do other work. For example, moving an electric charge through a material can cause the material to heat up, as in a burner on an electric stove.

You can see how a moving charge transfers energy when you get a shock from static electricity. As you walk across a rug, a charge builds up on your body. Once the charge is built up, it cannot move until you come in contact with something else. When you reach out to touch a doorknob, the charge has a path to follow. The electric potential energy of the charge moves the charge from you to the doorknob.

Why do you get a shock? Recall that the force of attraction or repulsion between charged particles is stronger when they are close together. As your hand gets closer to the doorknob, the electric potential of the static charge increases. At a certain point, the difference in electric potential between you and the doorknob is great enough to move the charge through the air to the doorknob. As the charge moves, some of its potential energy is changed into the heat, light, and sound of a spark.

CHECK YOUR READING What two factors determine whether a static charge will move?

 RESOURCE CENTER
CLASSZONE.COM
Find out more about lightning and lightning safety.

Lightning

The shock you get from a doorknob is a small-scale version of lightning. Lightning is a high-energy static discharge. This static electricity is caused by storm clouds. Lightning comes from the electric potential of millions of volts, which releases large amounts of energy in the form of light, heat, and sound. As you read about how lightning forms, follow the steps in the illustration on page 21.

1 **Charge Separation** Particles of moisture inside a cloud collide with the air and with each other, causing the particles to become electrically charged. Wind and gravity separate charges, carrying the heavier, negatively charged particles to the bottom of the cloud and the lighter, positively charged particles to the top of the cloud.

2 **Charge Buildup** Through induction, the negatively charged particles at the bottom of the cloud repel electrons in the ground, causing the surface of the ground to build up a positive charge.

3 **Static Discharge** When the electric potential, or voltage, created by the difference in charges is large enough, the negative charge moves from the cloud to the ground. The energy released by the discharge produces the flash of lightning and the sound of thunder.

E 20 Unit: Electricity and Magnetism

DIFFERENTIATE INSTRUCTION

 **More Reading Support**

C Does potential energy increase or decrease when a charge moves? *It decreases.*

D What is lightning? *high-energy static discharge*

Advanced Have students investigate why a television screen acquires a static charge when the television is turned on. *The picture on the screen is created by a beam of electrons. Excess electrons cause a negative static charge on the screen.*

 Challenge and Extension, p. 30

How Lightning Forms

Lightning is a type of static discharge. Storm clouds may develop very large charges, each with an electric potential of millions of volts.

1 Charge Separation

Collisions between particles in storm clouds separate charges. Negatively charged particles collect at the bottom of the cloud.

E
F

2 Charge Buildup

The negatively charged bottom part of the cloud induces a positive charge in the surface of the ground.

3 Static Discharge

The charge jumps through the air to the ground. Energy released by the discharge causes thunder and lightning.

READING VISUALS How is lightning like the shock you can get from a doorknob? How is it different?

Chapter 1: **Electricity 21** **E**

To help students interpret the visual of how lightning is formed, have students think about what happens in the visual illustrating static discharge. Ask:

- Do positively charged particles or negatively charged particles move when lightning strikes? *negative particles*
- What happens to some of the potential energy in the cloud? *It becomes light energy and sound.*

T This visual is also available as T6 in the Unit Transparency Book.

Integrate the Sciences

Nitrogen is essential to all organisms because it is part of proteins, DNA, and RNA. However, almost all nitrogen on Earth is in the form of atmospheric nitrogen, which is not usable by most organisms. Lightning helps to enrich soil by providing the energy needed to cause a chemical reaction between nitrogen and other chemicals in the air, to make compounds that organisms can use.

Develop Critical Thinking

EVALUATE The following safety precautions should be taken during a thunderstorm. Have students explain why each guideline should be observed, using what they know about electricity.

- Do not take shelter under an isolated tree. *The tree is positively charged and will attract lightning because of its height.*
- Move away from bodies of water. *Water conducts electricity.*
- Unplug appliances and use phones only in an emergency. *Electrical and phone wires conduct electricity.*

Ongoing Assessment

READING VISUALS *Answer: Both are discharges of static electricity. Lightning has much more charge and a much higher voltage.*

DIFFERENTIATE INSTRUCTION

? More Reading Support

E What causes particles in a cloud to become charged? *collisions between particles*

F What type of charge is at the bottom of clouds? *negative*

Advanced Ask students to explain why you can tell how far away a lightning strike is by comparing the time when you see the lightning and the time when you hear the thunder. *The speed of light is much greater than the speed of sound, so the farther away the lightning strike is, the more time between the lightning and the thunder.*

Have students who are interested in lightning read the following article:

 Challenge Reading, pp. 44–45

INVESTIGATE Conductors and Insulators

PURPOSE To determine what materials conduct electricity

TIPS *20 min.*

- You could assemble the battery, bulb, and wires ahead of time.
- Electrical tape can be used instead of duct tape.

WHAT DO YOU THINK? *Objects that allowed the bulb to light up are made from materials that are conductors. Those that did not allow the bulb to light up are insulators.*

CHALLENGE *The brightness of the bulb can be used to indicate how well different materials conduct a charge.*

 Datasheet, Conductors and Insulators, p. 31

Technology Resources

Customize this student lab as needed or look for an alternative. Print rubrics to assess student lab reports.

 Lab Generator CD-ROM

Metacognitive Strategy

Ask students to list questions that arose during the investigation. Have them write answers to questions that were answered during the investigation.

Ongoing Assessment

CHECK YOUR READING *Answer: A conductor allows an electric charge to pass through it easily, but an insulator does not.*

COMBINATION NOTES Make notes on the different ways materials can affect charge movement. Use sketches to help explain the concepts.

Materials affect charge movement.

After you walk across a carpet, a charge on your skin has no place to go until you touch or come very close to something. That is because an electric charge cannot move easily through air. However, a charge can move easily through the metal of a doorknob.

Conductors and Insulators

 G

A material that allows an electric charge to pass through it easily is called a **conductor.** Metals such as iron, steel, copper, and aluminum are good conductors. Most wire used to carry a charge is made of copper, which conducts very well.

 H

A material that does not easily allow a charge to pass through it is called an **insulator.** Plastic and rubber are good insulators. Many types of electric wire are covered with plastic, which insulates well. The plastic allows a charge to be conducted from one end of the wire to the other, but not through the sides of the wire. Insulators are also important in electrical safety, because they keep charges away from the body.

 CHECK YOUR READING What is the difference between a conductor and an insulator?

INVESTIGATE Conductors and Insulators

What materials conduct electricity?

PROCEDURE

1. Use tape to connect the battery, wires, and bulb holder as shown in the photograph. Make sure that the wires connected to the battery stay in full contact with the metal parts on either end. Test the bulb and the battery by touching the free ends of wire together. The bulb should light up.

2. Test each object in turn by touching it simultaneously with both free ends of wire. Make sure the ends of wire do not touch each other.

WHAT DO YOU THINK?

- Which objects allowed the light bulb to light up when the wires touched them? Which did not?
- How can you explain the difference between the two groups of objects?

CHALLENGE Do any of the materials you tested seem to conduct a charge better than other conductors? How could you use the setup you have to compare the degree of conducting ability of materials?

SKILL FOCUS Interpreting data

MATERIALS
- D cell (battery)
- 3 pieces of low-voltage wire
- duct tape
- flashlight bulb
- bulb holder
- objects of different materials

TIME 20 minutes

DIFFERENTIATE INSTRUCTION

 More Reading Support

G Is a copper penny an insulator or a conductor? *a conductor*

H Is plastic a conductor or an insulator? *an insulator*

Alternative Assessment Have students create a two-column chart that lists insulators and conductors. As they read this section and test the different materials in "Investigate Conductors and Insulators," have them fill in their charts.

Electrons can move freely in a material with low resistance, such as the copper wire in these power lines. Electrons cannot move freely in a material with high resistance, such as the ceramic insulator this worker is putting in place or his safety gloves.

History of Science
One of the pioneers in the study of electricity was Benjamin Franklin. His most famous experiment with charges occurred in the early 1750s, when he attached a key to a kite string to show the relationship between lightning and electricity. After his kite experiment, Franklin developed a lightning rod connected to bells that would ring when lightning was in the vicinity. The lightning rod quickly became a standard way to protect buildings from lightning strikes.

Teacher Demo
Show students the difference between a conductor and an insulator. Charge an electroscope with a rubber rod that has been rubbed with fur. Touch the electroscope with an insulator, such as a glass rod or a wooden dowel rod. The electroscope will not discharge. Touch the electroscope with a metal rod. The electroscope will discharge because the metal rod is a conductor.

Ongoing Assessment
Observe how differences in materials affect the movement of charges.

Ask: Would a penny or a dollar bill have greater resistance? *a dollar bill, because it is paper, which is made from wood*

 Answer: the amount, shape, and type of material

Resistance

Think about the difference between walking through air and walking through waist-deep water. The water resists your movement more than the air, so you have to work harder to walk. If you walked waist-deep in mud, you would have to work even harder.

Materials resist the movement of a charge in different amounts. Electrical **resistance** is the property of a material that determines how easily a charge can move through it. Electrical resistance is measured in units called **ohms.** The symbol for ohms is the Greek letter *omega* (Ω).

Most materials have some resistance. A good conductor such as copper, though, has low resistance. A good insulator, such as plastic or wood, has high resistance.

Like a thick drink in a straw, an electric charge moves more easily through a short, wide pathway than a long, narrow one.

Resistance depends on the amount and shape of the material as well as on the type of material itself. A wire that is thin has more resistance than a wire that is thick. Think of how you have to work harder to drink through a narrow straw than a wide one. A wire that is long has more resistance than a wire that is short. Again, think of how much harder it is to drink through a long straw than a short one.

CHECK YOUR READING What three factors affect how much resistance an object has?

Chapter 1: Electricity 23 **E**

DIFFERENTIATE INSTRUCTION

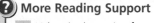 **More Reading Support**

I What is the unit of measure for resistance? *ohm*

J Which has more resistance, a thick or a thin wire? *thin*

Below Level Ask students to explain in their own words why resistance depends in part on the amount of a material. *Answers will vary but should reflect that more material means there is more surface area for an electric charge to move through.* To reinforce this notion with a visual example, draw a cross section of a large wire and a cross section of a smaller wire on the board. Ask: Which of these wires has less resistance? *the larger one*

Real World Example

One way of determining which type of wire to use is to consider its gauge. Gauge is a number that reflects the diameter of a single wire or the area of a cross section of a strand of wires. As gauge increases, the diameter of the wire decreases. If wires are made of the same material, resistance increases as gauge increases.

Teacher Demo

Bring to class a three-way lamp that contains a three-way light bulb. Turn on each level of brightness so that students can compare them. For each brightness, have students explain the relative resistance of the filaments.

Ongoing Assessment

 Answer: practically none at extremely low temperatures

By taking advantage of resistance, we can use an electric charge to do work. When a moving charge overcomes resistance, some of the charge's electrical energy changes into other forms of energy, such as light and heat. For example, the filament of a light bulb is often made of tungsten, a material with high resistance. When electricity moves through the tungsten, the filament gives off light, which is useful. However, the bulb also gives off heat. Because light bulbs are not usually used to produce heat, we think of the heat they produce as wasted energy.

A three-way light bulb has two filaments, each with a different level of resistance. The one with higher resistance produces brighter light. Both together give the brightest setting.

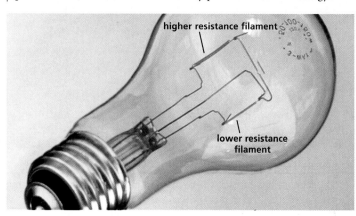

A material with low resistance is one that a charge can flow through with little loss of energy. Materials move electricity more efficiently when they have low resistances. Such materials waste less energy, so more is available to do work at the other end. That is why copper is used for electrical wiring. Even copper has some resistance, however, and using it wastes some energy.

Superconductors

Scientists have known for many years that some materials have practically no resistance at extremely low temperatures. Such materials are called superconductors, because they conduct even better than good conductors like copper. Superconductors have many uses. They can be used in power lines to increase efficiency and conserve energy, and in high-speed trains to reduce friction. Engineers are also testing superconducting materials for use in computers and other electronic devices. Superconductors would make computers work faster and might also be used to make better motors and generators.

Because superconductors must be kept extremely cold, they have not always been practical. Scientists are solving this problem by developing superconductors that will work at higher temperatures.

 How much resistance does a superconducting material have?

DIFFERENTIATE INSTRUCTION

? More Reading Support

K Why might superconductors not be practical? *They must be kept extremely cold.*

Below Level Ask: Why is it impractical to keep superconductors? *Even though the use of a superconductor decreases resistance and saves energy, the energy saved is much less than the energy used to keep the superconductor cold.*

Grounding

If a charge can pass through two different materials, it will pass through the one with the lower resistance. This is the principle behind an important electrical safety procedure—grounding. **Grounding** means providing a harmless, low-resistance path—a ground—for electricity to follow. In many cases, this path actually leads into the ground, that is, into the Earth.

Grounding is used to protect buildings from damage by lightning. Most buildings have some type of lightning rod, which is made from a material that is a good conductor. The rod is placed high up, so that it is closer to the lightning charge. The rod is connected to a conductor cable, and the cable is attached to a copper pole, which is driven into the ground.

Because of the rod's low resistance, lightning will strike the rod before it will strike the roof, where it might have caused a fire. Lightning hits the rod and passes harmlessly through the cable into the ground.

Grounding provides a path for electric current to travel into the ground, which can absorb the charge and make it harmless. The charge soon spreads out so that its voltage in any particular spot is low.

1 Lightning strikes the lightning rod, because the rod is the path of least resistance.

2 The rod conducts the charge to a conductor cable, which has low resistance.

3 The ground wire conducts the charge into the ground, where it spreads out and becomes harmless.

 What is a ground cable?

1.2 Review

KEY CONCEPTS

1. Explain what happens when you get a static electric shock as you touch a doorknob.

2. What is electric potential?

3. What three factors affect how much electrical resistance an object has?

4. How can a lightning rod protect a building from fire?

CRITICAL THINKING

5. **Infer** Object A has a positive charge. After Object A touches Object B, A still has a positive charge and the same amount of charge. What can you infer about the charge of B?

6. **Analyze** Why do lightning rods work better if they are placed high up, closer to the lightning charge?

CHALLENGE

7. **Apply** Could the same material be used as both a conductor and an insulator? Explain your answer.

Chapter 1: Electricity 25 **E**

CHAPTER INVESTIGATION

Focus

PURPOSE To model a lightning strike and the use of a lightning rod to control the path of lightning

OVERVIEW Students will model the buildup and discharge of a static charge. They will use wool to create a charge on a foam plate. They will transfer this charge to an aluminum pan. Then they will observe a static discharge as they bring other metal objects close to the pan. Students will find the following:

• Static discharge occurs when electric potential, which increases as distance decreases, is high enough.

• A low-resistance conductor can be used to control the path of the discharge.

Lab Preparation

• Ask students to bring in wool cloth and aluminum pie pans from home.

• Cut the wool cloth into conveniently sized pieces.

• Prior to the investigation have students read through the investigation and prepare their data tables. Or you may wish to copy and distribute datasheets and rubrics.

 UNIT RESOURCE BOOK, pp. 53–61

 SCIENCE TOOLKIT, F15

Lab Management

• Students should understand that the clay acts as an insulator, so that the charge does not pass to a student's hand.

• The surface on which the plate is placed must be nonconducting.

COOPERATIVE LEARNING Students should take turns doing each task, so that each student at some time charges the plate and observes the discharge.

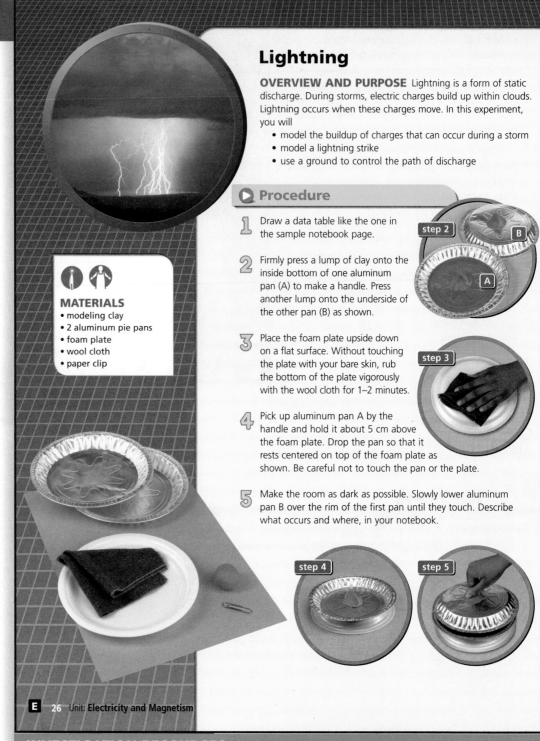

CHAPTER INVESTIGATION

Lightning

OVERVIEW AND PURPOSE Lightning is a form of static discharge. During storms, electric charges build up within clouds. Lightning occurs when these charges move. In this experiment, you will

• model the buildup of charges that can occur during a storm
• model a lightning strike
• use a ground to control the path of discharge

Procedure

1. Draw a data table like the one in the sample notebook page.

2. Firmly press a lump of clay onto the inside bottom of one aluminum pan (A) to make a handle. Press another lump onto the underside of the other pan (B) as shown.

   step 2

3. Place the foam plate upside down on a flat surface. Without touching the plate with your bare skin, rub the bottom of the plate vigorously with the wool cloth for 1–2 minutes.

   step 3

4. Pick up aluminum pan A by the handle and hold it about 5 cm above the foam plate. Drop the pan so that it rests centered on top of the foam plate as shown. Be careful not to touch the pan or the plate.

5. Make the room as dark as possible. Slowly lower aluminum pan B over the rim of the first pan until they touch. Describe what occurs and where, in your notebook.

   step 4   step 5

MATERIALS
• modeling clay
• 2 aluminum pie pans
• foam plate
• wool cloth
• paper clip

INVESTIGATION RESOURCES

 CHAPTER INVESTIGATION, Lightning
• Level A, pp. 53–56
• Level B, pp. 57–60
• Level C, p. 61

Advanced students should complete Levels B & C.

 Writing a Lab Report, D12–13

Technology Resources

Customize this student lab as needed or look for an alternative. Print rubrics to assess student lab reports.

 Lab Generator CD-ROM

6. Repeat steps 3–5 two more times, recording your observations in your notebook.

7. Open the paper clip partway, as shown. Repeat steps 3–4. Then, instead of using the second aluminum pan, slowly bring the pointed end of the paper clip toward the rim of the first pan until they touch. Record your observations.

8. Repeat step 7 two more times, touching the paper clip to the aluminum pan in different places.

Observe and Analyze | Write It Up

1. **RECORD OBSERVATIONS** Be sure your data table is complete. Draw pictures to show how the procedure varied between steps 5–6 and steps 7–8.

2. **ANALYZE** What did you observe in step 5 when the two aluminum pans touched? What do you think caused this to occur?

3. **COMPARE** How were your observations when you touched the aluminum pan with the paper clip different from those you made when you touched it with the other pan? How can you explain the difference?

Conclude | Write It Up

1. **ANALYZE** Use the observations recorded in your data table to answer the following question: When you used the paper clip, why were you able to control the point at which the static discharge occurred?

2. **INFER** What charges did the foam plate and aluminum pan have before you began the experiment? After you dropped the pan on the plate? After you touched the pan with the paper clip?

3. **IDENTIFY VARIABLES** What variables and controls affected the outcome of your experiment?

4. **IDENTIFY LIMITS** What limitations or sources of error could have affected your results?

5. **APPLY** In your experiment, what corresponds to storm clouds and lightning? How did the paper clip work like a lightning rod?

INVESTIGATE Further

CHALLENGE Where did the charge go when you touched the pie pan with the paper clip? Write a hypothesis to explain what happens in this situation and design an experiment to test your hypothesis.

Lightning
Observe and Analyze
Table 1. Observations of Static Discharge

Trial	Observations
With second aluminum pan	
1	
2	
3	
With paper clip	
4	
5	
6	

Conclude

Observe and Analyze | Write It Up

1. Sketches should show that in step 5 the aluminum pan was discharged with a second aluminum pan. In step 7, a paper clip discharged the pan.

2. A spark jumped from a random point on the outer edge of one pan to the other pan. A charge built up in the first pan and discharged to the second pan.

3. The spark jumped wherever the paper clip touched the pan instead of at some random point. The paper clip provided a path for the discharge.

Conclude | Write It Up

1. The paper clip gave the charge a path of least resistance to follow at a specific point on the pan.

2. The plate and pan initially had no charge. After the pan was dropped, they both had the same type of charge. The pan again had no net charge after being touched by the paper clip.

3. Variables: the size and shape of the metal object used to discharge the pan. Controls: the materials used to build up a charge, method of creating the charge, procedure used to discharge the charge.

4. accidentally touching the plate or the pans, the way that the wool was rubbed on the plate

5. The first aluminum pan is like the storm cloud because it is charged. The spark is like lightning. The paper clip grounded the discharge.

INVESTIGATE Further

CHALLENGE The charge passed onto the person holding the paper clip. Sample hypothesis: If the charge moved away from the paper clip, then the paper clip should not be attracted to a charged object, because it has no charge itself.

Post-Lab Discussion

• Ask: What could have been used to form a handle instead of the clay? *any item made of material that is a nonconductor*

• If students did not observe the expected results, have them reread the procedure. Ask: Where might you have made a mistake that affected your experiment? *Sample answer: Step 3 (could not generate a strong charge) or step 4 (accidentally touched the plate)*

▶ Set Learning Goals

Students will

- Describe electric current.
- Explain how current is related to voltage and resistance.
- Distinguish among different types of electric power cells.
- Perform an experiment to infer how electric current can be produced.

◀ 3-Minute Warm-Up

Display Transparency 5 or copy this exercise on the board:

Draw a diagram using the width of a horizontal arrow to show the amount of a charge. Adjust the width of the arrow to show what happens when the charge

- enters a material with low resistance. *The width of the arrow decreases slightly.*
- enters a material with high resistance. *The width of the arrow decreases a lot.*

T 3-Minute Warm-Up, p. T5

1.3 MOTIVATE

EXPLORE Current

PURPOSE To observe how resistance affects the flow of charge

TIP *20 min.* Check the setup ahead of class. If the bulb doesn't light, use a stronger battery or a thinner piece of graphite.

WHAT DO YOU THINK? *The bulb glows more dimly. As the wires are moved apart, the bulb receives less current.*

Teaching with Technology

Students can use a multimeter to check voltage and resistance for the setup in this exploration. (Directions for using a multimeter are discussed on p. 30.)

 KEY CONCEPT

Electric current is a flow of charge.

◀ BEFORE, you learned

- Charges move from higher to lower potential
- Materials can act as conductors or insulators
- Materials have different levels of resistance

▶ NOW, you will learn

- About electric current
- How current is related to voltage and resistance
- About different types of electric power cells

VOCABULARY

electric current p. 28
ampere p. 29
Ohm's law p. 29
electric cell p. 31

EXPLORE Current

How does resistance affect the flow of charge?

PROCEDURE

1. Tape the pencil lead flat on the posterboard.
2. Connect the wires, cell, bulb, and bulb holder as shown in the photograph.
3. Hold the wire ends against the pencil lead about a centimeter apart from each other. Observe the bulb.
4. Keeping the wire ends in contact with the lead, slowly move them apart. As you move the wire ends apart, observe the bulb.

WHAT DO YOU THINK?

- What happened to the bulb as you moved the wire ends apart?
- How might you explain your observation?

MATERIALS

- pencil lead
- posterboard
- electrical tape
- 3 lengths of wire
- D cell battery
- flashlight bulb
- bulb holder

VOCABULARY
Don't forget to make a four square diagram for the term *electric current*.

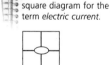

Electric charge can flow continuously.

Static charges cannot make your television play. For that you need a different type of electricity. You have learned that a static charge contains a specific, limited amount of charge. You have also learned that a static charge can move and always moves from higher to lower potential. However, suppose that, instead of one charge, an electrical pathway received a continuous supply of charge and the difference in potential between the two ends of the pathway stayed the same. Then, you would have a continuous flow of charge. Another name for a flow of charge is **electric current**. Electric current is the form of electricity used to supply energy in homes, schools, and other buildings.

E 28 Unit: **Electricity and Magnetism**

RESOURCES FOR DIFFERENTIATED INSTRUCTION

Below Level

UNIT RESOURCE BOOK
- Reading Study Guide A, pp. 35–36
- Decoding Support, p. 48

 AUDIO CDS

R Additional INVESTIGATION,
Making a Coin Battery, A, B, & C, pp. 62–70;
Teacher Instructions, pp. 206–207

Advanced

UNIT RESOURCE BOOK
Challenge and Extension, p. 41

English Learners

UNIT RESOURCE BOOK
Spanish Reading Study Guide, pp. 39–40

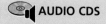 **AUDIO CDS**

- Audio Readings in Spanish
- Audio Readings (English)

Current, Voltage, and Resistance

A

Electric current obeys the same rules as moving static charges. Charge can flow only if it has a path to follow, that is, a material to conduct it. Also, charge can flow only from a point of higher potential to one of lower potential. However, one concept that does not apply to a moving static charge applies to current. Charge that flows steadily has a certain rate of flow. This rate can be measured. The standard unit of measure for current is the **ampere,** or amp. An amp is the amount of charge that flows past a given point per unit of time. One amp equals one coulomb per second. The number of amps—or amperage—of a flowing charge is determined by both voltage and resistance.

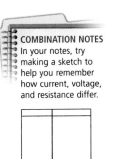

COMBINATION NOTES
In your notes, try making a sketch to help you remember how current, voltage, and resistance differ.

Electric current, or amperage, can be compared to the flow of water through a pipe. Electric potential, or voltage, is like pressure pushing the water through the pipe. Resistance, or ohms, is like the diameter of the pipe, which controls how much water can flow through. Water pressure and pipe size together determine the rate of water flow. Similarly, voltage and resistance together determine the rate of flow of electric charge.

How Potential Affects Current

Current increases with potential, just as water flow increases with water pressure.

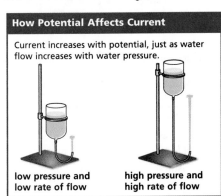

low pressure and low rate of flow

high pressure and high rate of flow

How Resistance Affects Current

Current decreases as resistance increases, just as water flow decreases as resistance to flow increases.

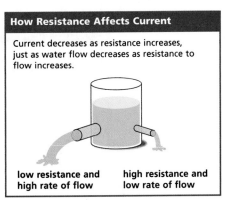

low resistance and high rate of flow

high resistance and low rate of flow

Ohm's Law

B

You now have three important measurements for the study of electricity: volts, ohms, and amps. The scientist for whom the ohm is named discovered a mathematical relationship among these three measurements. The relationship, called **Ohm's law,** is expressed in the formula below.

 SIMULATION
CLASSZONE.COM

See Ohm's law in action.

$$\text{Current} = \frac{\text{Voltage}}{\text{Resistance}} \qquad I = \frac{V}{R}$$

I is current measured in amps (A), *V* is voltage measured in volts (V), and *R* is resistance measured in ohms (Ω).

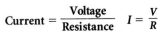 What two values do you need to know to calculate the amperage of electric current?

Chapter 1: Electricity **29** **E**

DIFFERENTIATE INSTRUCTION

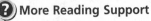 **More Reading Support**

A Which unit is used to measure current? *ampere (amp)*

B Which law relates current, voltage, and resistance? *Ohm's law*

English Learners English learners may be unfamiliar with the use of similes in writing. The second paragraph above uses *like* in two different comparisons. "Electric potential, or voltage, is like pressure pushing the water though the pipe. Resistance, or ohms, is like the diameter of the pipe, . . ." Explain how using *like* helps to make comparisons in writing. Ask students to write simple comparisons using *like*. For example, "The sun is like a giant heat lamp."

Teach Difficult Concepts

Because most light bulbs and appliances come on immediately when a switch is flipped, students may think that electrons move almost instantly from the plug to the bulb or appliance.

Explain to students that while electrons move when current is produced, they actually move along the wire very slowly. It is the energy that moves quickly. Compare current to what happens with energy transfer through mechanical waves. When you talk, the air particles next to your mouth move. Those air particles do not rush across the room at the speed of sound to hit a student's ear. Rather, they move slowly, but hit other air particles that hit more air particles. Very quickly, the air particles near the student's ear receive the same disturbance and hit the eardrum. The energy of the sound travels faster than the particles that create the sound.

Technology Resources

Visit **ClassZone.com** for background on common student misconceptions.

MISCONCEPTION DATABASE

Ongoing Assessment

Describe electric current.

Ask: Why can't you operate a stereo from a static charge? *You need a continuous supply of fresh charge at a constant potential.*

Explain how current is related to voltage and resistance.

Ask: If resistance increases and voltage decreases, what happens to the current? *It decreases.*

CHECK YOUR READING *Answer: voltage and resistance*

Integrate the Sciences

Current can cause discomfort or even damage to the human body.

- At .001 amperes a person will feel a faint tingle.

- A person will feel a painful shock at .006 amperes.

- A person holding a charge source that is producing current ranging from 0.05 to .15 amperes cannot let go of the source.

- A person experiences difficulty breathing at exposure to current ranging from 0.05 to 0.15 amperes.

- A current ranging from 1.0 to 4.3 amperes can stop the rhythmic pumping of the heart.

- Cardiac arrest occurs at 10 amperes.

Mathematics Connection

Tell students to assume that the resistance of skin may be as low as 1000 ohms on a day with high humidity but 100 times as much on a dry day. Ask: If you touch the poles of a 1.5-volt battery on a dry day, how much current might move through your hand? *1.5 × 10⁻⁵ amps* How much current might be there on a day with high humidity? *1.5 × 10⁻³ amps*

Although the practice problems involve determining current if voltage and resistance are known, students should realize that the formula for Ohm's law can be rearranged to solve for a different variable in the equation when that variable is the unknown.

Developing Algebra Skills

- Math Support, p. 49
- Math Practice, p. 50

▶ **Practice the Math** *Answers:*

1. $I = \dfrac{V}{R} = \dfrac{220\ volts}{55\ ohms} = 4\ amps$

2. $I = \dfrac{V}{R} = \dfrac{12\ volts}{24\ ohms} = 0.5\ amp$

Ongoing Assessment

CHECK YOUR READING *Answer: resistance*

You have read that current is affected by both voltage and resistance. Using Ohm's law, you can calculate exactly how much it is affected and determine the exact amount of current in amps. Use the formula for current to solve the sample problem below.

Calculating Current

▶ **Sample Problem**

What is the current in an electrical pathway with an electric potential of 120 volts and a resistance of 60 ohms?

What do you know?	voltage = 120 V, resistance = 60 Ω
What do you want to find out?	current
Write the formula:	$I = \dfrac{V}{R}$
Substitute into the formula:	$I = \dfrac{120\ V}{60\ \Omega}$
Calculate and simplify:	$I = 2\ A$
Check that your units agree:	Unit is amps. Unit of current is amps. Units agree.
Answer:	2 A

▶ **Practice the Math**

1. What is the current in an electrical pathway in which the voltage is 220 V and the resistance is 55 Ω?

2. An electrical pathway has a voltage of 12 volts and a resistance of 24 ohms. What is the current?

READING TiP

The terms *voltmeter, ohmmeter, ammeter,* and *multimeter* are all made by adding a prefix to the word *meter*.

? C
? D

Measuring Electricity

Volts, ohms, and amps can all be measured using specific electrical instruments. Volts can be measured with a voltmeter. Ohms can be measured with an ohmmeter. Amps can be measured with an ammeter. These three instruments are often combined in a single electrical instrument called a multimeter.

To use a multimeter, set the dial on the type of unit you wish to measure. For example, the multimeter in the photograph is being used to test the voltage of a 9-volt battery. The dial is set on volts in the 0–20 range. The meter shows that the battery's charge has an electric potential of more than 9 volts, which means that the battery is good. A dead battery would have a lower voltage.

CHECK YOUR READING What does an ohmmeter measure?

DIFFERENTIATE INSTRUCTION

? **More Reading Support**

C What type of meter measures current?
ammeter

D What does a multimeter measure?
volts, ohms, and amps

Inclusion To enable students with learning disabilities to visualize the effect of electrical energy in their lives, have them locate on the Internet satellite photographs of the United States taken at night during the massive blackout on August 14, 2003. For comparison, have students also find photographs showing the same area before the blackout. Advise students to enter the date and the term *satellite* in a search engine. Be sure to find sites with the actual photographs of the blackout rather than ones that have been touched up.

INVESTIGATE Electric Cells

How can you produce electric current?

PROCEDURE

① Insert the paper clip and the penny into the lemon, as shown in the photograph. The penny and paper clip should go about 3 cm into the lemon. They should be close, but not touching.

② On the multimeter, go to the DC volts (V⎓) section of the dial and select the 0–2000 millivolt range (2000 m).

③ Touch one of the leads of the multimeter to the paper clip. Touch the other lead to the penny. Observe what is shown on the display of the multimeter.

WHAT DO YOU THINK?

• What did you observe on the display of the multimeter?
• How can you explain the reading on the multimeter?

CHALLENGE Repeat this experiment using different combinations of fruits or vegetables and metal objects. Which combinations work best?

SKILL FOCUS
Inferring

MATERIALS
• paper clip
• penny
• large lemon
• multimeter
For Challenge
• additional fruits or vegetables
• metal objects

TIME
20 minutes

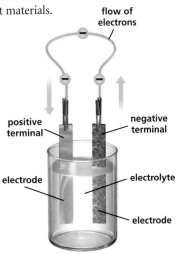

Electric cells supply electric current.

Electric current can be used in many ways. Two basic types of device have been developed for producing current. One type produces electric current using magnets. You will learn more about this technology in Chapter 3. The other type is the **electric cell,** which produces electric current using the chemical or physical properties of different materials.

Electrochemical Cells

An electrochemical cell is an electric cell that produces current by means of chemical reactions. As you can see in the diagram, an electrochemical cell contains two strips made of different materials. The strips are called electrodes. The electrodes are suspended in a third material called the electrolyte, which interacts chemically with the electrodes to separate charges and produce a flow of electrons from the negative terminal to the positive terminal.

Batteries are made using electrochemical cells. Technically, a battery is two or more cells connected to each other. However, single cells, such as C cells and D cells, are often referred to as batteries.

flow of electrons

positive terminal

negative terminal

electrode

electrolyte

electrode

Chapter 1: **Electricity** 31 **E**

INVESTIGATE Electric Cells

PURPOSE To infer how electric current can be produced

TIPS *20 min.*

• Demonstrate how to use the multimeter.

• Have sandpaper available. If pennies are corroded, results will be better if students sand the corrosion off the coin.

WHAT DO YOU THINK? *The meter shows a steady voltage in the circuit, usually between 0.03 V and 0.05 V. The lemon, the penny, and the paper clip interact to produce current.*

CHALLENGE *The most current will come from acidic fruits and relatively active metals.*

Ⓡ Datasheet, Electric Cells, p. 42

Technology Resources

Customize this student lab as needed or look for an alternative. Print rubrics to assess student lab reports.

Lab Generator CD-ROM

Teaching with Technology

If you have probeware, you may wish to adapt this activity so a current and voltage probe system can be used instead of a multimeter.

DIFFERENTIATE INSTRUCTION

? More Reading Support

E How do electrochemical cells produce current? *by chemical reactions*

F What do the electrodes react with in an electrochemical cell? *the electrolyte*

Additional Investigation To reinforce Section 1.3 learning goals, use the following full-period investigation:

Ⓡ **Additional INVESTIGATION,** Making a Coin Battery, A, B, & C, pp. 62–70, 206–207

Advanced Have students research the details of electrochemical cells. Have them investigate how different metals and electrolytes interact and use the information to design an electrochemical cell that could produce a relatively high voltage.

Ⓡ Challenge and Extension, p. 41

Chapter 1 **31** **E**

Real World Example

Alkaline cells are commonly used dry cells. These cells differ from carbon-zinc cells in that they use potassium hydroxide instead of ammonium chloride in the electrolyte. Potassium hydroxide is a base (alkaline) material, so these cells are referred to as alkaline batteries. These cells have a much longer shelf life than zinc-carbon cells, and they perform better while being used and in cold weather. They do not use ammonium ions, which corrode zinc and cause the cells to be more likely to leak. Alkaline cells also do not produce any gaseous products.

Integrate the Sciences

The heart beats in a regular pattern because it receives a consistent pattern of electrical signals. If the pattern becomes irregular or is interrupted, a pacemaker can be surgically implanted to provide regular signals. Because surgery is required to change the battery in a pacemaker, the batteries must last a long time. Pacemaker batteries normally last from 4 to 8 years.

Ongoing Assessment

Distinguish among different types of electric power cells.

Ask: Are rechargeable batteries primary cells or storage cells? *storage cells*

Answer: They contain a solid paste electrolyte.

Answer: storage cell

 **RESOURCE CENTER** CLASSZONE.COM

Learn more about electrochemical cells.

 G

Primary Cells The electrochemical cell shown on page 31 is called a wet cell, because the electrolyte is a liquid. Most household batteries in use today have a solid paste electrolyte and so are called dry cells. Both wet cells and dry cells are primary cells. Primary cells produce electric current through chemical reactions that continue until one or more of the chemicals is used up.

The primary cell page 33 is a typical zinc-carbon dry cell. It has a negative electrode made of zinc. The zinc electrode is made in the shape of a can and has a terminal—in this case, a wide disk of exposed metal—on the bottom of the cell. The positive electrode consists of a carbon rod and particles of carbon and manganese dioxide. The particles are suspended in an electrolyte paste. The positive electrode has a terminal—a smaller disk of exposed metal—at the top of the rod. A paper separator prevents the two electrodes from coming into contact inside the cell.

When the two terminals of the cell are connected—for example, when you turn on your flashlight—a chemical reaction between the zinc and the electrolyte produces electrons and positive zinc ions. The electrons flow through the wires connecting the cell to the flashlight bulb, causing the bulb to light up. The electrons then travel through the carbon rod and combine with the manganese dioxide. When the zinc and manganese dioxide stop reacting, the cell dies.

 CHECK YOUR READING Why are most household batteries called dry cells?

 H

Storage Cells Some batteries produce current through chemical reactions that can be reversed inside the battery. These batteries are called storage cells, secondary cells, or rechargeable batteries. A car battery like the lead-acid battery shown on page 33 is rechargeable. The battery has a negative electrode of lead and a positive electrode of lead peroxide. As the battery produces current, both electrodes change chemically into lead sulfate, and the electrolyte changes into water.

When storage cells are producing current, they are said to be discharging. Whenever a car engine is started, the battery discharges to operate the ignition motor. A car's battery can also be used when the car is not running to operate the lights or other appliances. If the battery is used too long in discharge mode, it will run down completely.

While a car is running, however, the battery is continually being charged. A device called an alternator, which is run by the car's engine, produces current. When electrons flow into the battery in the reverse direction from discharging, the chemical reactions that produce current are reversed. The ability of the battery to produce current is renewed.

 CHECK YOUR READING What kind of battery can be charged by reversing chemical reactions?

E 32 Unit: Electricity and Magnetism

DIFFERENTIATE INSTRUCTION

? More Reading Support

G What type of primary cell contains a liquid electrolyte? *a wet cell*

H Which type of cell is in a car battery? *storage cell*

Alternative Assessment Batteries contain active electrolytes. Divide the class into three groups and have each group prepare a presentation about what safety precautions should be observed for common dry cells, alkaline batteries, and lead–acid storage batteries in cars.

Batteries

Both primary cells and storage cells produce electricity through chemical reactions.

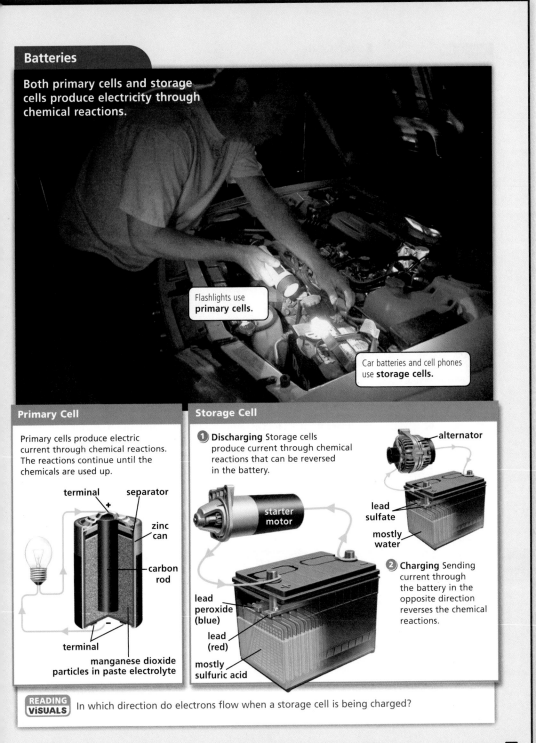

Flashlights use **primary cells.**

Car batteries and cell phones use **storage cells.**

Primary Cell

Primary cells produce electric current through chemical reactions. The reactions continue until the chemicals are used up.

terminal
+
separator
zinc can
carbon rod
terminal
−
manganese dioxide particles in paste electrolyte

Storage Cell

① **Discharging** Storage cells produce current through chemical reactions that can be reversed in the battery.

alternator
starter motor
lead sulfate
mostly water

② **Charging** Sending current through the battery in the opposite direction reverses the chemical reactions.

lead peroxide (blue)
lead (red)
mostly sulfuric acid

READING VISUALS In which direction do electrons flow when a storage cell is being charged?

Teach from Visuals

To help students interpret the diagrams of a primary cell and a storage cell, ask:

- Why is a paper separator needed for the primary cell that is shown? *The electrolyte is active and might react with the zinc can if the paper separator were not present.*

- What is likely to happen when an alternator is not working correctly? *The battery will not recharge, and it will stop producing current.*

Ongoing Assessment

READING VISUALS *Answer: from the positive terminal to the negative terminal*

DIFFERENTIATE INSTRUCTION

Advanced Have interested students research why dry cells containing mercury were popular and why they have been outlawed in the United States. *The voltage remains consistent throughout the battery life. The mercury causes a disposal problem because it pollutes the environment.* Tell students that one use of mercury batteries was to power small pieces of equipment in the space program. Ask: What can you infer about a mercury battery that would make it suitable for this use? *A small mercury battery produces adequate and consistent current. It can produce current for a long period of time.*

Ongoing Assessment

 CHECK YOUR READING *Answer: from the Sun*

Reinforce (the **BIG** idea)

Have students relate the section to the Big Idea.

 Reinforcing Key Concepts, p. 43

1.3 ASSESS & RETEACH

Assess

A Section 1.3 Quiz, p. 5

Reteach

Have students draw a concept map that shows that both primary cells and storage cells are types of electrochemical cells. Have them list the following terms under the appropriate type of cell: *can be reversed, cannot be reversed, liquid electrolyte, paste electrolyte.*

Technology Resources

Have students visit **ClassZone.com** for reteaching of Key Concepts.

 CONTENT REVIEW

 CONTENT REVIEW CD-ROM

READING TIP
The word *solar* comes from the Latin word *sol,* which means the Sun.

Solar Cells

Some materials, such as silicon, can absorb energy from the Sun or other sources of light and then give off electrons, producing electric current. Electric cells made from such materials are called solar cells.

Solar cells are often used to make streetlights come on automatically at night. Current from the cell operates a switch that keeps the lights turned off. When it gets dark, the current stops, the switch closes, and the streetlights come on.

This NASA research aircraft is powered only by the solar cells on its upper surface.

Many houses and other buildings now get at least some of their power from solar cells. Sunlight provides an unlimited source of free, environmentally safe energy. However, it is not always easy or cheap to use that energy. It must be collected and stored because solar cells do not work at night or when sunlight is blocked by clouds or buildings.

 CHECK YOUR READING Where do solar cells get their energy?

1.3 Review

KEY CONCEPTS

1. How is electric current different from a static charge that moves?
2. How can Ohm's law be used to calculate the electrical resistance of a piece of wire?
3. How do rechargeable batteries work differently from nonrechargeable ones?

CRITICAL THINKING

4. **Infer** Electrical outlets in a house maintain a steady voltage, even when the amount of resistance on them changes. How is this possible?
5. **Analyze** Why don't solar cells eventually run down as electrochemical cells do?

CHALLENGE

6. **Apply** Several kinds of electric cells are discussed in this section. Which do you think would be the most practical source of electrical energy on a long trek through the desert? Explain your reasoning.

ANSWERS

1. Electric current is a continuous flow of charge; static charge changes position but does not flow continuously.

2. Run current through the wire; measure voltage and amperage. Divide voltage by amperage to get resistance.

3. Rechargeable batteries use chemical reactions that can be reversed by sending current through the battery.

4. The amperage also changes.

5. Solar cells obtain their energy from the Sun rather

than from chemical reactions, so there are no reactants to be used up.

6. solar cells; because sunlight is abundant in the desert, solar cells would not stop producing power over time and do not require a current to recharge

MATH TUTORIAL
CLASSZONE.COM
Click on Math Tutorial
for more help with
equations.

SKILL: USING VARIABLES

Which Formula Is Best?

A rock band needs an amplifier, and an amplifier needs a volume control. A volume control works by controlling the amount of resistance in an electrical pathway. When resistance goes down, the current—and the volume—go up. Ohm's law expresses the relationship among voltage (V), resistance (R), and amperage (I). If you know the values of two variables, you can use Ohm's law to find the third. The law can be written in three ways, depending on which variable you wish to find.

$$I = \frac{V}{R} \qquad R = \frac{V}{I} \qquad V = IR$$

A simple way to remember these three versions of the formula is to use the pyramid diagram below. Cover up the variable you are looking for. The visible part of the diagram will give you the correct formula to use.

Example

What is the voltage of a battery that produces a current of 1 amp through a wire with a resistance of 9 ohms?

(1) You want to find voltage, so cover up the V in the pyramid diagram. To find V, the correct formula to use is V = IR.

(2) Insert the known values into the formula. $V = 1\,A \cdot 9\,\Omega$

(3) Solve the equation to find the missing variable. $1 \cdot 9 = 9$

ANSWER 9 volts

Answer the following questions.

1. What is the voltage of a battery that sends 3 amps of current through a wire with a resistance of 4 ohms?

2. What is the resistance of a wire through which 2 amps of current flow at 220 volts?

3. What is the amperage of a current flowing at 120 volts through a wire with a resistance of 5 ohms?

CHALLENGE Dimmer switches also work by varying resistance. A club owner likes the way the lights look at 1/3 normal current. The normal current is 15 amps. The voltage is constant at 110. How much resistance will he need?

A volume control works by changing the amount of resistance to the flow of current.

ANSWERS

1. V = IR = 3 amps · 4 ohms = 12 V

2. $R = \frac{V}{I} = \frac{220\ V}{2\ amps} = 110\ ohms$

3. $I = \frac{V}{R} = \frac{120\ V}{5\ ohms} = 24\ amps$

CHALLENGE $R = \frac{V}{I} = \frac{110\ V}{5\ amps} = 22\ ohms$

MATH IN SCIENCE
Math Skills Practice for Science

Set Learning Goal

To use inverse operations to manipulate variables in an equation

Present the Science

The voltage between the two terminals in a house outlet is kept at a constant average of 120 volts. According to Ohm's law, plugging in items with different resistances will produce different currents through those items.

Develop Algebra Skills

Point out that if one form of the Ohm's law equation is known, other forms can be derived by using inverse operations. One form of Ohm's law, $I = \frac{V}{R}$ is introduced on p. 29. To solve for V, multiply both sides of the equation by R. Ask: How would you solve for R? *Multiply both sides by R, then divide both by I.*

DIFFERENTIATION TIP For students with learning disabilities, provide a pyramid on pieces of tagboard. Have them cut out the pyramid, then cut the pieces of the pyramid apart. Students can move the variable they are solving for to the left of the pyramid so they can see the rest of the equation to the right.

Close

Ask: Current used to operate stoves, dryers, and certain other appliances is 220 volts. If the resistance in the wires is 10 ohms, what is the amperage?

$$I = \frac{V}{R} = \frac{220V}{10\ ohms} = 22\ amps$$

• Math Support, p. 51
• Math Practice, p. 52

Technology Resources

Students can visit **ClassZone.com** for practice in using variables.

 MATH TUTORIAL

BACK TO

the BIG idea

Have students look at the photograph on pp. 6–7. Ask them to use the photograph to summarize what they have learned about the difference between static electricity and electric current. *Sample answer: The dragon must be powered by electric current because it stays lit. A current can be supplied by batteries.*

○ KEY CONCEPTS SUMMARY

SECTION 1.1

Ask: What particle is shown on the left? *an electron*

Ask: How does the type of charge affect the way that two charged particles interact? *Two charged particles attract if the charges differ. They repel if the charges are alike.*

SECTION 1.2

Ask: What are two factors that affect charge movement? *electric potential and resistance*

Ask: Should a grounding wire be made of high-resistance or low-resistance material? *low resistance*

SECTION 1.3

Ask: What type of apparatus for producing electric current is shown? *an electrochemical cell*

Ask: Identify the electrodes and the electrolyte in the diagram. *The metal strips suspended in the liquid are the electrodes, and the liquid is the electrolyte.*

Review Concepts

- Big Idea Flow Chart, p. T1
- Chapter Outline, pp. T7–T8

---

1 Chapter Review

the BIG idea

Moving electric charges transfer energy.

 CONTENT REVIEW
CLASSZONE.COM

○ KEY CONCEPTS SUMMARY

1.1 Materials can become electrically charged.

Electric charge is a property of matter.

Electrons have a negative charge. Protons have a positive charge. Unlike charges attract. Like charges repel.

Static charges are caused by the movement of electrons, resulting in an imbalance of positive and negative charges.

VOCABULARY
electric charge p. 10
electric field p. 10
static charge p. 11
induction p. 13

1.2 Charges can move from one place to another.

Charge movement is affected by
- electric potential, measured in volts
- resistance, measured in ohms

A conductor has low resistance.
An insulator has high resistance.
A ground is the path of least resistance.

VOCABULARY
electric potential p. 19
volt p. 19
conductor p. 22
insulator p. 22
resistance p. 23
ohm p. 23
grounding p. 25

1.3 Electric current is a flow of charge.

Electric current is measured in amperes, or amps.
Ohm's law states that current equals voltage divided by resistance.
Electrochemical cells produce electric current through chemical reactions.

VOCABULARY
electric current p. 28
ampere p. 29
Ohm's law p. 29
electric cell p. 31

Technology Resources

Have students visit **ClassZone.com** or use the CD-ROM for a cumulative review of concepts.

 CONTENT REVIEW

 CONTENT REVIEW CD-ROM

Engage students in a whole-class interactive review of Key Concepts. Edit content as you wish.

POWER PRESENTATIONS

Reviewing Vocabulary

Copy the chart below, and write each term's definition. Use the meanings of the underlined roots to help you.

Word	Root	Definition
EXAMPLE current	to run	continuous flow of charge
1. static charge	standing	
2. induction	into + to lead	
3. electric cell	chamber	
4. conductor	with + to lead	
5. insulator	island	
6. resistance	to stop	
7. electric potential	power	
8. grounding	surface of Earth	

Write a vocabulary term to match each clue.

9. In honor of scientist Alessandro Volta (1745–1827)

10. In honor of the scientist who discovered the relationship among voltage, resistance, and current

11. The amount of charge that flows past a given point in a unit of time.

Reviewing Key Concepts

Multiple Choice *Choose the letter of the best answer.*

12. An electric charge is a
   a. kind of liquid
   b. reversible chemical reaction
   c. type of matter
   d. force acting at a distance

13. A static charge is different from electric current in that a static charge
   a. never moves
   b. can either move or not move
   c. moves only when resistance is low enough
   d. moves only when voltage is high enough

14. Charging by induction means charging
   a. with battery power
   b. by direct contact
   c. at a distance
   d. using solar power

15. Electric potential refers to
   a. the amount of energy a charge has
   b. the number of electrons that make up a charge
   c. whether a charge is positive or negative
   d. whether or not a charge can move

16. A superconductor is a material that, when very cold, has no
   a. amperage
   b. resistance
   c. electric charge
   d. electric potential

17. Ohm's law says that when resistance goes up, current
   a. increases   c. stays the same
   b. decreases   d. matches voltage

18. Electrochemical cells include
   a. all materials that build up a charge
   b. primary cells and storage cells
   c. batteries and solar cells
   d. storage cells and lightning rods

Short Answer *Write a short answer to each question.*

19. What determines whether a charge you get when walking across a rug is positive or negative?

20. What is the difference between resistance and insulation?

21. What is one disadvantage of solar cells?

Reviewing Vocabulary

1. *buildup of electric charge in an object caused by many particles with the same charge*

2. *buildup of a charge without direct contact*

3. *a device that produces electric current using chemical or physical properties of different materials*

4. *a material that allows an electric charge to pass through it*

5. *a material that resists passage of an electric charge through it*

6. *the property that determines how much a material resists the flow of electric charge*

7. *the amount of electric potential energy per unit charge at a point in an electric field*

8. *providing a harmless low-resistance path for an electric charge to flow into the ground*

9. *volt*

10. *ohm and Ohm's law*

11. *ampere*

Reviewing Key Concepts

12. *d*

13. *c*

14. *c*

15. *a*

16. *b*

17. *b*

18. *b*

19. *the materials that make up the carpet and your shoes*

20. *Resistance is the degree to which an object resists the passage of an electric charge through it. If the resistance of a particular object is high enough to prevent a charge from flowing through it, the object is an insulator.*

21. *They don't provide current at night or when an obstruction keeps sunlight from hitting them.*

ASSESSMENT RESOURCES

UNIT ASSESSMENT BOOK
- Chapter Test A, pp. 6–9
- Chapter Test B, pp. 10–13
- Chapter Test C, pp. 14–17
- Alternative Assessment, pp. 18–19

SPANISH ASSESSMENT BOOK
Spanish Chapter Test, pp. 301–304

Technology Resources

Edit test items and answer choices.

 Test Generator CD-ROM

Visit **ClassZone.com** to extend test practice.

 Test Practice

Thinking Critically

22. from the negative terminal to the positive terminal

23. The chemicals involved in the reaction will be used up, and the flow of charge will decrease.

24. whether the chemical reaction can be reversed by reversing the flow of current

25. PVC plastics

26. copper

27. germanium

Using Math in Science

28. $\frac{240}{10} = 24$ ohms

29. $\frac{240}{8} = 30$ amps

30. $1.2 \cdot 40 = 48$ volts

31. $\frac{400}{2000} = 0.2$ amp

the BIG idea

32. Moving electric charges in the form of electric current transfer energy to the light bulbs. The resistance of the filaments causes electrical energy to be converted into light.

33. The diagrams and paragraph should demonstrate an understanding of the difference between static charges and electric current. Sample answer: Static charges contain specific, limited amount of charge; they build up and discharge at once. Circuits provide a steady continuous current, with a potential difference that is maintained.

UNIT PROJECTS

Give students the appropriate Unit Project worksheets from the URB for their projects. Both directions and rubrics can be used as a guide.

R Unit Projects, pp. 5–10

Thinking Critically

Use the diagram of an electrochemical cell below to answer the next three questions.

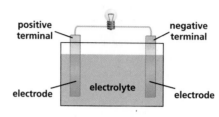

positive terminal   negative terminal

electrode   electrolyte   electrode

22. **ANALYZE** In which direction do electrons flow between the two terminals?

23. **PREDICT** What changes will occur in the cell as it discharges?

24. **ANALYZE** What determines whether the cell is rechargeable or not?

Use the graph below to answer the next three questions.

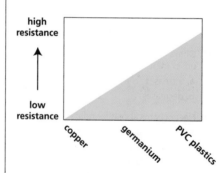

high resistance

low resistance

copper   germanium   PVC plastics

25. **INFER** Which material could you probably use as an insulator?

26. **INFER** Which material could be used in a lightning rod?

27. **APPLY** Materials that conduct electrons under some—but not all—conditions are known as semiconductors. Which material is probably a semiconductor?

Using Math in Science

Use the formula for Ohm's law to answer the next four questions.

$$I = \frac{V}{R}$$

28. An electrical pathway has a voltage of 240 volts and a current of 10 amperes. What is the resistance?

29. A 240-volt air conditioner has a resistance of 8 ohms. What is the current?

30. An electrical pathway has a current of 1.2 amperes and resistance of 40 ohms. What is the voltage?

31. An electrical pathway has a voltage of 400 volts and resistance of 2000 ohms. What is the current?

the BIG idea

32. **INFER** Look back at the photograph on pages 6 and 7. Based on what you have learned in this chapter, describe what you think is happening to keep the dragon lit.

33. **COMPARE AND CONTRAST** Draw two simple diagrams to compare and contrast static charges and electric current. Add labels and captions to make your comparison clear. Then write a paragraph summarizing the comparison.

UNIT PROJECTS

If you are doing a unit project, make a folder for your project. Include in your folder a list of the resources you will need, the date on which the project is due, and a schedule to keep track of your progress. Begin gathering data.

MONITOR AND RETEACH

If students have trouble applying the concepts in items 22–24, have them refer to the diagrams on pp. 31 and 33. Ask: How do the diagrams on pp. 31 and 33 correspond to the diagram for the Thinking Critically problems? *both are diagrams of an electrochemical cell* Have students discuss the flow of electrons and the chemical reaction that occur in the cell.

Students may benefit from summarizing one or more sections of the chapter.

R Summarizing the Chapter, pp. 71–72

Standardized Test Practice

For practice on your state test, go to . . .
TEST PRACTICE
CLASSZONE.COM

Interpreting Diagrams

Use the illustration below to answer the following questions. Assume that the balloons start off with no net charge.

1. What will happen if a negatively charged rod is brought near one of the balloons without touching it?

   a. The balloons will move toward each other.

   b. The balloons will move away from each other.

   c. Electrons on the balloons will move toward the rod.

   d. Electrons on the balloons will move away from the rod.

2. What will happen if a positively charged rod is brought near one of the balloons without touching it?

   a. The balloons will move toward each other.

   b. The balloons will move away from each other.

   c. Electrons on the balloons will move toward the rod.

   d. Electrons on the balloons will move away from the rod.

3. In the previous question, the effect of the rod on the balloons is an example of

   a. charging by contact   **c.** induction

   b. charge polarization   **d.** conduction

4. What will happen if a negatively charged rod is brought near one of the balloons and the balloons are then separated?

   a. The balloon farthest from the rod will become positively charged.

   b. The balloon farthest from the rod will become negatively charged.

   c. Both balloons will become positively charged.

   d. Both balloons will have no net charge.

5. In the previous question, separating the balloons prevents

   a. charging by contact

   b. charge polarization

   c. induction

   d. discharge

6. What will happen if a negatively charged rod is brought near one of the balloons, then taken away, and the balloons are then separated?

   a. The balloon farthest from the rod will become positively charged.

   b. The balloon farthest from the rod will become negatively charged.

   c. Both balloons will become positively charged.

   d. Both balloons will have no net charge.

Extended Response

Answer the two questions below in detail. Include some of the terms from the word box. Underline each term that you use in your answers.

| charge separation | recharging | resistance |
| source of current | static charge | induce |

7. Describe the events leading up to and including a bolt of lightning striking Earth from a storm cloud.

8. Explain the advantages and disadvantages of storage cells over other types of electric cells.

Interpreting Diagrams

| 1. d | 3. c | 5. d |
| 2. c | 4. b | 6. d |

Extended Response

7. RUBRIC

4 points for a response that correctly answers the question and uses the following terms accurately:

   • charge separation
   • induce
   • static charge
   • resistance

Sample: Collisions between particles in a storm cloud cause <u>charge separation</u>. Positively charged particles move to the top of the cloud; negatively charged particles move to the bottom. The bottom of the cloud <u>induces</u> a positive charge in the ground. When the difference in electric potential becomes great enough, the <u>static charge</u> overcomes the <u>resistance</u> of the air, and the charge moves from the cloud to the ground.

3 points correctly answers the question and uses three terms accurately

2 points correctly answers the question and uses two terms accurately

1 point correctly answers the question or uses one term accurately

8. RUBRIC

3 points for a response that compares and contrasts and uses the following terms accurately:

   • source of current
   • recharging

Sample: Storage cells can be recharged repeatedly and so can last a lot longer than primary cells. Storage cells work at night or other times when sunlight is not available, which is not true of solar cells. However, storage cells do run down. They also need a <u>source of current</u> and can't be used while they are <u>recharging</u>.

2 points compares and contrasts and uses one term accurately

1 point compares or contrasts without using either of the terms

---

METACOGNITIVE ACTIVITY

Have students answer the following questions in their **Science Notebook:**

1. What did you find the most challenging to understand about electricity?

2. What questions do you still have about electricity?

3. How did you determine which resources to include for your Unit Project?

2 Circuits and Electronics

Physical Science
UNIFYING PRINCIPLES

PRINCIPLE 1

Matter is made of particles too small to see.

PRINCIPLE 2

Matter changes form and moves from place to place.

PRINCIPLE 3

Energy changes from one form to another, but it cannot be created or destroyed.

PRINCIPLE 4

Physical forces affect the movement of all matter on Earth and throughout the universe.

Unit: Electricity and Magnetism
BIG IDEAS

CHAPTER 1
Electricity

Charged particles transfer electric energy.

CHAPTER 2
Circuits and Electronics

Circuits control the flow of electric charge.

CHAPTER 3
Magnetism

Current can produce magnetism, and magnetism can produce current.

CHAPTER 2
KEY CONCEPTS

SECTION **2.1**

Charge needs a continuous path to flow.

1. Electric charge flows in a loop.
2. Current follows the path of least resistance.
3. Safety devices control current.

SECTION **2.2**

Circuits make electric current useful.

1. Circuits are constructed for specific purposes.
2. Circuits can have multiple paths.
3. Circuits convert electrical energy into other forms of energy.

SECTION **2.3**

Electronic technology is based on circuits.

1. Electronics use coded information.
2. Computer circuits process digital information.
3. Computers can be linked with other computers.

[T] The Big Idea Flow Chart is available on p. T9 in the **UNIT TRANSPARENCY BOOK.**

Previewing Content

SECTION

 2.1 **Charge needs a continuous path to flow.** pp. 43–50

1. Electric charge flows in a loop.

A **circuit** is a closed path through which a charge can flow. It has at least four basic parts.

Parts of a Circuit	
Voltage source	Provides electric potential Example: battery
Conductor	Provides connecting path Example: copper wire
Switch	Closes or opens the path Example: light switch
Electrical device	Changes electrical energy into other energy forms Example: resistor

2. Current follows the path of least resistance.

Conductors, such as many metals, have a lower resistance than the materials around them, and provide a path through which charge will flow. Insulators have high resistance.

- A **short circuit** is a path that allows current to go where it is not intended, and may be dangerous.
- In a grounded circuit a third wire leads stray current safely into the ground.

3. Safety devices control current.

Fuses and circuit breakers open a circuit when the level of current becomes dangerously high, thereby stopping the flow of charge.

- Fuses have to be replaced when they blow.
- Circuit breakers can be used over and over again.
- A ground-fault circuit interrupter (GFCI) outlet can be reset repeatedly. It breaks a circuit when it detects a change in current.

SECTION

 2.2 **Circuits make electric current useful.** pp. 51–56

1. Circuits are constructed for specific purposes.

Circuits are designed to be used for specific purposes, such as lighting bulbs, moving motors, and performing calculations.

2. Circuits can have multiple paths.

In a **series circuit,** current follows a single path.

- Every element in a series circuit must be functional, or the whole circuit stops working.
- Every resistor added to a series circuit decreases the current available to each resistor.
- The voltages of batteries in series add together.

In a **parallel circuit,** current follows multiple paths.

- If an element in a parallel circuit is not functional, the other elements still work.
- Every device on a parallel circuit gets full current.
- The voltages of batteries in parallel do not add together.

Series Circuit	Parallel Circuit
Each device in a series circuit is wired on a single path. 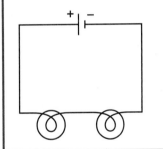	Each device in a parallel circuit has its own connection to the voltage source. 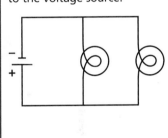

3. Circuits convert electrical energy into other forms of energy.

Circuit elements can convert electrical energy into other forms of energy. They make electrical energy useful.

Common Misconceptions

HOW TO MAKE A COMPLETE CIRCUIT Students may hold the misconception that a circuit is complete if there is a wire connecting a battery to a bulb. There must be a complete conductive path from the battery to the bulb and back.

 This misconception is addressed on p. 44.

MISCONCEPTION DATABASE

CLASSZONE.COM Background on student misconceptions

HOW CURRENT IS DISTRIBUTED Students may hold the misconception that lights added to a series circuit dim progressively from one to the next instead of universally across the circuit. Adding light bulbs in a series circuit causes all the lights to dim due to diminishing current available to any one bulb.

 This misconception is addressed on p. 52.

Previewing Content

SECTION

2.3 Electronic technology is based on circuits. pp. 57–67

1. Electronics use coded information.

An **electronic** device uses electric current to represent coded information. Many electronic devices use a **binary code.**

- **Digital** electronic devices, such as computers, use a binary code consisting of 1s and 0s.
- **Analog** signals, such as sound waves, can be converted to digital information that computers recognize.

2. Computer circuits process digital information.

Integrated circuits are highly complex, tiny circuits. They are usually built on chips made of silicon. They provide processing in computers, cars, calculators, and other devices. **Computer** hardware is the physical equipment required to do computer work. Computer software is the set of instructions and languages required to run the equipment. Personal computers have parts that perform four basic functions: input, storage, processing, and output.

3. Computers can be linked with other computers.

Computers can be linked together in networks. The largest computer network is the Internet, a decentralized network devised to keep going even if many links are not functioning.

The World Wide Web is an international Internet-based environment in which Web sites can be posted and viewed.

Previewing Labs

Lab Generator CD-ROM
Edit these Pupil Edition labs and generate alternative labs.

EXPLORE (the BIG idea)

Will the Flashlight Still Work? p. 41
Students rearrange batteries in a flashlight to observe that a circuit must be complete and its parts arranged properly for it to work.

TIME 10 minutes
MATERIALS flashlight with batteries, piece of paper

What's Inside a Calculator? p. 41
Students explore the circuit board in a calculator to think about the function of different parts.

TIME 10 minutes
MATERIALS hand-held calculator, small screwdriver

Internet Activity: Circuits, p. 41
Students use the Internet to build a virtual circuit, including a switch.

TIME 20 minutes
MATERIALS computer with Internet access

SECTION 2.1

EXPLORE Circuits, p. 43
Students use a battery, light bulb, and aluminum foil to complete a simple circuit.

TIME 10 minutes
MATERIALS strips of aluminum foil, electrical tape, D battery, light bulb

INVESTIGATE Fuses, p. 48
Students model a fuse using steel wool and a battery in a simple circuit

TIME 15 minutes
MATERIALS 2 pieces of insulated wire (30 cm in length) with alligator clips, single strand of steel wool, glass jar, 15 cm tape, 6-volt battery

---

SECTION 2.2

INVESTIGATE Circuits, p. 54
Students experiment with series and parallel arrangements of batteries to determine which causes bulbs to burn brighter.

TIME 15 minutes
MATERIALS 4 insulated wires (30 cm in length) with alligator clips, small light bulb in a holder, 2 batteries in holders

---

SECTION 2.3

EXPLORE Codes, p. 57
Students write the name of their street using numbers instead of letters to model information coding.

TIME 10 minutes
MATERIALS notebook, small piece of paper

INVESTIGATE Digital Information, p. 59
Students exchange codes to recreate a simple model of a digital image.

TIME 30 minutes
MATERIALS graph paper, plain paper

CHAPTER INVESTIGATION
Design an Electronic Communication Device, pp. 66–67
Students design, build, and test a battery-powered Morse code communicator to be marketed as a toy.

 Morse Code Chart, p. 113

TIME 40 minutes
MATERIALS 2 batteries, light bulb in holder, 50 cm copper wire (22 gauge), 2 insulated wire leads (30 cm in length) with alligator clips, 2 craft sticks, toothpick, paper clip, 20 cm × 20 cm piece of cardboard, clothespin, 10 cm × 10 cm piece of aluminum foil, rubber band, scissors, 10 cm electrical tape, wire cutters, Morse Code Chart

---

 Additional INVESTIGATION, Wire a Room, A, B, & C, pp. 123–131; Teacher Instructions, pp. 206–207

Previewing Chapter Resources

	INTEGRATED TECHNOLOGY	**LABS AND ACTIVITIES**

CHAPTER 2

Circuits and Electronics

 CLASSZONE.COM
- eEdition Plus
- EasyPlanner Plus
- Misconception Database
- Content Review
- Test Practice
- Visualization
- Resource Center
- Internet Activity: Circuits
- Math Tutorial

 SCILINKS.ORG

 CD-ROMS
- eEdition
- EasyPlanner
- Power Presentations
- Content Review
- Lab Generator
- Test Generator

 AUDIO CDS
- Audio Readings
- Audio Readings in Spanish

 EXPLORE the Big Idea, p. 41
- Will the Flashlight Still Work?
- What's Inside a Calculator?
- Internet Activity: Circuits

 UNIT RESOURCE BOOK
Unit Projects, pp. 5–10

 Lab Generator CD-ROM
Generate customized labs.

SECTION

 2.1

Charge needs a continuous path to flow.
pp. 43–50

Time: 2 periods (1 block)
 Lesson Plan, pp. 73–74

 RESOURCE CENTER, Electrical Safety

 UNIT TRANSPARENCY BOOK
- Big Idea Flow Chart, p. T9
- Daily Vocabulary Scaffolding, p. T10
- Note-Taking Model, p. T11
- 3-Minute Warm-Up, p. T12

 EXPLORE Circuits, p. 43
- INVESTIGATE Fuses, p. 48
- Science on the Job, p. 50

 UNIT RESOURCE BOOK
- Datasheet, Fuses, p. 82
- Additional INVESTIGATION, Wire a Room, A, B, & C, pp. 123–131

SECTION

 2.2

Circuits make electric current useful.
pp. 51–56

Time: 2 periods (1 block)
 Lesson Plan, pp. 84–85

 MATH TUTORIAL

UNIT TRANSPARENCY BOOK
- Daily Vocabulary Scaffolding, p. T10
- 3-Minute Warm-Up, p. T12

 INVESTIGATE Circuits, p. 54
- Math in Science, p. 56

 UNIT RESOURCE BOOK
- Datasheet, Circuits, p. 93
- Math Support, p. 111
- Math Practice, p. 112

SECTION

 2.3

Electronic technology is based on circuits.
pp. 57–67

Time: 4 periods (2 blocks)
 Lesson Plan, pp. 95–96

 • **RESOURCE CENTER,** Electronics
• **VISUALIZATION,** Hard Drive

UNIT TRANSPARENCY BOOK
- Big Idea Flow Chart, p. T9
- Daily Vocabulary Scaffolding, p. T10
- 3-Minute Warm-Up, p. T13
- "Analog and Digital Signals" Visual, p. T14
- Chapter Outline, pp. T15–T16

 EXPLORE Codes, p. 57
- INVESTIGATE Digital Information, p. 59
- CHAPTER INVESTIGATION, Design an Electronic Communication Device, pp. 66–67

 UNIT RESOURCE BOOK
- Datasheet, Digital Information, p. 104
- Morse Code Chart, p. 113
- CHAPTER INVESTIGATION, Design an Electronic Communication Device, A, B, & C, pp. 114–122

KEY TO ICONS

 CD/CD-ROM

 Teacher Edition

 UNIT TRANSPARENCY BOOK

 SPANISH ASSESSMENT BOOK

 INTERNET

 Pupil Edition

 UNIT RESOURCE BOOK

UNIT ASSESSMENT BOOK

 SCIENCE TOOLKIT

READING AND REINFORCEMENT

ASSESSMENT

STANDARDS

- Frame Game, B26–27
- Outline, C43
- Daily Vocabulary Scaffolding, H1–8

 UNIT RESOURCE BOOK
- Vocabulary Practice, pp. 108–109
- Decoding Support, p. 110
- Summarizing the Chapter, pp. 132–133

 Audio Readings CD
Listen to Pupil Edition.

Audio Readings in Spanish CD
Listen to Pupil Edition in Spanish.

- Chapter Review, pp. 69–70
- Standardized Test Practice, p. 71

 UNIT ASSESSMENT BOOK
- Diagnostic Test, pp. 20–21
- Chapter Test, A, B, & C, pp. 25–36
- Alternative Assessment, pp. 37–38

 Spanish Chapter Test, pp. 305–308

 Test Generator CD-ROM
Generate customized tests.

 Lab Generator CD-ROM
Rubrics for Labs

National Standards
A.1–8, A.9.a–g, B.3.a, B.3.d, E.1–5

See p. 40 for the standards.

 UNIT RESOURCE BOOK
- Reading Study Guide, A & B, pp. 75–78
- Spanish Reading Study Guide, pp. 79–80
- Challenge and Extension, p. 81
- Reinforcing Key Concepts, p. 83

 Ongoing Assessment, pp. 43–49

 Section 2.1 Review, p. 49

 UNIT ASSESSMENT BOOK
Section 2.1 Quiz, p. 22

National Standards
A.2–7, A.9.a–b, A.9.e–f, B.3.a

 UNIT RESOURCE BOOK
- Reading Study Guide, A & B, pp. 86–89
- Spanish Reading Study Guide, pp. 90–91
- Challenge and Extension, p. 92
- Reinforcing Key Concepts, p. 94

 Ongoing Assessment, pp. 51–55

 Section 2.2 Review, p. 55

UNIT ASSESSMENT BOOK
Section 2.2 Quiz, p. 23

National Standards
A.2–8, A.9.a–f, B.3.a, B.3.d

 UNIT RESOURCE BOOK
- Reading Study Guide, A & B, pp. 97–100
- Spanish Reading Study Guide, pp. 101–102
- Challenge and Extension, p. 103
- Reinforcing Key Concepts, p. 105
- Challenge Reading, pp. 106–107

 Ongoing Assessment, pp. 57–64

 Section 2.3 Review, p. 65

UNIT ASSESSMENT BOOK
Section 2.3 Quiz, p. 24

National Standards
A.1–8, A.9.a–g, E.1–5

Previewing Resources for Differentiated Instruction

CHAPTER INVESTIGATION

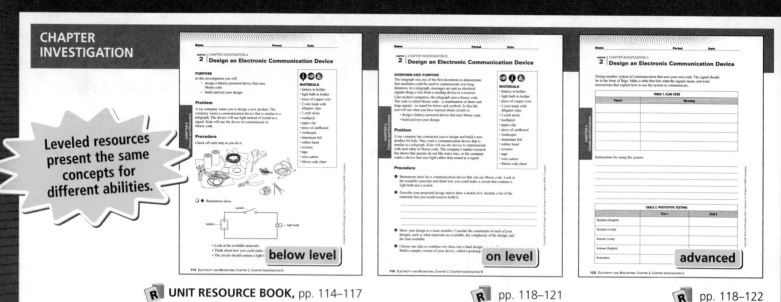

> Leveled resources present the same concepts for different abilities.

below level

on level

advanced

READING STUDY GUIDE

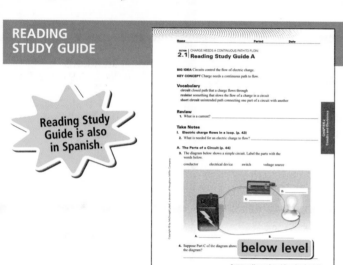

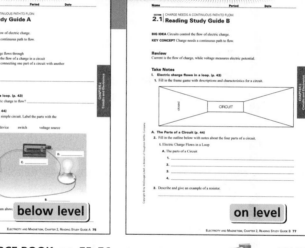

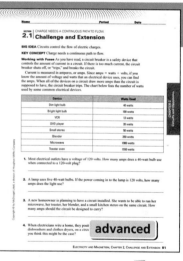

> Reading Study Guide is also in Spanish.

below level

on level

advanced

CHAPTER TEST

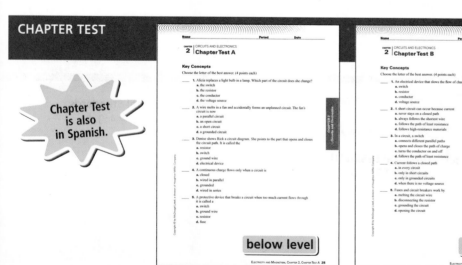

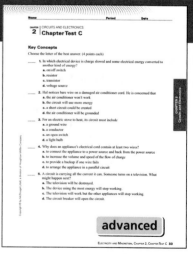

> Chapter Test is also in Spanish.

below level

on level

advanced

TECHNOLOGY

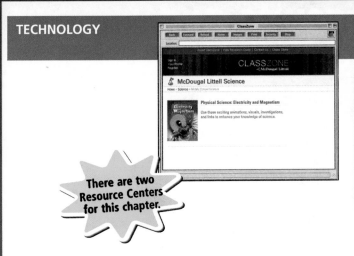

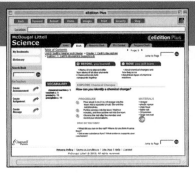

There are two Resource Centers for this chapter.

CLASSZONE.COM

CD/CD-ROMS

CLASSZONE.COM

VISUAL CONTENT

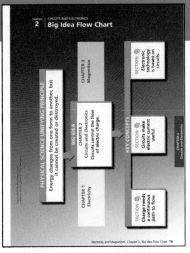

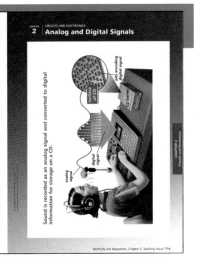

UNIT TRANSPARENCY BOOK, p. T9

p. T11

p. T14

MORE SUPPORT

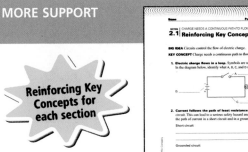

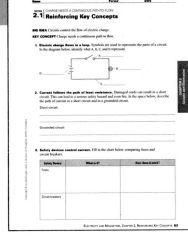

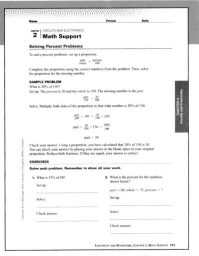

Reinforcing Key Concepts for each section

UNIT RESOURCE BOOK, p. 83

pp. 108–109

p. 111

INTRODUCE

the **BIG** idea

Have students look at the photograph of the student handling electronic equipment and discuss how the question in the box links to the Big Idea:

- How would you describe the equipment that the student is holding?

- Look closely at the equipment. Can you see patterns on the circuit boards? What might these patterns have to do with charges that move from place to place?

National Science Education Standards

Content

B.3.a Energy is a property of many substances that is often associated with electricity. Energy is transferred in many ways.

B.3.d Circuits transfer electrical energy. Heat, light, sound, and chemical changes are produced.

Process

A.1–8 Identify questions that can be answered through scientific investigations; design and conduct an investigation; use tools; use evidence; think critically between evidence and explanation; recognize different explanations and predictions; communicate procedures and explanations; use mathematics.

A.9.a–g Understand scientific inquiry by using different investigations, methods, mathematics, technology, and explanations based on logic and evidence. Data often results in new investigations.

E.1–5 Identify a problem; design, implement, and evaluate a solution or product; communicate technological design.

CHAPTER

2 Circuits and Electronics

the **BIG** idea

Circuits control the flow of electric charge.

Key Concepts

SECTION
2.1 Charge needs a continuous path to flow.
Learn how circuits are used to control the flow of charge.

SECTION
2.2 Circuits make electric current useful.
Learn about series circuits and parallel circuits.

SECTION
2.3 Electronic technology is based on circuits.
Learn about computers and other electronic devices.

Internet Preview

CLASSZONE.COM

Chapter 2 online resources:
Content Review, Simulation, Visualization, two Resource Centers, Math Tutorial, Test Practice

E 40 Unit: **Electricity and Magnetism**

How can circuits control the flow of charge?

INTERNET PREVIEW

CLASSZONE.COM For student use with the following pages:

Review and Practice
- Content Review, pp. 42, 68
- Math Tutorial: Percents and Proportions, p. 56
- Test Practice, p. 71

Activities and Resources
- Internet Activity: Circuits, p. 41
- Resource Centers: Electrical Safety, p. 46; Electronics, p. 57
- Visualization: Hard Drive, p. 63

NSTA *SCi*
scilinks.org *LINKS*

Electronic Circuits
Code: MDL066

EXPLORE (the BIG idea)

Will the Flashlight Still Work?

Experiment with a flashlight to find out if it will work in any of the following arrangements: with one of the batteries facing the wrong way, with a piece of paper between the batteries, or with one battery removed. In each case, switch on the flashlight and observe.

Observe and Think
When did the flashlight work? Why do you think it worked or did not work in each case?

What's Inside a Calculator?

Use a small screwdriver to open a simple calculator. Look at the circuit board inside.

Observe and Think How do you think the metal lines relate to the buttons on the front of the calculator? to the display? What is the source of electrical energy? How is it connected to the rest of the circuit?

Internet Activity: Circuits

Go to **ClassZone.com** to build a virtual circuit. See if you can complete the circuit and light the bulb.

Observe and Think
What parts are necessary to light the bulb? What happened when you opened the switch? closed the switch?

NSTA
scilinks.org
SCI LINKS

Electronic Circuits **Code: MDL066**

EXPLORE (the BIG idea)

These inquiry-based activities are appropriate for use at home or as a supplement to classroom instruction.

Will the Flashlight Still Work?

PURPOSE To observe an incomplete circuit. Students rearrange batteries in a flashlight to test different arrangements.

TIP *10 min.* Make sure students put the batteries in the correct position and remove the paper at the end of the activity.

Answer: The flashlight did not work in any of the arrangements. None was a complete circuit.

REVISIT after p. 44.

What's Inside a Calculator?

PURPOSE To infer the function of a circuit board and hypothesize what the parts do.

TIP *10 min.* Ask students not to touch the inside of the calculator. Tell students to keep track of the small screws so that the calculator can be reassembled.

Answer: Metal lines connect the circuit to the buttons and the display. The source of electrical energy is the battery. It is connected to the rest of the circuit with a metal strip.

REVISIT after p. 57.

Internet Activity: Circuits

PURPOSE To build a virtual circuit.

TIP *20 min.* Before they begin building a virtual circuit, students can try different parts of the simulation to get a feel for what the individual parts do.

Answer: The battery and the wire are necessary to light the bulb. The light bulb turned off when the switch was opened and on when the switch was closed.

REVISIT after p. 45.

TEACHING WITH TECHNOLOGY

Graphics Software Have students use graphics software to draw schematic diagrams of the circuits they design in "Investigate Circuits" on p. 54.

Digital Camera You might want to take photographs of some students' circuits during the Chapter Investigation on pp. 66–67.

PREPARE

◐ CONCEPT REVIEW
Activate Prior Knowledge

- Remind students that electric charge can build up in an object, then move suddenly out of that object to a region of lesser charge. Ask students to think of a situation in nature like this. *lightning*

- Have students think about what unseen activity happens when they plug a device or an appliance into an electrical wall socket and turn it on. Prompt students to compare this with lightning.

- Have students list materials that make good conductors of electricity and those that make good insulators.

▶ TAKING NOTES

Outline

Students may need to be reminded that successive lines should be indented. Indenting is part of the visual enhancement that helps the reader quickly organize details and sub-topics with a main idea or concept.

Vocabulary Strategy

Suggest that students, when using the frame game diagram, write the vocabulary term with larger lettering in the center, or with a contrasting color, or any other way that makes it stand out.

Vocabulary and Note-Taking Resources

- Vocabulary Practice, pp. 108–109
- Decoding Support, p. 110

- Note-Taking Model, p. T11
- Daily Vocabulary Scaffolding, p. T10

- Frame Game, B26–27
- Outline, C43
- Daily Vocabulary Scaffolding, H1–8

◐ CONCEPT REVIEW

- Energy can change from one form to another.
- Energy can move from one place to another.
- Current is the flow of charge through a conductor.

◐ VOCABULARY REVIEW

electric current p. 28
electric potential p. 18
conductor p. 22
resistance p. 23
ampere p. 29

CONTENT REVIEW
CLASSZONE.COM
Review concepts and vocabulary.

▶ TAKING NOTES

OUTLINE

As you read, copy the headings on your paper in the form of an outline. Then add notes in your own words that summarize what you read.

VOCABULARY STRATEGY

Write each new vocabulary term in the center of a **frame game** diagram. Decide what information to frame it with. Use examples, descriptions, parts, sentences that use the term in context, or pictures. You can change the frame to fit each term.

See the Note-Taking Handbook on pages R45–R51.

SCIENCE NOTEBOOK

I. ELECTRIC CHARGE FLOWS IN A LOOP.
  A. THE PARTS OF A CIRCUIT
    1. voltage source
    2. connection
    3. electrical device
    4. switch

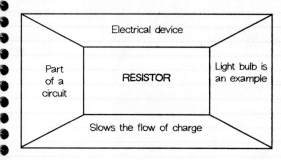

Electrical device | Part of a circuit | **RESISTOR** | Light bulb is an example | Slows the flow of charge

CHECK READINESS

Administer the Diagnostic Test to determine students' readiness for new science content and their mastery of requisite math skills.

 Diagnostic Test, pp. 20–21

Technology Resources

Students needing content and math skills should visit **ClassZone.com**.

- **CONTENT REVIEW**
- **MATH TUTORIAL**

 CONTENT REVIEW CD-ROM

KEY CONCEPT
Charge needs a continuous path to flow.

◀ **BEFORE,** you learned
- Current is the flow of charge
- Voltage is a measure of electric potential
- Materials affect the movement of charge

▶ **NOW,** you will learn
- About the parts of a circuit
- How a circuit functions
- How safety devices stop current

VOCABULARY

circuit p. 43
resistor p. 44
short circuit p. 46

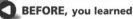
EXPLORE Circuits

How can you light the bulb?

PROCEDURE

1. Tape one end of a strip of foil to the negative terminal, or the flat end, of the battery. Tape the other end of the foil to the tip at the base of the light bulb, as shown.

2. Tape the second strip of foil to the positive terminal, or the raised end, of the battery.

3. Find a way to make the bulb light.

WHAT DO YOU THINK?
- How did you make the bulb light?
- Can you find other arrangements that make the bulb light?

MATERIALS
- 2 strips of aluminum foil
- electrical tape
- D cell (battery)
- light bulb

Electric charge flows in a loop.

In the last chapter, you read that current is electric charge that flows from one place to another. Charge does not flow continuously through a material unless the material forms a closed path, or loop. A **circuit** is a closed path through which a continuous charge can flow. The path is provided by a low-resistance material, or conductor, usually wire. Circuits are designed to do specific jobs, such as light a bulb.

Circuits can be found all around you and serve many different purposes. In this chapter, you will read about simple circuits, such as the ones in flashlights, and more complex circuits, such as the ones that run toys, cameras, computers, and more.

 CHECK YOUR READING How are circuits related to current?

RESOURCES FOR DIFFERENTIATED INSTRUCTION

Below Level
UNIT RESOURCE BOOK
- Reading Study Guide A, pp. 75–76
- Decoding Support, p. 110

 **AUDIO CDS**

R **Additional INVESTIGATION,**
Wire a Room, A, B, & C, pp. 123–131;
Teacher Instruction, pp. 206–207

Advanced
UNIT RESOURCE BOOK
Challenge and Extension, p. 81

English Learners
UNIT RESOURCE BOOK
Spanish Reading Study Guide, pp. 79–80

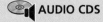 **AUDIO CDS**
- Audio Readings in Spanish
- Audio Readings (English)

2.1 FOCUS

▶ **Set Learning Goals**
Students will
- Describe the parts of a circuit.
- Explain how a circuit functions.
- Explain how electrical safety devices stop current.
- Model a fuse in an experiment using a simple circuit.

◀ **3-Minute Warm-Up**

Display Transparency 12 or copy this exercise on the board:

Match the definitions to the terms.

Definitions

1. the flow of charge *e*
2. a measure of electric potential *c*
3. a path for current *a*

Terms

a. conductor   d. charge
b. insulator   e. current
c. voltage

T 3-Minute Warm-Up, p. T12

2.1 MOTIVATE

EXPLORE Circuits

PURPOSE To introduce the idea that completing (closing) a circuit allows charge to flow

TIP *10 min.* Students may want to fold the foil into a narrower strip. Make sure the aluminum foil is in direct contact with the battery terminals. Alert students that the foil will get warm if it is left connected.

WHAT DO YOU THINK? *By touching the aluminum foil to the metal part of the light bulb. As long as metal is touching metal and the circuit is complete, the bulb will light.*

Ongoing Assessment

CHECK YOUR READING *Answer: Circuits provide a closed path for current.*

Address Misconceptions

IDENTIFY Ask: How do you complete a circuit? If students answer by connecting a wire from a battery to a bulb, then they may not realize that there must be a complete conductive path from the battery to the bulb and back.

CORRECT Have students examine a standard home battery and describe the two ends. Point out the plus or minus sign at each end, explaining that it indicates charge. Remind students that current flows around a circuit *from* one charge *to* the opposite charge.

REASSESS Ask students to imagine they have an incomplete light-bulb circuit from which the battery is missing. Ask: How would you connect a battery to complete the circuit? *Connect one end of the circuit wire to the battery's negative terminal and the other end to the positive terminal.*

Technology Resources

Visit **ClassZone.com** for background on common student misconceptions.

MISCONCEPTION DATABASE

Teach Difficult Concepts

Help students review voltage and current. Use the analogy of a person squeezing a tube of toothpaste to force it out of the tube. Compare voltage to the squeezing force applied by the person's hand to the tube and current to the surge of toothpaste.

EXPLORE (the BIG idea)

Revisit "Will the Flashlight Still Work?" on p. 41. Have students explain their results.

Ongoing Assessment

Describe the parts of a circuit.

Ask: Suppose an incomplete circuit has three elements: a conductor, a resistor, and a switch. What does the circuit lack? *a voltage source*

READING VISUALS *Answer: Yes; the circuit would still be complete.*

The Parts of a Circuit

The illustration below shows a simple circuit. Circuits typically contain the following parts. Some circuits contain many of each part.

REMINDER
Remember, a battery consists of two or more cells.

① **Voltage Source** The voltage source in a circuit provides the electric potential for charge to flow through the circuit. Batteries are often the voltage sources in a circuit. A power plant may also be a voltage source. When you plug an appliance into an outlet, a circuit is formed that goes all the way to a power plant and back.

② **Conductor** A circuit must be a closed path in order for charge to flow. That means that there must be a conductor, such as wire, that forms a connection from the voltage source to the electrical device and back.

?A

③ **Switch** A switch is a part of a circuit designed to break the closed path of charge. When a switch is open, it produces a gap in the circuit so that the charge cannot flow.

?B

④ **Electrical Device** An electrical device is any part of the circuit that changes electrical energy into another form of energy. A **resistor** is an electrical device that slows the flow of charge in a circuit. When the charge is slowed, some energy is converted to light or heat. A light bulb is an example of a resistor.

Circuit Parts

The parts of a basic circuit include a voltage source, conductor, switch, and one or more electrical devices.

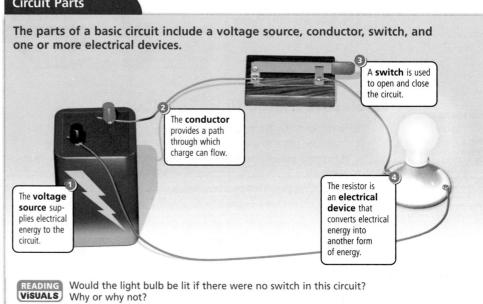

③ A **switch** is used to open and close the circuit.

② The **conductor** provides a path through which charge can flow.

① The **voltage source** supplies electrical energy to the circuit.

④ The resistor is an **electrical device** that converts electrical energy into another form of energy.

READING VISUALS Would the light bulb be lit if there were no switch in this circuit? Why or why not?

DIFFERENTIATE INSTRUCTION

? More Reading Support

A How does an open switch affect a circuit? *It stops the flow of charge.*

B How does a resistor affect the flow of charge? *It slows the flow.*

English Learners This section uses the word "or" several times to signal a definition. For example, "Tape the second strip of foil to the positive terminal, or the raised end, of the battery." English learners may mistakenly read such sentences in terms of "either/or." Help students understand that in these cases "or" indicates a definition or an alternate term.

Below Level Copy the four parts of a circuit on the board, or have students list them on an index card for reference.

Open and Closed Circuits

Current in a circuit is similar to water running through a hose. The flow of charge differs from the flow of water in an important way, however. The water does not require a closed path to flow. If you cut the hose, the water continues to flow. If you cut a wire, the charge stops flowing.

Batteries have connections at both ends so that charge can follow a closed path to and from the battery. The cords that you see on appliances might look like single cords but actually contain at least two wires. The wires connect the device to a power plant and back to make a closed path.

Switches work by opening and closing the circuit. A switch that is on closes the circuit and allows charge to flow through the electrical devices. A switch that is off opens the circuit and stops the current.

REMINDER
Current requires a closed loop.

CHECK YOUR READING How are switches used to control the flow of charge through a circuit?

Standard symbols are used to represent the parts of a circuit. Some common symbols are shown in the circuit diagrams below. The diagrams represent the circuit shown on page 44 with the switch in both open and closed positions. Electricians and architects use diagrams such as these to plan the wiring of a building.

Circuit Diagrams

Symbols are used to represent the parts of a circuit. The circuit diagrams below show the circuit from page 44 in both an open and closed position.

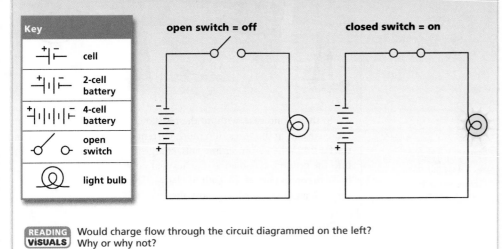

Key
- cell
- 2-cell battery
- 4-cell battery
- open switch
- light bulb

open switch = off   closed switch = on

READING VISUALS Would charge flow through the circuit diagrammed on the left? Why or why not?

DIFFERENTIATE INSTRUCTION

? More Reading Support

C What happens to the flow of charge in a circuit if you cut a wire? *It stops.*

D What happens when you close a switch? *You turn on the circuit.*

Inclusion Students with visual impairments might find this activity helpful. Set up a simple circuit with a battery, switch, and light bulb. Make a photocopy of the diagrams on this page. Trace the diagrams (but not the key) with a pointed pen, pressing heavily into a soft backing, to create a raised pattern on the back of the sheet. Have students explore the physical setup and then use their fingers to feel the raised circuit that was traced on the back of the paper.

Teach from Visuals

Have students study the "Circuit Parts" and the "Circuit Diagrams" visuals on pp. 44–45.

- Ask: What is different about these two ways of drawing a circuit? *The diagrams on p. 45 uses special symbols instead of a realistic drawing.*

- Ask: If the circuit diagram was flipped so that the bulb was on the left and the battery was on the right, would it still represent the circuit in the drawing? *Yes*

Teacher Demo

Use this demonstration to help students understand the distinction between a closed switch and an open one. Turn off the room's light. Ask:

- Is the switch open or closed? *open*

- Is there current in the circuit that goes to the lights? *no*

Now turn on the light. Ask:

- Is the switch open or closed? *closed*

- Is there current in the circuit that goes to the lights? *yes*

EXPLORE (the **BIG** idea)

Revisit "Internet Activity: Circuits" on p. 41. Have students describe the circuit they built.

Ongoing Assessment

Explain how a circuit functions.

Ask: How does the design of a battery allow it to fit into a circuit and complete the path? *The battery has connections at both ends so that charge can follow a closed path to and from the battery.*

CHECK YOUR READING *Answer: They open and close the circuit.*

READING VISUALS *Answer: No; the circuit is not complete.*

Integrate the Sciences

Multiple sclerosis (MS) is caused by the "short circuiting" of electrical impulses along nerve pathways. Messages between the human brain and other body parts travel along nerve paths that are similar to wires in circuits. These nerve paths are covered and protected with myelin, which acts as insulation. In MS, the myelin begins to break down, electrical impulses leak out of the nerve path, and a "short circuit" results. People with MS may have difficulty with movement, balance, sight, and other bodily functions.

Teach from Visuals

To help students interpret the two illustrations of wiring in a lamp, ask:

- In which lamp is the wiring damaged? *lamp 2*

- How can you tell? *Frayed ends are unintentionally connecting the input and output wires to create a short circuit.*

Ongoing Assessment

CHECK YOUR READING *Answer: Charge flows where it is not meant to, and it could flow through a person if the person touches it or start a fire if the wires become overheated.*

OUTLINE
Add this heading to your outline, along with supporting ideas.

. I. Main idea
   A. Supporting idea
     1. Detail
     2. Detail
   B. Supporting idea

E

RESOURCE CENTER
CLASSZONE.COM

Explore resources on electrical safety.

Current follows the path of least resistance.

Since current can follow only a closed path, why are damaged cords so dangerous? And why are people warned to stay away from fallen power lines? Although current follows a closed path, the path does not have to be made of wire. A person can become a part of the circuit, too. Charge flowing through a person is dangerous and sometimes deadly.

Current follows the path of least resistance. Materials with low resistance, such as certain metals, are good conductors. Charge will flow through a copper wire but not the plastic coating that covers it because the copper is a good conductor and plastic is not. Water is also a good conductor when mixed with salt from a person's skin. That is why it is dangerous to use electrical devices near water.

Short Circuits

F

A **short circuit** is an unintended path connecting one part of a circuit with another. The current in a short circuit follows a closed path, but the path is not the one it was intended to follow. The illustration below shows a functioning circuit and a short circuit.

1 **Functioning Circuit** The charge flows through one wire, through the light bulb, and then back through the second wire to the outlet.

2 **Short Circuit** The cord has been damaged and the two wires inside have formed a connection. Now the path of least resistance is through one wire and back through the second wire.

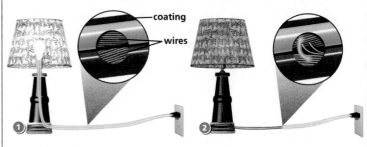

In the second case, without the resistance from the lamp, there is more current in the wires. Too much current can overheat the wires and start a fire. When a power line falls, charge flows along the wire and into the ground. If someone touches that power line, the person's body becomes part of the path of charge. That much charge flowing through a human body is almost always deadly.

CHECK YOUR READING Why are short circuits dangerous?

DIFFERENTIATE INSTRUCTION

? More Reading Support

E In an electrical circuit, what follows the path of least resistance? *current*

F What do you call an unintended path connecting one part of a circuit with another? *short circuit*

Below Level Review the concepts of conductors and insulators. Relate these concepts to a short circuit. Help students understand the critical role of proper insulation in preventing short circuits. Ask: What might happen if you drove a nail into an electric cord? Why would this be very dangerous? *The nail could create a short circuit, because it is made of metal and is a good conductor. A short circuit would be dangerous because the nail could conduct a high current to another low-resistance material outside of the cord.*

Grounding a Circuit

?
G

Recall that when lightning strikes a lightning rod, charge flows into the ground through a highly conductive metal rod rather than through a person or a building. In other words, the current follows the path of least resistance. The third prong on some electrical plugs performs a similar function. A circuit that connects stray current safely to the ground is known as a grounded circuit. Because the third prong grounds the circuit, it is sometimes called the ground.

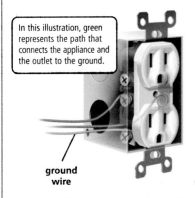

In this illustration, green represents the path that connects the appliance and the outlet to the ground.

Orange is used in this illustration to represent the path that connects the appliance's circuit to a power source and back.

ground wire

connects to ground wire

Normally, charge flows through one prong, along a wire to an appliance, then back along a second wire to the second prong. If there is a short circuit, the charge might flow dangerously to the outside of the shell of the appliance. If there is a ground wire, the current will flow along the third wire and safely into the ground, along either a buried rod or a cold water pipe.

 CHECK YOUR READING   What is the purpose of a ground wire?

Safety devices control current.

?
H

Suppose your living room wiring consists of a circuit that supplies current to a television and several lights. One hot evening, you turn on an air conditioner in the living room window. The wires that supply current to the room are suddenly carrying more current than before. The lights in the room become dim. Too much current in a circuit is dangerous. How do you know if there is too much current in a wire?

Fortunately, people have been using electric current for over a hundred years. An understanding of how charge flows has led to the development of safety devices. These safety devices are built into circuits to prevent dangerous situations from occurring.

⚠ **SAFETY TIPS**

- Never go near a fallen power line.

- Never touch an electrical appliance when you are in the shower or bathtub.

- Always dry your hands thoroughly before using an electrical appliance.

- Never use an electrical cord that is damaged in any way.

- Never bend or cut a ground prong in order to make a grounded plug fit into an ungrounded outlet.

Teach from Visuals

To help students understand the illustration of the outlet and plug, ask: What has the artist done to help viewers understand how a ground wire works? *removed the wall behind the plug box to show the wires and made the plug transparent*

Real World Example

If you were to see the actual wires inside the coating, you would find that the current-bearing hot wires are red and the neutral wires are gray or white. This illustration uses orange to indicate the entire pathway for the circuit to and from the outlet and green for the entire grounding pathway.

Teacher Demo

Show a three-prong plug and a two-prong plug to the class. Ask: Why do some plugs have a grounding plug and others do not? *Appliances that draw large current or have a metal case need greater safety protection than appliances that draw a small amount of current or are encased in plastic.* Hold up the two-prong plug. Ask: If you plug this into a grounded outlet, will the appliance be grounded? *No, because the cord does not have a ground wire.*

Ongoing Assessment

 Answer: to conduct stray current safely to the ground

DIFFERENTIATE INSTRUCTION

?
More Reading Support

G What is a grounded circuit? *an electric circuit that conduct stray current safely to the ground*

H What happens if there is to much current in a wire? *it is dangerous*

Additional Investigation To reinforce Section 2.1 learning goals, use the following full-period investigation:

R **Additional INVESTIGATION,** Wire a Room, A, B, & C, pp. 123–131, 206–207 (Advanced students should complete Levels B and C.)

Teach from Visuals

Ask: How do the pictures of a new fuse and a blown fuse differ? *The new fuse shows the metal strip from end-to-end of the tube; the metal strip of the blown fuse does not reach the other end.*

INVESTIGATE Fuses

PURPOSE To model a fuse by using steel wool to open a simple circuit

TIP *15 min.*

SAFETY Students should not complete the circuit with the clips until the steel wool strand is inside the jar. Resistance can produce enough heat in the steel wool to cause burns.

INCLUSION Carefully monitor the proceedings of any ADD or ADHD students during this investigation or pair them with students who follow directions very carefully. Make sure they do not complete the circuit while the steel wool strand is touching skin or flammable materials. (It should be inside the jar.)

WHAT DO YOU THINK? *The circuit was opened because the current melted the steel wool strand. Opening the circuit stopped the current.*

CHALLENGE *Both a fuse and the steel wool melt and open a circuit when there is too much current in them. In a home circuit, there would be other devices in a circuit that would stop working when a fuse was blown.*

 Datasheet, Fuses, p. 82

Technology Resources

Customize this student lab as needed or look for an alternative. Print rubrics to assess student lab reports.

 Lab Generator CD-ROM

Ongoing Assessment

Identify electrical safety devices and explain how they stop current.

Ask: Why does a normal level of current leave a fuse intact? *A fuse is designed with metal that is solid at normal current. It melts only if there is enough current to heat it up.*

How Fuses Work

If you turn on an air conditioner in a room full of other electrical appliances that are already on, the circuit could overheat. But if the circuit contains a fuse, the fuse will automatically shut off the current. A fuse is a safety device that opens a circuit when there is too much current in it. Fuses are typically found in older homes and buildings. They are also found in cars and electrical appliances like air conditioners.

A fuse consists of a thin strip of metal that is inserted into the circuit. The charge in the closed circuit flows through the fuse. If too much charge flows through the fuse, the metal strip melts. When the strip has melted and the circuit is open, the fuse is blown. The photographs on the left show a new fuse and a blown fuse. As you can see, charge cannot flow across the melted strip. It has broken the circuit and stopped the current.

How much current is too much? That varies. The electrician who installs a circuit knows how much current the wiring can handle. He or she uses that knowledge to choose the right kind of fuse. Fuses are measured in amperes, or amps. Remember that amperage is a measure of current. If a fuse has blown, it must be replaced with a fuse of the same amperage. But a fuse should be replaced only after the problem that caused it to blow has been fixed.

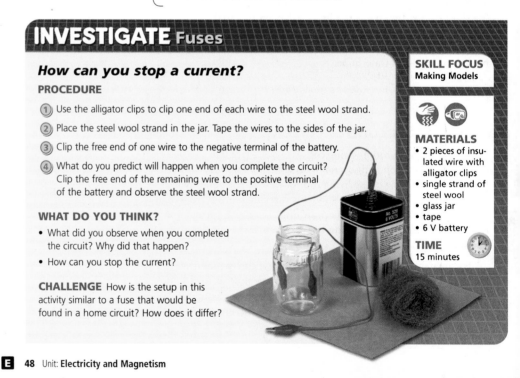

new fuse

blown fuse

INVESTIGATE Fuses

How can you stop a current?

PROCEDURE

1. Use the alligator clips to clip one end of each wire to the steel wool strand.
2. Place the steel wool strand in the jar. Tape the wires to the sides of the jar.
3. Clip the free end of one wire to the negative terminal of the battery.
4. What do you predict will happen when you complete the circuit? Clip the free end of the remaining wire to the positive terminal of the battery and observe the steel wool strand.

WHAT DO YOU THINK?

- What did you observe when you completed the circuit? Why did that happen?
- How can you stop the current?

CHALLENGE How is the setup in this activity similar to a fuse that would be found in a home circuit? How does it differ?

SKILL FOCUS
Making Models

MATERIALS
- 2 pieces of insulated wire with alligator clips
- single strand of steel wool
- glass jar
- tape
- 6 V battery

TIME
15 minutes

DIFFERENTIATE INSTRUCTION

 More Reading Support

I What happens to a fuse when too much current flows in a circuit? *Metal in the fuse melts and breaks the circuit so that current can no longer flow in it.*

Advanced Have students borrow the instruction manual for a car. Prompt them to find information about fuses used in the car. Students might find where fuses are used, what the fuse ratings are (in amps), and how often particular fuses should be replaced during regular maintenance. Encourage participants to connect their findings with the concepts presented in this chapter, and share their ideas with the class.

 Challenge and Extension, p. 81

Other Safety Devices

Most modern homes do not use fuses. Instead, they use safety devices called circuit breakers. Circuit breakers, unlike fuses, do not have to be replaced every time they open the circuit. Like fuses, circuit breakers automatically open the circuit when too much charge flows through it. If the circuit becomes overloaded or there is a short circuit, the wire and the breaker grow hot. That makes a piece of metal inside the breaker expand. As it expands, it presses against a switch. The switch is then flipped to the off position and the current is stopped. Once the problem is solved, power can be restored manually by simply flipping the switch back. The illustration on the right shows a circuit breaker.

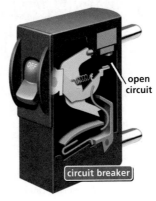

open circuit

circuit breaker

 CHECK YOUR READING How are circuit breakers similar to fuses?

The photograph at the bottom right shows another safety device—a ground-fault circuit interrupter (GFCI) outlet. Sometimes a little current leaks out of an outlet or an appliance. Often it is so small you do not notice it. But if you happen to have wet hands, touching even a small current can be very dangerous.

GFCI outlets are required in places where exposure to water is common, such as in kitchens and bathrooms. A tiny circuit inside the GFCI outlet monitors the current going out and coming in. If some of the current starts to flow through an unintended path, there will be less current coming in to the GFCI. If that happens, a circuit breaker inside the GFCI outlet opens the circuit and stops the current. To close the circuit again, you push "Reset."

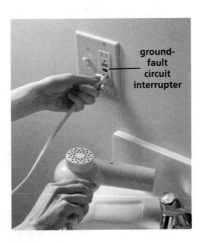

ground-fault circuit interrupter

2.1 Review

KEY CONCEPTS

1. Describe three parts of a circuit and explain what each part does.
2. Explain the function of a ground wire.
3. What do fuses and circuit breakers have in common?

CRITICAL THINKING

4. **Apply** Suppose you have built a circuit for a class project. You are using a flat piece of wood for its base. How could you make a switch out of a paperclip and two nails?
5. **Communicate** Draw a diagram of a short circuit. Use the symbols for the parts of a circuit.

CHALLENGE

6. **Evaluate** A fuse in a home has blown and the owner wants to replace it with a fuse that can carry more current. Why might the owner's decision lead to a dangerous situation?

Chapter 2: **Circuits and Electronics** 49 **E**

ANSWERS

1. *Sample answer: The voltage source provides the electric potential for charge to flow in a circuit. The conductor provides a path on which charge can flow. A switch turns the circuit on and off.*

2. *It prevents current from flowing where it could be*

dangerous and instead sends it into the ground.

3. *Both are safety devices; both stop the current.*

4. *You could attach the paper clip to one nail. To close the switch, bend it down so that it connects with the other nail.*

5. *Diagrams should include a voltage source and connection.*

6. *It might allow too much current in the circuit, which could heat wires and start a fire.*

Set Learning Goal

To understand how electricians use their knowledge about electricity

Present the Science

Electricians have special meters that enable them to safely test current, voltage, and resistance in circuits. A multimeter is one such device. You can use a multimeter to test for continuity (whether a circuit is open or closed) and to measure the resistance of parts of a circuit.

Discussion Questions

Tell students to think of themselves in the role of an electrician modernizing an electrical system in an old building. Ask:

- How would you evaluate the old system? *test the function with meters, lamps, etc.*

- Which safety devices would you upgrade or add? *add new wire to carry more current, install circuit breakers instead of fuses, replace wires that had worn insulation*

- How would you test the new system? *load up with some appliances, see if breakers work and wires stay cool; use meters to make sure ground wires work*

Close

Ask: Why is it important that a person studying to be an electrician learn his or her coursework thoroughly? *The coursework would include safety information that an electrician must know. If students studying to be electricians failed to learn critical safety information, they could seriously injure themselves or others.*

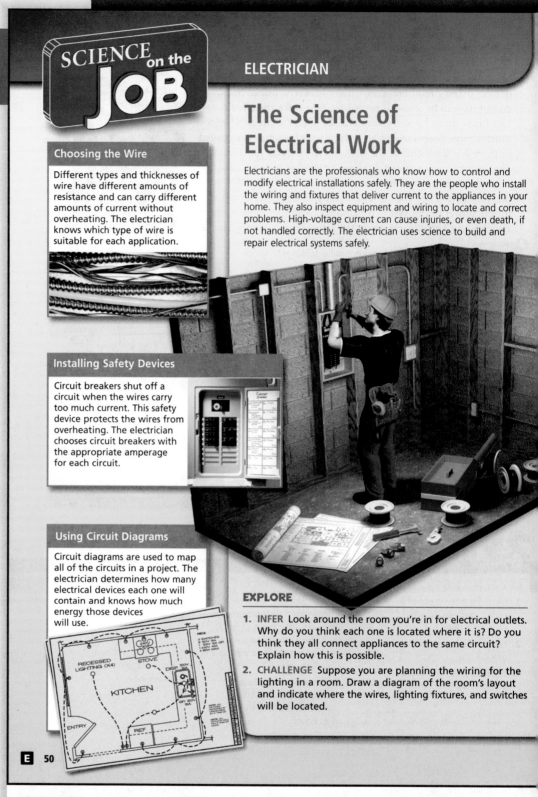

SCIENCE on the JOB

ELECTRICIAN

The Science of Electrical Work

Electricians are the professionals who know how to control and modify electrical installations safely. They are the people who install the wiring and fixtures that deliver current to the appliances in your home. They also inspect equipment and wiring to locate and correct problems. High-voltage current can cause injuries, or even death, if not handled correctly. The electrician uses science to build and repair electrical systems safely.

Choosing the Wire

Different types and thicknesses of wire have different amounts of resistance and can carry different amounts of current without overheating. The electrician knows which type of wire is suitable for each application.

Installing Safety Devices

Circuit breakers shut off a circuit when the wires carry too much current. This safety device protects the wires from overheating. The electrician chooses circuit breakers with the appropriate amperage for each circuit.

Using Circuit Diagrams

Circuit diagrams are used to map all of the circuits in a project. The electrician determines how many electrical devices each one will contain and knows how much energy those devices will use.

E 50

EXPLORE

1. **INFER** Look around the room you're in for electrical outlets. Why do you think each one is located where it is? Do you think they all connect appliances to the same circuit? Explain how this is possible.

2. **CHALLENGE** Suppose you are planning the wiring for the lighting in a room. Draw a diagram of the room's layout and indicate where the wires, lighting fixtures, and switches will be located.

EXPLORE

1. *INFER Sample answer: The outlets are placed low so that cords can reach them, and there are more of them in the front of the room where the computer and overhead projector are. They could all be on one circuit if they each form their own complete connection to and from the power source.*

2. *CHALLENGE Students' diagrams should include wires, fixtures, and switches drawn in reasonable places.*

2.2 Circuits make electric current useful.

◀ **BEFORE, you learned**

- Charge flows in a closed circuit
- Circuits have a voltage source, conductor, and one or more electrical devices
- Current follows the path of least resistance

▶ **NOW, you will learn**

- How circuits are designed for specific purposes
- How a series circuit differs from a parallel circuit
- How electrical appliances use circuits

VOCABULARY

series circuit p. 52
parallel circuit p. 53

THINK ABOUT

How does it work?

You know what a telephone does. But did you ever stop to think about how the circuits and other electrical parts inside of it work together to make it happen?

This photo shows an old telephone that has been taken apart to reveal its circuits. As you can see, there are a lot of different parts. Each one has a function. Pick two or three of the parts. What do you think each part does? How do you think it works? How might it relate to the other parts inside the telephone?

OUTLINE
Remember to include this heading in your outline.

I. Main idea
  A. Supporting idea
    1. Detail
    2. Detail
  B. Supporting idea

Circuits are constructed for specific purposes.

How many things around you right now use electric current? Current is used to transfer energy to so many things because it is easy to store, distribute, and turn off and on. Each device that uses current is a part of at least one circuit—the circuit that supplies its voltage.

Most electrical appliances have many circuits inside of them that are designed to carry out specific functions. Those circuits may be designed to light bulbs, move motor parts, or calculate. Each of those circuits may have thousands—or even millions—of parts. The functions that a circuit can perform depend on how those parts are set up within the circuit.

CHECK YOUR READING Why is the design of a circuit important?

RESOURCES FOR DIFFERENTIATED INSTRUCTION

Below Level

UNIT RESOURCE BOOK
- Reading Study Guide A, pp. 86–87
- Decoding Support, p. 110

 AUDIO CDS

Advanced

UNIT RESOURCE BOOK
Challenge and Extension, p. 92

English Learners

UNIT RESOURCE BOOK
Spanish Reading Study Guide, pp. 90–91

AUDIO CDS

- Audio Readings in Spanish
- Audio Readings (English)

▶ **Set Learning Goals**
Students will

- Identify how circuits are designed for specific purposes.
- Contrast series and parallel circuits.
- Explain how electrical appliances use circuits.
- Experiment with parallel and series circuits.

◀ **3-Minute Warm-Up**

Display Transparency 12 or copy this exercise on the board:

Decide if these statements are true. If not, correct them.

1. When a light is switched off, the circuit is closed. *open*

2. A fuse is a safely device that prevents electrical fires. *true*

3. Electric current flows in the most resistant path. *least resistant*

T 3-Minute Warm-Up, p. T12

2.2 MOTIVATE

THINK ABOUT

PURPOSE To show that electrical devices include many circuits

DISCUSS Have volunteers explain what they see in the telephone circuitry and offer their ideas about the functions of the circuits. If possible, bring in the parts of an old telephone to pass around. Speculate about how different functional circuits might be linked together inside the telephone.

Ongoing Assessment

Identify how circuits are designed for specific purposes.

Ask: What makes an electrical appliance appropriate for its job? *the combination of many circuits*

CHECK YOUR READING *Answer: The design determines the circuit's function.*

Teach from Visuals

To help students understand the relationship between the realistic rendering of the series circuit and the schematic diagram, ask:

• How are the two pictures of the series circuit similar? *Both pictures show the same circuit.*

• How are they different? *One is a real photograph while the other is a diagram with symbols.*

Address Misconceptions

IDENTIFY Ask: Suppose you have a series circuit with three identical light bulbs. You know that each bulb converts some of the electrical energy into light and heat. How do you think the brightness of the third bulb will compare to the brightness of the first bulb? If students answer that it will be dimmer, they may hold the misconception that diminution—or reduction—of current in a series circuit is progressive per resistor, rather than universal across the circuit.

CORRECT Remind students that a circuit is a closed loop. Any change in resistance anywhere in the circuit affects current everywhere in the closed loop. The effect of a change in resistance is not limited to just part of the circuit.

REASSESS Again have students imagine a closed circuit with three light bulbs that are lit. Ask: What will happen to the brightness of the three bulbs when I add a fourth bulb? *All the light bulbs will dim.*

Ongoing Assessment

CHECK YOUR READING *Answer: If one light bulb burns out, the rest of the bulbs turn off. The bulbs in a series circuit dim with each added bulb.*

Circuits can have multiple paths.

Even a simple circuit can contain several parts. When you flip the light switch in your classroom, how many different lights go on? If you count each light bulb or each fluorescent tube, there might be as many as ten or twelve light bulbs. There is more than one way those light bulbs could be connected in one circuit. Next, you will read about two simple ways that circuits can be constructed.

Series Circuits

READING TiP
The word *series* means a number of things arranged one after another.

 A

A **series circuit** is a circuit in which current follows a single path. That means that all of the parts in a series circuit are part of the same path. The photograph and diagram below show a series circuit. The charge coming from the D cell flows first through one light bulb, and then through the next one.

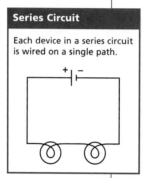

Series Circuit

Each device in a series circuit is wired on a single path.

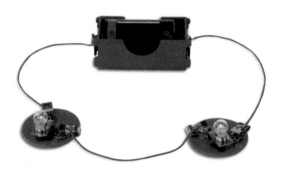

 B

A series circuit uses a minimal amount of wire. However, a disadvantage of a series circuit is that all of the elements must be in working order for the circuit to function. If one of the light bulbs burns out, the circuit will be broken and the other bulb will be dark, too. Series circuits have another disadvantage. Light bulbs and other resistors convert some energy into heat and light. The more light bulbs that are added to a series circuit, the less current there is available, and the dimmer all of the bulbs become.

CHECK YOUR READING Give two disadvantages of a series circuit.

If voltage sources are arranged in series, the voltages will add together. Sometimes batteries are arranged in series to add voltage to a circuit. For example, the circuits in flashlights are usually series circuits. The charge flows through one battery, through the next, through the bulb, and back to the first battery. The flashlight is brighter than it would be if its circuit contained only a single battery.

DIFFERENTIATE INSTRUCTION

? **More Reading Support**

A In which type of circuit does current follow a single path? *series circuit*

B What happens when one bulb out of several goes out in a series circuit? *They all go out.*

English Learners Help students recognize sentences that indicate a cause-effect relationship. Have them make a two-column chart like the one below. Give them several cause-effect sentences and have them put each sentence part under the correct heading. (The examples are from this page.)

Cause	Effect
If one of the light bulbs burns out,	the circuit will be broken and the other bulb will be dark, too.
The more light bulbs that are added to a series circuit,	the less current there is available, and the dimmer all of the bulbs become.

Parallel Circuits

C

A **parallel circuit** is a circuit in which current follows more than one path. Each path is called a branch. The current divides among all possible branches, so that the voltage is the same across each branch. The photograph and diagram below show a simple parallel circuit.

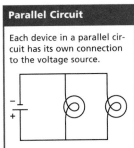

Parallel Circuit

Each device in a parallel circuit has its own connection to the voltage source.

Parallel circuits require more wire than do series circuits. On the other hand, there is more than one path on which the charge may flow. If one bulb burns out, the other bulb will continue to glow. As you add more and more light bulbs to a series circuit, each bulb in the circuit grows dimmer and dimmer. Because each bulb you add in a parallel circuit has its own branch to the power source, the bulbs burn at their brightest.

A flashlight contains batteries wired in a series circuit. Batteries can be wired in parallel, too. If the two positive terminals are connected to each other and the two negative terminals are connected to each other, charge will flow from both batteries. Adding batteries in parallel will not increase the voltage supplied to the circuit, but the batteries will last longer.

The circuits in most businesses and homes are connected in parallel. Look at the illustration of the kitchen and its wiring. This is a parallel circuit, so even if one electrical device is switched off, the others can still be used. The circuits within many electrical devices are combinations of series circuits and parallel circuits. For example, a parallel circuit may have branches that contain several elements arranged in series.

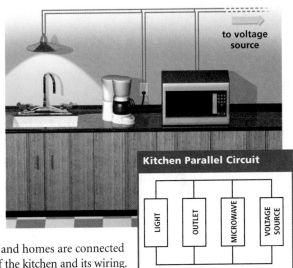

to voltage source

Kitchen Parallel Circuit

LIGHT | OUTLET | MICROWAVE | VOLTAGE SOURCE

Why are the circuits in buildings and homes arranged in parallel?

DIFFERENTIATE INSTRUCTION

INVESTIGATE Circuits

PURPOSE To experiment with series and parallel circuits in order to determine which arrangement makes light bulbs burn brighter

TIPS *15 min.*

- The clips may get warm from carrying current. Have students disconnect the wires when they are finished.
- Students with motor disabilities can be paired with students who can perform tasks requiring manual dexterity, such as clipping wires to battery terminals.

WHAT DO YOU THINK? *The first circuit; series; the voltages added together.*

CHALLENGE *The diagram or sketch should show four light bulbs connected in a parallel circuit.*

 Datasheet, Circuits, p. 93

Technology Resources

Customize this student lab as needed or look for an alternative. Print rubrics to assess student lab reports.

 Lab Generator CD-ROM

Teaching with Technology

Have students use graphics software to draw a schematic diagram of the circuits they create in the investigation. Have them include a key like that on p. 45.

Integrate the Sciences

Like electrical appliances that convert electrical energy into other forms of energy, the human body converts chemical energy into other forms of energy. For example, muscle cells break down glucose (blood sugar) molecules, releasing chemical energy. The muscle tissue converts this energy into energy of motion so the person can push, pull, lift, walk, and so on.

Ongoing Assessment

CHECK YOUR READING *Answer: heat, motion, sound*

INVESTIGATE Circuits

How can you produce brighter light?

PROCEDURE

1. Clip one end of a wire to the light bulb and the other end to the negative terminal of one battery to form a connection.
2. Use another wire to connect the positive terminal of the battery with the negative terminal of a second battery, as shown in the photograph.
3. Use a third wire to connect the positive terminal of the second battery to the light bulb. Observe the light bulb.
4. Remove the wires. Find a way to reconnect the wires to produce the other type of circuit.

WHAT DO YOU THINK?

- Which circuit produced brighter light? What type of circuit was it?
- Why did the light bulb glow brighter in that circuit?

CHALLENGE Suppose you wanted to construct a new circuit consisting of four light bulbs and only one battery. How would you arrange the light bulbs so that they glow at their brightest? Your answer should be in the form of either a diagram or a sketch of the circuit.

Circuits convert electrical energy into other forms of energy.

We use electrical energy to do many things besides lighting a string of light bulbs. For example, a circuit in a space heater converts electrical energy into heat. A circuit in a fan converts electrical energy into motion. A circuit in a bell converts electrical energy into sound. That bell might also be on a circuit that makes it ring at certain times, letting you know when class is over.

Branches, switches, and other elements in circuits allow for such control of current that our calculators and computers can use circuits to calculate and process information for us. All of these things are possible because voltage is electric potential that can be converted into energy in a circuit.

CHECK YOUR READING Name three types of energy that electrical energy can be converted into.

DIFFERENTIATE INSTRUCTION

? More Reading Support

D What kind of energy do circuits carry? *electrical energy*

Advanced Challenge students to find out what type of energy a microwave oven converts electrical energy into and how that type of energy cooks food. For example, one way is by dispersing the energy so that it reaches all surfaces of the food. Students could contrast the way a microwave oven cooks food with the way a conventional electrical oven does.

 Challenge and Extension, p. 92

A toaster is an example of an electrical appliance containing a circuit that converts energy from one form to another. In a toaster, electrical energy is converted into heat. Voltage is supplied to the toaster by plugging it into a wall outlet, which completes the circuit from a power plant. The outlet is wired in parallel with other outlets, so the appliance will always be connected to the same amount of voltage.

① When you push the handle down, a piece of metal connects to contact points on a circuit board that act as a switch and run current through the circuit.

② Charge flows through a resistor in the circuit called a heating element. The heating element is made up of a type of wire that has a very high resistance. As charge flows through the heating element, electrical energy is converted into heat.

③ The holder in the toaster is loaded onto a timed spring. After a certain amount of time passes, the spring is released, the toast pops up, and the circuit is opened. The toaster shuts off automatically, and your toast is done.

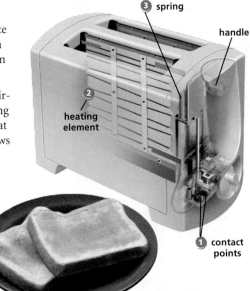

③ spring

handle

② heating element

① contact points

CHECK YOUR READING Summarize the way a circuit in a toaster works. (Remember that a summary includes only the most important information.)

2.2 Review

KEY CONCEPTS

1. Explain how a circuit can perform a specific function.

2. How are series circuits and parallel circuits similar? How do they differ?

3. Describe three electrical appliances that use circuits to convert electrical energy into other forms of energy.

CRITICAL THINKING

4. **Analyze** Why are the batteries of flashlights often arranged in series and not in parallel?

5. **Infer** You walk past a string of small lights around a window frame. Only two of the bulbs are burned out. What can you tell about the string of lights?

⚠ CHALLENGE

6. **Apply** Explain how the circuit in a space heater converts electrical energy into heat. Draw a diagram of the circuit, using the standard symbols for circuit diagrams.

Chapter 2: **Circuits and Electronics 55** **E**

Ongoing Assessment

Explain how electrical appliances use circuits.

Ask: What is the main kind of resistor in a toaster? Describe the energy conversion it performs. *A heating element converts electrical energy into heat energy.*

CHECK YOUR READING *Answer: The circuit closes when you push down the handle. Charge flows through a heating element. The heating element heats the toast.*

Reinforce (the **BIG** idea)

Have students relate the section to the Big Idea.

Ⓡ Reinforcing Key Concepts, p. 94

2.2 ASSESS & RETEACH

Assess

Ⓐ Section 2.2 Quiz, p. 23

Reteach

Use the board for this activity. In the middle of the board, draw a vertical line. Draw a battery and two light bulbs on opposite sides of the line. Have volunteers go to the board and draw lines to connect the elements in the left half in a series circuit and on the right in a parallel circuit. Ask them to identify resistors, conductors, and a voltage source.

Technology Resources

Have students visit **ClassZone.com** for reteaching of Key Concepts.

🌐 **CONTENT REVIEW**

💿 **CONTENT REVIEW CD-ROM**

Set Learning Goal

To calculate voltage drop in a circuit by using percentages

Present the Science

Voltage drop is directly proportional to the resistance of a device: the higher a device's resistance, the greater the voltage drop when current passes through it. In a series circuit, the voltage drop of all the devices add up to the voltage supplied by the voltage source.

Develop Number Sense

Make sure students understand the difference between proportions and equations. A proportion is an equation in which the items are ratios. What this means is that when two ratios are equal to each other, a proportion is formed. For example, $a/b = c/d$. A good way to remember the difference is that proportions include ratios.

DIFFERENTIATION TIP For students with perceptual difficulties, work out the proportion on graph paper, magnifying the dimensions. Place parts of the proportion so that numerator and denominator positions—and the line separating each pair—are very clear.

Close

Direct students' attention to the lights in the picture and have them read the caption. Ask: How could an electrician decide whether the voltage drop in this light display is safe or not? *The electrician could determine the rated voltage for the wires and then figure out what 5 percent of those values would be. He or she could test the wires with special equipment and compare the test values with the calculated percentage.*

- Math Support, p. 111
- Math Practice, p. 112

Technology Resources

Students can visit **ClassZone.com** for practice with percents and proportions.

 MATH TUTORIAL

MATH in SCIENCE

MATH TUTORIAL
CLASSZONE.COM
Click on Math Tutorial for more help with percents and proportions.

SKILL: SOLVING PERCENT PROBLEMS

Voltage Drop

A voltage drop occurs when current passes through a wire or an electrical device. The higher the resistance of a wire, the greater the voltage drop. Too much voltage drop can cause the device to overheat.

The National Electric Code—a document of guidelines for electricians—states that the voltage drop across a wire should be no more than 5 percent of the voltage from the voltage source. To find 5 percent of a number, you can set up the calculation as a proportion.

Example

The lighting in a hotel includes many fixtures that will be arranged on long wires. The electrician needs to know the maximum voltage drop allowed in order to choose the proper wire. The circuit will use a voltage source of 120 V. What is 5% of 120?

(1) Write the problem as the proportion.

$$\frac{\text{voltage drop}}{\text{voltage}} = \frac{\text{percent}}{100}$$

(2) Substitute.

$$\frac{\text{voltage drop}}{120} = \frac{5}{100}$$

(3) Calculate and simplify.

$$\frac{\text{voltage drop}}{120} \cdot 120 = \frac{5}{100} \cdot 120$$

$$\text{voltage drop} = 6$$

ANSWER The maximum voltage drop in the wire is 6 V.

Use the proportion to answer the following questions.

1. If the voltage source is increased to 277 V, what is the maximum voltage drop in the wire?

2. To be on the safe side, the electrician decided to find a wire with a voltage drop that is 3 percent of the voltage from the voltage source. What is the voltage drop in the wire?

CHALLENGE A student wants to hang a string of lights outside and connect it to an extension cord. The voltage drop across the extension cord is 3.1 V. The outlet supplies 240 V. Does the voltage drop in the extension cord meet the code guidelines?

The many lights in this spectacular display in Kobe, Japan, produce a large voltage drop. The appropriate type of wire must be used to supply its current.

ANSWERS

1. *voltage drop/voltage = percent/100; voltage drop/277 = 5/100; voltage drop = 5/100 · 277; voltage drop = 13.85 V*
2. *voltage drop/voltage = percent/100; voltage drop/120 = 3/100; voltage drop = (3/100) · 120; voltage drop = 3.6 V*
CHALLENGE *3.1/240 = percent/100; percent = (3.1/240) · 100; percent = 1.3 Yes; 1.3 percent is less than 5 percent.*

KEY CONCEPT

2.3 Electronic technology is based on circuits.

◀ **BEFORE,** you learned
- Charge flows in a closed loop
- Circuits are designed for specific purposes
- Electrical appliances use circuits

▶ **NOW,** you will learn
- How information can be coded
- How computer circuits use digital information
- How computers work

VOCABULARY

electronic p. 57
binary code p. 58
digital p. 58
analog p. 60
computer p. 61

EXPLORE Codes

How can information be coded?

PROCEDURE

① Write the numbers 1 to 26 in your notebook. Below each number, write a letter of the alphabet. This will serve as your key.

② On a separate piece of paper, write the name of the street you live on using numbers instead of words. For each letter of the word, use the number that is directly above it on your key.

③ Exchange messages with a partner and use your key to decode your partner's information.

MATERIALS
- notebook
- small piece of paper

WHAT DO YOU THINK?
- How can information be coded?
- Under what types of circumstances would information need to be coded?

Electronics use coded information.

A code is a system of symbols used to send a message. Language is a code, for example. The symbols used in written language are lines and shapes. The words on this page represent meanings coded into the form of letters. As you read, your brain decodes the lines and shapes that make up each word, and you understand the message that is encoded.

RESOURCE CENTER
CLASSZONE.COM

Find out more about electronics.

An **electronic** device is a device that uses electric current to represent coded information. In electronics, the signals are variations in the current. Examples of electronic devices include computers, calculators, CD players, game systems, and more.

CHECK YOUR READING Describe the signals used in electronic devices.

Chapter 2: Circuits and Electronics **57** E

▶ **Set Learning Goals**

Students will
- Learn how electronics use coded information.
- Explain what digital information is.
- Understand how the parts of a computer work together.
- Make a model of a digital image in an experiment.

◀ **3-Minute Warm-Up**

Display Transparency 13 or copy this exercise on the board:

Draw simple circuit diagrams for
- a series circuit
- a parallel circuit

Use standard symbols or label the parts.

T 3-Minute Warm-Up, p. T13

2.3 MOTIVATE

EXPLORE Codes

PURPOSE To encode information according to a specific set of rules

TIPS *10 min.* Leave a space between each number. Graph paper may be useful to students with visual impairments or messy handwriting.

WHAT DO YOU THINK? *by using symbols to represent letters; when sending confidential information*

EXPLORE the BIG idea

Revisit "What's Inside a Calculator?" on p. 41. Have students explain their inferences.

Ongoing Assessment

CHECK YOUR READING *Answer: They are variations in an electric current.*

RESOURCES FOR DIFFERENTIATED INSTRUCTION

Below Level

UNIT RESOURCE BOOK
- Reading Study Guide A, pp. 97–98
- Decoding Support, p. 110

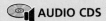 **AUDIO CDS**

Advanced

UNIT RESOURCE BOOK
- Challenge and Extension, p. 103
- Challenge Reading, pp. 106–107

English Learners

UNIT RESOURCE BOOK
Spanish Reading Study Guide, pp. 101–102

 AUDIO CDS

- Audio Readings in Spanish
- Audio Readings (English)

Teach from Visuals

To help students understand the binary decision tree, ask:

- Where in the diagram are the questions asked? *in the word bubbles to the left*

- Where are the questions answered? *above or beside the arrows*

- Where are directions given? *in the word bubbles to the right, or for the last question, in the word bubble at the bottom*

Develop Critical Thinking

INFER Have students infer a plausible explanation for the development of computer technology based on binary code, rather than some other type. Ask: Why do you suppose computer scientists developed computer technology based on binary code, rather than on some other type of number code? *Binary code is the simplest possible code to use in circuits because they require only off and on states.*

Ongoing Assessment

Learn how electronics use coded information.

Ask: What kind of information does a computer understand? *information that is binary, or digital information*

CHECK YOUR READING *Answer: Make it a series of binary questions.*

Binary Code

The English alphabet contains only 26 letters, yet there is no limit to the number of messages that can be expressed with it. That is because the message is conveyed not only by the letters that are chosen but also by the order in which they are placed.

Many electronic devices use a coding system with only two choices, as compared with the 26 in the alphabet. A coding system consisting of two choices is a **binary code.** As with a language, complex messages can be sent using binary code. In electronics, the two choices are whether an electric current is on or off. Switches in electronic circuits turn the current on and off. The result is a message represented in pulses of current.

It may be hard to imagine how something as complex as a computer game can be expressed with pulses of current. But it is a matter of breaking down information into smaller and smaller steps. You may have played the game 20 questions. In that game, you receive a message by asking someone only yes-or-no questions. The player answering the questions conveys the message only in yes's and no's, a binary code.

The diagram on the left shows how a decision-making process can be written in simple steps. The diagram is similar to a computer program, which tells a computer what to do. Each step of the process has been broken down into a binary question. If you determine exactly what you mean by *cold* and *hot*, then anyone using this program—or even a computer—would arrive at the same conclusion for a given set of conditions.

? A

Diagram:
- Should I wear a jacket?
- Is it raining? — yes → Wear a raincoat.
- no
- Is it cold outside? — yes → Wear a warm coat.
- no
- Is it hot outside? — yes → Wear no jacket.
- no
- Wear a light jacket.

CHECK YOUR READING How can a process be broken down into simple steps?

Digital Information

You can think of the yes-or-no choices in a binary system as being represented by the numbers 0 and 1. Information that is represented as numbers, or digits, is called **digital** information. In electronics, a circuit that is off is represented by 0, and a circuit that is on is represented by 1.

? B

Digital information is represented in long streams of digits. Each 0 or 1 is also known as a bit, which is short for *binary digit*. A group of 8 bits is known as a byte. You might have heard the term *gigabyte* in reference to the amount of information that can be stored on a computer. One gigabyte is equal to about 1 billion bytes. That's 8 billion 0s and 1s!

DIFFERENTIATE INSTRUCTION

? More Reading Support

A How many choices does a binary code allow? *two*

B What word describes information represented by digits? *digital*

English Learners Have students write the definitions of *electronic, binary code, digital, analog,* and *computer* in their Science Word Dictionaries. Help students recognize when prepositional phrases begin a sentence. For example, "Below each number, write a letter of the alphabet" and "On a separate piece of paper, write the name of the street you live on" ("Explore Codes," p. 57). List common prepositions on a poster or on the board for students to reference.

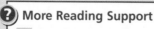

Computers, digital cameras, CD players, DVD players, and other devices use digital information. Digital information is used in electronic devices more and more. There are at least two reasons for this:

- Digital information can be copied many times without losing its quality. The 1s are always copied as 1s, and the 0s are always copied as 0s.
- Digital information can be processed, or worked with, on computers.

For example, a photograph taken on a digital camera can be input to a computer in the form of digital information. Once the photograph is on a computer, the user can modify it, copy it, store it, and send it.

Many portable devices such as game systems and MP3 players can also be used with computers. Because computers and the devices use the same type of information, computers can be used to add games, music, and other programs to the devices. The photograph at right of a watch shows an example of a portable device that uses digital information.

This watch also functions as an MP3 player—it can store songs as digital files.

CHECK YOUR READING Why is digital information often used in electronic devices?

INVESTIGATE Digital Information

How can you save a drawing in 1s and 0s?

PROCEDURE

1. Draw a 10-square by 10-square grid on a piece of graph paper.

2. Fill in some of the squares of the grid to draw a picture or pattern. Look at the example shown, but draw your own picture.

3. Starting in the upper left-hand corner of your grid, write 0 for every blank square and 1 for every filled-in square. Write a continuous series of 1s and 0s for all rows.

4. Exchange coded information with a partner who has not seen your picture. Draw a new grid in your notebook and fill it in using your partner's information.

WHAT DO YOU THINK?

- How were you able to reproduce your partner's picture?
- How is this activity similar to saving an image on a computer?

CHALLENGE Suppose you used three colored markers in your drawing—red, yellow, and green. How could you represent your color drawing using only 1s and 0s?

SKILL
Making models

MATERIALS
- graph paper
- plain paper

TIME
30 minutes

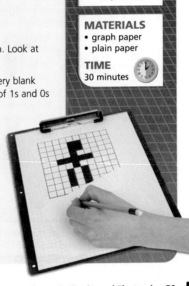

Chapter 2: **Circuits and Electronics** 59 **E**

INVESTIGATE Digital Information

PURPOSE To make a simple model of a computer graphic to demonstrate how information can be stored digitally.

TIPS *30 min.*

- Use graph paper with a large cell size.
- Some possibilities for the image include smiley faces, simple flowers, sunbursts, numbers, capital letters, and familiar symbols such as dollar and cent signs.

WHAT DO YOU THINK? *By translating the filled squares as 1s and the blank squares as 0s; the computer translates an image into a stream of 1s and 0s, or digital information.*

CHALLENGE *Use more than one digit to represent each color.*

 Datasheet, Digital Information, p. 104

Technology Resources

Customize this student lab as needed or look for an alternative. Print rubrics to assess student lab reports.

 Lab Generator CD-ROM

Ongoing Assessment

Explain what digital information is.

Ask: How are the two states of current in a digital circuit represented? *with 1s and 0s*

CHECK YOUR READING *Answer: because it can be used with computers and it can be copied many times without losing its quality*

Teach Difficult Concepts

Have students graph clock motion to illustrate the difference between analog and digital values. The hands of a clock move in a continuous circle over and over again. You can represent this movement on a line graph by showing a continuing wave with repeating high and low values. The line graph is an analog representation that connects each moment in time with the next.

Alternatively, you can represent a digital clock on a bar graph by plotting hours (one o'clock, two o'clock, etc.) or smaller time intervals as bars in repeating patterns. This bar graph would not connect one moment of time with the next. It is either one o'clock or two o'clock, for example. To help students with this concept, try the demo.

Teacher Demo

Compare a digital clock with one that has a second hand, both showing the correct time. The digital clock shows when each minute has passed, but nothing in between, even though time is passing. However, the analog clock shows a continuous movement to represent the passage of time.

Teach from Visuals

To help students understand microscopic pits in the illustration of analog and digital signals, ask: What do the pits represent? *a stream of 1s*

 This visual is also available as T14 in the Unit Transparency Book.

Real World Connection

Telephone exchanges convert analog speech into digital code before sending the signals. At the receiving end, the binary code is translated back into analog sound and delivered to the listener. Digital code allows more information to be sent along existing telephone lines.

Ongoing Assessment

READING VISUALS *Answer: the curved line; the numbers*

Analog to Digital

Some electronic devices use a system of coding electric current that differs from the digital code. Those electronics use analog information. **Analog** information is information that is represented in a continuous but varying form.

For example, a microphone records sound waves as analog information. The analog signal that is produced varies in strength as the sound wave varies in strength, as shown below. In order for the signal to be burned onto a CD, it is converted into digital information.

① The sound waves are recorded in the microphone as an analog electrical signal.

② The signal is sent through a computer circuit that measures, or samples, each part of the wave. The signal is sampled many thousands of times every second.

③ Each measurement of the wave is converted into a stream of digits. Microscopic pits representing the stream of digits are burned onto the CD. A stereo converts the signal from a digital back to analog form, making it possible for people to hear what was recorded.

Analog and Digital Signals

Sound is recorded as an analog signal and converted to digital information for storage on a CD.

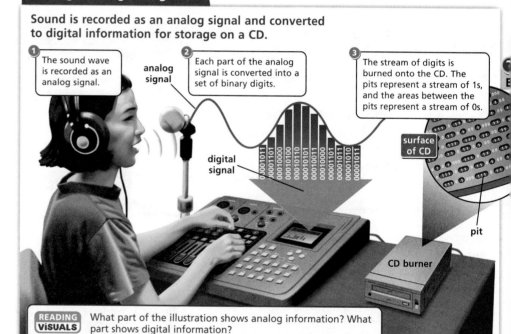

① The sound wave is recorded as an analog signal.

analog signal

② Each part of the analog signal is converted into a set of binary digits.

③ The stream of digits is burned onto the CD. The pits represent a stream of 1s, and the areas between the pits represent a stream of 0s.

digital signal

surface of CD

pit

CD burner

READING VISUALS What part of the illustration shows analog information? What part shows digital information?

DIFFERENTIATE INSTRUCTION

? More Reading Support

D Which type of information is represented in a continuous but varying form? *analog*

E Which type of information consists of a stream of digits? *digital*

Alternative Assessment Encourage students to devise their own demonstrations to illustrate the concept of analog versus digital. Have participants write a short paragraph interpreting the demonstration.

Advanced Have students who are interested in thought-communication devices read the following article:

 Challenge Reading, pp. 106–107

Computer circuits process digital information.

A **computer** is an electronic device that processes digital information. Computers have been important in science and industry for a long time. Scientists use computers to gather, store, process, and share scientific data. As computers continue to get faster, smaller, and less expensive, they are turning up in many places.

Suppose you get a ride to the store. If the car you're riding in is a newer car, it probably has a computer inside it. At the store, you buy a battery, and the clerk records the sale on a register that is connected to a computer. You put the battery in your camera and take a picture, and the camera has a computer inside it.

VOCABULARY
Remember to make a frame game diagram for *computer*.

Integrated Circuits

The first digital computer weighed 30 tons and took up a whole room. After 60 years of development, computers the size of a postage stamp are able to complete the same tasks in less time. New technology in computer circuits has led to very small and powerful computers.

Computers process information on circuits that contain many switches, branches, and other elements that allow for a very fine control of current. An integrated circuit is a miniature electronic circuit. Tiny switches, called transistors, in these circuits turn off and on rapidly, signaling the stream of digits that represent information. Over a million of these switches may be on one small integrated circuit!

CHECK YOUR READING How do integrated circuits signal digital information?

Most integrated circuits are made from silicon, an element that is very abundant in Earth's crust. When silicon is treated with certain chemicals, it becomes a good semiconductor. A semiconductor is a material that is more conductive than an insulator but less conductive than a conductor. Silicon is a useful material in computers because the flow of current in it can be finely controlled.

Microscopic circuits are etched onto treated silicon with chemicals or lasers. Transistors and other circuit parts are constructed layer by layer on the silicon. A small, complex circuit on a single piece of silicon is known as a silicon chip, or microchip.

This integrated circuit is smaller than the common ant, *Camponotus pennsylvanicus,* which ranges in length from 6 to 17 mm.

History of Science

The proliferation of computers has depended on their shrinking over the years. Circuit parts have shrunk from tubes and wires to transistors on integrated circuits. Computer size has followed suit—from main frames, to minicomputers, to personal computers, to handheld computers. Miniaturization has made it possible to computerize cars, pacemakers, and ovens, for example.

Integrate the Sciences

Fortunately for consumers of computers, silicon is the second most abundant element in Earth's crust. Almost 28 percent of the crust is silicon. Only oxygen is more abundant. While silicon never occurs in nature in its pure form, it is very common in chemical compounds in rock, sand, clay, soil, and sea water, and even in body tissues of many plants and animals.

Ongoing Assessment

CHECK YOUR READING *Answer: Tiny switches (transistors) turn on and off.*

DIFFERENTIATE INSTRUCTION

More Reading Support

F What are the tiny switches that carry the digital signal called?
transistors

English Learners Ask students to think of other contexts for the word *chip.* From their examples, have students discuss a reasonable definition for *chip,* such as "a thin, rigid piece of material." Have students connect their definition to integrated circuits.

Teach from Visuals

Make sure students understand that the illustration is spread across two pages. Ask: What do the numbers in the big illustration at the bottom of pp. 62 and 63 correspond to on the pages? *the numbered paragraphs on p. 63*

Metacognitive Strategy

Have students write and illustrate an analogy to computer hardware and software. For example, computer hardware is like the cover and pages of a book while the software is the actual story.

Develop Critical Thinking

PROVIDE EXAMPLES Have students investigate the types of software that are on computers at home, in the library, or at school. Ask students to think of games they have played on the computer. Discuss with students what some of the more common software programs are. *Sample answers might include spreadsheet software, graphic design software, and word processing software.*

Ongoing Assessment

CHECK YOUR READING *Answer: The physical parts of the computer are the hardware; the computer programs and instructions and languages are the software.*

OUTLINE
Use an outline to take notes about personal computers.

I. Main idea
  A. Supporting idea
    1. Detail
    2. Detail
  B. Supporting idea

Personal Computers

When you think of a computer, you probably think of a monitor, mouse, and keyboard—a personal computer (PC). All of the physical parts of a computer and its accessories are together known as hardware. Software refers to the instructions, or programs, and languages that control the hardware. The hardware, software, and user of a computer all work together to complete tasks.

CHECK YOUR READING What is the difference between hardware and software?

Computers have two kinds of memory. As the user is working, information is saved on the computer's random-access memory, or RAM. RAM is a computer's short-term memory. Most computers have enough RAM to store billions of bits. Another type of memory is called read-only memory, or ROM. ROM is a computer's long-term memory, containing the programs to start and run the computer. ROM can save information even after a computer is turned off.

The illustration below shows how a photograph is scanned, modified, and printed using a personal computer. The steps fall into four main functions—input, storage, processing, and output.

 G

How a PC Works

Digital information can move through input, processing, storage, and output devices.

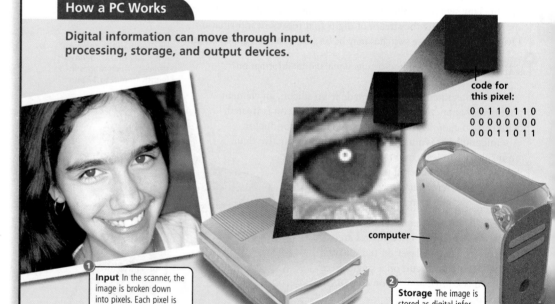

code for this pixel:
0 0 1 1 0 1 1 0
0 0 0 0 0 0 0 0
0 0 0 1 1 0 1 1

computer—

1 Input In the scanner, the image is broken down into pixels. Each pixel is translated into a series of digits representing a color.

2 Storage The image is stored as digital information on the hard drive in the computer.

E 62 Unit: Electricity and Magnetism

DIFFERENTIATE INSTRUCTION

 More Reading Support

G What are the four main functions of a computer? *input, storage, processing, output*

Below Level Write the terms from the chart below in random order on the board. Have students copy the terms and write *H* for hardware or *S* for software after each. Discuss why they made the choices they did.

Hardware	Software
• mouse	• the computer's instructions for displaying colors
• key on keyboard	• word-processing program
• integrated circuit	• spreadsheet program
• computer screen	• a search engine

1 Input The user scans the photograph on a scanner. Each small area, or pixel, of the photograph is converted into a stream of digits. The digital information representing the photograph is sent to the main computer circuit, which is called the central processing unit, or CPU.

2 Storage The user saves the photograph on a magnetic storage device called the hard drive. Small areas of the hard drive are magnetized in one of two directions. The magnetized areas oriented in one direction represent 1s, and the areas oriented in the opposite direction represent 0s, as a way to store the digital information.

3 Processing The photograph is converted back into pixels on the monitor, or screen, for the user to see. The computer below has a software program installed for altering photographs. The user adds more input to the computer with the mouse and the keyboard to improve the photograph.

4 Output The user sends the improved photograph to a printer. The printer converts the digital information back to pixels, and the photograph is printed.

See how hard drives store information.

CHECK YOUR READING During which one of the four main computer functions is information converted into digital information?

monitor

printer

4 Output Digital information is translated back into pixels and the photograph is printed.

3 Processing The image is altered using software on the computer.

READING VISUALS How has the photograph of the girl been altered on the computer?

Chapter 2: **Circuits and Electronics 63** **E**

Teach from Visuals

To verify that students interpret the Internet network map correctly, ask: Where is the center of the Web? *The Web is a network does not have a center.*

History of Science

Though the Internet was in existence long before the early 1990s, there was no universal language for posting and browsing Web sites. A British computer engineer, Tim Berners-Lee, conceived of the World Wide Web (WWW) in 1989. In 1990, he wrote the computer language HTML, which is used to structure text on Web sites, and the first Web browser and Web server, special programs that allow users to post and access Web sites.

Ongoing Assessment

CHECK YOUR READING *Answer: Earlier networks behaved like a series circuit, so if a link to one computer was broken, the entire network of links went down. If some links go down on the Internet, however, the others still work.*

Computers can be linked with other computers.

You may have been at a computer lab or a library and had to wait for a printer to print something for you. Offices, libraries, and schools often have several computers that are all connected to the same printer. A group of computers that are linked together is known as a network. Computers can also be linked with other computers to share information. The largest network of computers is the Internet.

The Origin of the Internet

People have been using computer networks to share information on university campuses and military bases for decades. The computers within those networks were connected over telephone systems from one location to another. But those networks behaved like a series circuit. If the link to one computer was broken, the whole network of links went down.

The network that we now call the Internet is different. The United States Department of Defense formed the Internet by linking computers on college campuses across the country. Many extra links were formed, producing a huge web of connected computers. That way, if some links are broken, others still work.

 CHECK YOUR READING How does the Internet differ from earlier networks?

The Internet Today

The Internet now spans the world. E-mail uses the Internet. E-mail has added to the ways in which people can "meet," communicate, conduct business, and share stories. The Internet can also be used to work on tasks that require massive computing power. For example, millions of computers linked together, along with their combined information, might one day be used to develop a cure for cancer or model the workings of a human mind.

This map shows a representation of Internet traffic in the early 1990s. A map of Internet traffic now would be even more full of lines.

DIFFERENTIATE INSTRUCTION

More Reading Support

J What is a computer network? *a group of computers that are linked together*

K What is the largest network of computers? *the Internet*

Inclusion There are various hardware and software programs available for students with disabilities. Some examples include screen magnification and enhancement software that provides higher levels of magnification, contrast, and color enhancement; screen reading software that takes information shown on a computer screen and translates it into spoken words using a speech synthesizer; and voice input, which is a keyboard and mouse alternative for typing words and sentences into a word processor, and operating program controls such as menus and buttons.

When you think of the Internet, you might think of the World Wide Web, or the Web. The Web consists of all of the information that can be accessed on the Internet. This information is stored on millions of host computers all over the world. The files that you locate are called Web pages. Each Web page has an address that begins with *www*, which stands for World Wide Web. The system allows you to search or surf through all of the information that is available on it. You might use the Web to research a project. Millions of people use the Web every day to find information, to shop, or for entertainment.

You may have heard of the Bronze Age or the Iron Age in your history class. Digital information and the Internet have had such a strong impact on the way we do things that some people refer to the era we live in as the Information Age.

Internet Usage

Host Computers (in millions) vs. Year

SOURCE: *Internet Software Consortium* (http://www.isc.org)

2.3 Review

KEY CONCEPTS

1. Describe an example of coded information.

2. What is digital information? Give three examples of devices that use digital information.

3. Give an example of each of the following in terms of computers: input, storage, processing, and output.

CRITICAL THINKING

4. **Compare** Morse code uses a signal of dots and dashes to convey messages. How is Morse code similar to digital code?

5. **Infer** The word *integrated* means "brought together to form a whole." How does that definition apply to an integrated circuit?

CHALLENGE

6. **Predict** Computers as we know them did not exist 50 years ago, and now they are used for many purposes. How do you think people will use computers 50 years from now? Write a paragraph describing what you think the computers of the future will be like and how they will be used.

Chapter 2: **Circuits and Electronics** 65 **E**

Teach from Visuals

To help students interpret the graph, ask: What does the increasing steepness of the curve mean? *Every year, more and more host computers contribute to the Internet.*

Reinforce (the **BIG** idea)

Have students relate the section to the Big Idea.

 Reinforcing Key Concepts, p. 105

2.3 ASSESS & RETEACH

Assess

 Section 2.3 Quiz, p. 24

Reteach

Have students make a drawing or diagram that represents all of the following concepts:

- computer
- binary code
- digital

Instruct them to include appropriate captions and labels. Then have students share and discuss their work.

Technology Resources

Have students visit **ClassZone.com** for reteaching of Key Concepts.

 CONTENT REVIEW

 CONTENT REVIEW CD-ROM

ANSWERS

1. Sample answer: Each part of a photograph is represented as a 1 or a 0.

2. information that is represented as numbers; examples may include computers, digital cameras, CD or DVD players, MP3 players.

3. Answers might include keystrokes, CD, calculations, communication.

4. Both use binary code to send a message.

5. Many circuit parts have been brought together to form an integrated circuit.

6. Answers should include a description of computers and some applications.

Chapter 2 **65** **E**

Focus

PURPOSE To design a battery-powered communicator to be marketed as a toy

OVERVIEW Students will design a simple device capable of visual signaling in Morse code. They will

- build a prototype,
- test it,
- and write up an evaluation.

Lab Preparation

- Copy and distribute the Morse Code Chart to students.
- Prior to the investigation have students read through the investigation and prepare their data tables. Or you may wish to copy and distribute datasheets and rubrics.

 UNIT RESOURCE BOOK, pp. 113–122

 SCIENCE TOOLKIT, F13

Lab Management

- Assign students to teams of three to five people.
- Have students make careful notes of all prototype trials. If time is short, have them write up their prototype evaluations as homework.

INCLUSION If any participants have hearing-impairments, have their team summarize the brainstorming ideas on paper or on the board. Invite students with hearing impairments to be physically involved in the experiment by letting them make the long and short signals of light.

Teaching with Technology

Use a digital camera to photograph the circuits that students produce. Post the pictures on a bulletin board, or make color photocopies to distribute to the class.

CHAPTER INVESTIGATION

Design an Electronic Communication Device

OVERVIEW AND PURPOSE

The telegraph was one of the first inventions to demonstrate that machines could be used to communicate over long distances. In a telegraph, messages are sent as electrical signals along a wire from a sending device to a receiver.

Like modern computers, the telegraph uses a binary code. The code is called Morse code—a combination of short and long signals—to stand for letters and symbols. In this lab, you will use what you have learned about circuits to

- design a battery-powered device that uses Morse code
- build and test your design

MATERIALS

- 2 batteries
- light bulb in holder
- piece of copper wire
- 2 wire leads with alligator clips
- 2 craft sticks
- toothpick
- paper clip
- piece of cardboard
- clothespin
- aluminum foil
- rubber band
- scissors
- tape
- wire cutters
- Morse code chart

▶ Problem

A toy company has contracted you to design and build a new product for kids. They want a communication device that is similar to a telegraph. Kids will use the device to communicate with each other in Morse code. The company's market research has shown that parents do not like noisy toys, so the company wants a device that uses light rather than sound as a signal.

▶ Procedure

1. Brainstorm ideas for a communication device that can use Morse code. Look at the available materials and think how you could make a circuit that contains a light bulb and a switch.

2. Describe your proposed design and/or draw a sketch of it in your **Science Notebook**. Include a list of the materials that you would need to build it.

INVESTIGATION RESOURCES

 CHAPTER INVESTIGATION, Design an Electronic Communication Device
- Morse Code Chart, p. 113
- Level A, pp. 114–117
- Level B, pp. 118–121
- Level C, p. 122
Advanced students should complete Levels B & C.

 Writing a Lab Report, D12–13

Technology Resources

Customize this student lab as needed or look for an alternative. Print rubrics to assess student lab reports.

Lab Generator CD-ROM

3 Show your design to a team member. Consider the constraints of each of your designs, such as what materials are available, the complexity of the design, and the time available.

4 Choose one idea or combine two ideas into a final design to test with your group. Build a sample version of your device, called a prototype.

5 Test your device by writing a short question. Translate the question into Morse code. Make long and short flashes of light on your device to send your message. Another person on your team should write down the message received in Morse code, translate the message, and send an answer.

6 Complete at least two trials. Each time, record the question in English, the question in code, the answer in code, and the answer in English.

7 Write a brief evaluation of how well the signal worked. Use the following criteria for your evaluation for each trial.

- What errors, if any, occurred while you were sending the signal?
- What errors, if any, occurred while you were receiving the signal?
- Did the translated answer make sense? Why or why not?

▶ Observe and Analyze

1. **MODEL** Draw a sketch of your final design. Label the parts. Next to your sketch, draw a circuit diagram of your device.

2. **INFER** How do the parts of your circuit allow you to control the flow of current?

3. **COMPARE** How is the signal that is used in your system similar to the digital information used by computers to process information? How does the signal differ?

4. **APPLY** A small sheet of instructions will be packaged with the device. Write a paragraph for the user that explains how to use it. Keep in mind that the user will probably be a child.

▶ Conclude

1. **EVALUATE** What problems, if any, did you encounter when testing your device? How might you improve upon the design?

2. **IDENTIFY LIMITS** What are the limitations of your design? You might consider its estimated costs, where and how kids will be able to use it, and the chances of the device breaking.

3. **APPLY** How might you modify your design so that it could be used by someone with limited vision?

4. **SYNTHESIZE** Write down the steps that you have used to develop this new product. Your first step was to brainstorm an idea.

▶ INVESTIGATE Further

CHALLENGE Design another system of communication that uses your own code. The signal should be in the form of flags. Make a table that lists what the signals mean and write instructions that explain how to use the system to communicate.

Design an Electronic Communication Device
Observe and Analyze

Table 1. Prototype Testing

	Trial 1	Trial 2
Question (English)		
Question (code)		
Answer (code)		
Answer (English)		
Evaluation		

Conclude

Chapter 2: **Circuits and Electronics** 67 **E**

▶ Observe and Analyze

1. *Drawings should be labeled and include circuit diagrams.*

2. *The switch allows you to start and stop the flow of current.*

3. *Both are binary codes. The signal is different from digital information in that it is conveyed in long and short flashes of light instead of 0s and 1s.*

4. *The instructions should be simple and in logical order. If students need help, refer them back to the procedure and their datasheets.*

▶ Conclude

1. *Improvements should reflect problems encountered.*

2. *Answers should indicate an understanding of the intended audience (children) and technological constraints. Sample answer: using light signals to communicate with Morse code takes some time; if you make a mistake, you have to go back.*

3. *Sample answer: A buzzer or bell could be wired into the circuit instead of a light bulb.*

4. *Brainstorm an idea, describe or sketch a design, compare design limitations, choose a design, build a prototype, test the device, improve on the design.*

▶ INVESTIGATE Further

CHALLENGE Answer: Accept any signaling system that is logical, well thought out, and presented neatly and legibly.

Post-Lab Discussion

- Ask: What do you think was the hardest part of the process for designing the toy? Why? *Sample answers: getting started; realizing from instructions that purpose was to build a circuit with light bulbs; recognizing that light flashes were coded information.*

- Invite students to think of a different signal that could be used to communicate Morse code to people with visual impairments. Have them design a follow-up experiment that tests a device that uses sound signals.

the BIG idea

Have students look at the photograph on pp. 40–41. Ask how they know that the student is handling circuit boards and how those circuits might be used. *Circuit boards are parts of computers and other digital devices.*

◄ KEY CONCEPTS SUMMARY

SECTION 2.1
Ask: Which device interrupts the path of an electric charge? *the switch*

Ask: Is the circuit open or closed? *closed*

SECTION 2.2
Ask: What does a parallel circuit have that a similar series circuit does not have? *separate branches from the devices to the voltage source in the circuit*

SECTION 2.3
Ask: What type of information do computer circuits use? *digital*

Review Concepts

- Big Idea Flow Chart, p. T9
- Chapter Outline, pp. T15–T16

---

2 Chapter Review

the BIG idea
Circuits control the flow of electric charge.

CONTENT REVIEW
CLASSZONE.COM

◄ KEY CONCEPTS SUMMARY

2.1 **Charge needs a continuous path to flow.**

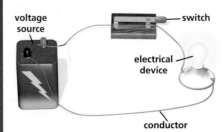
voltage source
switch
electrical device
conductor

Charge flows in a closed path. Circuits provide a closed path for current. Circuit parts include voltage sources, switches, conductors, and electrical devices such as resistors.

VOCABULARY
circuit p. 43
resistor p. 44
short circuit p. 46

2.2 **Circuits make electric current useful.**

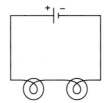
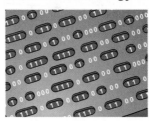

Each device in a **series circuit** is wired on a single path.

Each device on a **parallel circuit** has its own connection to the voltage source.

VOCABULARY
series circuit p. 52
parallel circuit p. 53

2.3 **Electronic technology is based on circuits.**

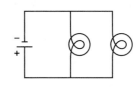

Electronic devices use electrical signals to represent coded information. Computers process information in digital code which uses 1s and 0s to represent the information.

VOCABULARY
electronic p. 57
binary code p. 58
digital p. 58
analog p. 60
computer p. 61

Technology Resources

Have students visit **ClassZone.com** or use the CD-ROM for a cumulative review of concepts.

CONTENT REVIEW

CONTENT REVIEW CD-ROM

Engage students in a whole-class interactive review of Key Concepts. Edit content as you wish.

POWER PRESENTATIONS

Reviewing Vocabulary

Draw a Venn diagram for each of the term pairs below. Write the terms above the ovals. In the center, write characteristics that the terms have in common. In the circles write the ways in which they differ. A sample diagram has been completed for you.

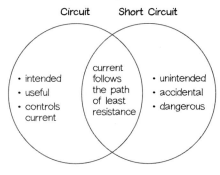

Circuit   Short Circuit

- intended
- useful
- controls current

current follows the path of least resistance

- unintended
- accidental
- dangerous

1. resistor; switch

2. series circuit; parallel circuit

3. digital; analog

4. digital; binary code

5. electronic; computer

Reviewing Key Concepts

Multiple Choice *Choose the letter of the best answer.*

6. Current always follows
   a. a path made of wire
   b. a path containing an electrical device
   c. a closed path
   d. an open circuit

7. When you open a switch in a circuit, you
   a. form a closed path for current
   b. reverse the current
   c. turn off its electrical devices
   d. turn on its electrical devices

8. Which one of the following parts of a circuit changes electrical energy into another form of energy?
   a. resistor
   b. conductor
   c. base
   d. voltage source

9. A circuit breaker is a safety device that
   a. must be replaced after each use
   b. has a wire that melts
   c. supplies voltage to a circuit
   d. stops the current

10. What happens when more than one voltage source is added to a circuit in series?
    a. The voltages are added together.
    b. The voltages cancel each other out.
    c. The voltages are multiplied together.
    d. The voltage of each source decreases.

11. Which of the following is an electronic device?
    a. flashlight
    b. calculator
    c. lamp
    d. electric fan

12. Which word describes the code used in digital technology?
    a. binary
    b. analog
    c. alphabetical
    d. Morse

13. Computers process information that has been
    a. broken down into simple steps
    b. converted into heat
    c. represented as a wave
    d. coded as an analog signal

Short Answer *Write a short answer to each question.*

14. How can hardware, software, and the user of a computer work together to complete a task?

15. Describe three parts of personal computer and explain the main function of each.

Chapter 2: **Circuits and Electronics** 69 **E**

Reviewing Vocabulary

1. *Resistor: slows current, produces heat and light, light bulb is an example. Switch: opens and closes a circuit, turns electrical devices on and off. Shared: are circuit parts, control the flow of current*

2. *Series circuit: each part wired in same path, uses minimum amount of wire; all elements must be in working order for circuit to function; resistors in it convert some energy into light and heat. Parallel circuit: each part has its own path to voltage source, it requires more wire, light bulbs burn brighter then they do in a series circuit. Shared: they are ways to wire a circuit.*

3. *Digital: a series of 1s and 0s, binary. Analog: continuous signal. Shared: types of electrical signal.*

4. *Digital: uses numbers. Binary code: uses two choices. Shared: type of code.*

5. *Electronic: uses electrical signals. Computer: carries out programs. Shared: use coded information.*

Reviewing Key Concepts

6. *c*

7. *c*

8. *a*

9. *d*

10. *a*

11. *b*

12. *a*

13. *a*

14. *A user adds input to a computer by using hardware, such as a keyboard or mouse. Software allows a user to operate programs on a computer.*

15. *Sample answer: A computer contains a hard drive, which is the main storage device. The monitor allows a user to see a document. The printer allows a user to print a document.*

Thinking Critically

16. *A–C will glow. D and E will not glow because they do not form a closed path from one battery terminal to the other.*

17. *A, B, or C; wire, light bulb*

18. *series*

19. *The other bulbs would go out.*

20. *All of the bulbs would become dimmer.*

21. *Diagrams should show a battery and three bulbs in a parallel circuit using more wire. (Accept any number between 2 and 4.)*

22. *The switch should be placed near the battery, not on a branch to a single bulb.*

23. *1011; by finding a pattern*

24. *Accept any three—heat, light, sound, and motion.*

25. *The CD uses digital information.*

26. *integrated circuit, or microprocessor*

27. *Even if one link goes down, the others can function.*

the BIG idea

28. *Answers should reflect information learned in the chapter.*

29. *Answers should discuss movement of current and devices that convert electrical energy into other forms of energy.*

30. *Sample answer: Charge needs a continuous path to flow. Charge flows in a loop, or circuit. The parts of a circuit are a voltage source, conductor, switch, and electrical device. Charge flows through closed circuits. Switches turn a circuit's electrical devices off and on by opening and closing the circuit.*

UNIT PROJECTS

Check to make sure students are working on their projects. Check schedules and work in progress.

 Unit Projects, pp. 5–10

Thinking Critically

Use the illustrations below to answer the next two questions.

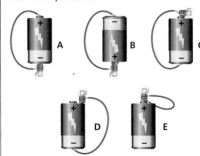

16. **PREDICT** In which arrangement(s) above will the light bulb glow? For each arrangement in which you do not think the bulb will not glow, explain your reasoning.

17. **APPLY** Which arrangement could be used as a battery tester? List the materials that you would use to make a battery tester.

Use the diagram below to answer the next five questions.

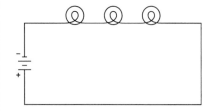

18. Is this a series circuit or a parallel circuit?

19. Explain what would happen if you unscrewed one of the bulbs in the circuit.

20. Explain what would happen if you wired three more bulbs into the circuit.

21. Draw and label a diagram of the same elements wired in the other type of circuit. Does your sketch involve more or fewer pieces of wire? How many?

22. Imagine you want to install a switch into your circuit. Where would you add the switch? Explain your answer.

 70 Unit: **Electricity and Magnetism**

---

23. **ANALYZE** Look for a pattern in the digital codes below, representing the numbers 1–10. What is the code for the number 11? How do you know?

0001; 0010; 0011; 0100; 0101; 0110; 0111; 1000; 1001; 1010

24. **APPLY** A computer circuit contains millions of switches that use temperature-dependent materials to operate lights, sounds, and a fan. How many different types of energy is current converted to in the computer circuit? Explain.

25. **ANALYZE** A music recording studio makes a copy of a CD that is itself a copy of another CD. Explain why the quality of the copied CDs is the same as the original CD.

26. **INFER** A new watch can be programmed to perform specific tasks. Describe what type of circuit the watch might contain.

27. **SYNTHESIZE** Explain how the Internet is like a worldwide parallel circuit.

the BIG idea

28. **ANALYZE** Look back at the photograph on pages 40–41. Think about the answer you gave to the question. How has your understanding of circuits changed?

29. **SYNTHESIZE** Explain how the following statement relates to electric circuits: "Energy can change from one form to another and can move from one place to another."

30. **SUMMARIZE** Write a paragraph summarizing how circuits control current. Using the heading at the top of page 43 as your topic sentence. Then give an example from each red and blue heading on pages 43–45.

UNIT PROJECTS

If you need to do an experiment for your unit project, gather the materials. Be sure to allow enough time to observe results before the project is due.

MONITOR AND RETEACH

If students have trouble applying the concepts in items 18–22, review the differences between series and parallel circuits. Have students make their own schematic sketches of the two types of circuits.

Students may benefit from summarizing one or more sections of the chapter.

Summarizing the Chapter, pp. 132–133

Standardized Test Practice

Interpreting Diagrams

The four circuit diagrams below use the standard symbols for the parts of a circuit.

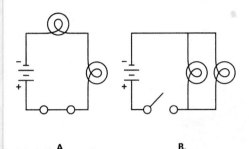 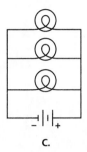 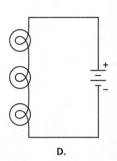

A. **B.** **C.** **D.**

Study the diagrams and answer the questions that follow.

1. Which diagram shows a series circuit, with one voltage source and two light bulbs?

  a. A   **c.** C

  b. B   **d.** D

2. Which diagram shows a parallel circuit powered by a battery, with three light bulbs?

  a. A   **c.** C

  b. B   **d.** D

3. The light bulbs in these diagrams limit the flow of charge and give off heat and light. Under which category of circuit parts do light bulbs belong?

  a. switches   **c.** resistors

  b. conductors   **d.** voltage sources

4. In which diagram would the light bulbs be dark?

  a. A   **c.** C

  b. B   **d.** D

5. If all light bulbs and voltage sources were equal, how would the light from each of the bulbs in diagram C compare to the light from each of the bulbs in diagram A?

  a. The bulbs in diagram C would give less light than the bulbs in diagram A.

  b. The bulbs in diagram C would give more light than the bulbs in diagram A.

  c. The bulbs in diagram C would give the same amount of light as the bulbs in diagram A.

  d. It cannot be determined which bulbs would give more light.

Extended Response

Answer the two questions below in detail. Include some of the terms from the word box. Underline each term you use in your answer.

flow of charge	electric current	binary code
open circuit	digital	signal

6. What are two types of safety devices designed to control electric current and prevent dangerous accidents? How does each work?

7. Explain how an electronic circuit differs from an electric circuit. What role do electronic circuits play in computer operations?

METACOGNITIVE ACTIVITY

Have students answer the following questions in their **Science Notebook:**

1. What did you find most surprising about the uses of circuits and control of electric charge?

2. Which topics in this chapter would you like to learn more about?

3. How have you solved a problem while working on your Unit Project?

Interpreting Diagrams

1. a   *3. c*   *5. b*

2. c   *4. b*

Extended Response

6. RUBRIC

4 points for a response including two of the following safety features and an explanation of how each works:

  • fuse   • GFCI outlet

  • circuit breaker

Sample: Fuses and circuit breakers are common safety devices in homes. Both stop the <u>*flow of charge*</u> *when wires become dangerously hot. A fuse has a strip of metal that melts if it gets hot enough; that* <u>*opens*</u> *the* <u>*circuit*</u>*. A circuit breaker flips to the off position when a metal switch heats up, expands, and pushes the breaker switch off.*

3 points for a response that includes two safety features and an explanation for how one of them works

2 points for a response that contains one safety feature and an explanation of how it works

1 point for a response that names one or two safety features but gives no explanation of how they work

7. RUBRIC

4 points for a response that correctly answers the question and uses the following terms accurately:

  • binary code   • signal

  • digital   • electric current

Sample: Electronic circuits carry a <u>*signal*</u> *coded in* <u>*electric current,*</u> *but electric circuits do not necessarily carry such a signal. The electric impulses in an electronic circuit are interpreted as* <u>*binary code*</u> *in* <u>*digital*</u> *processing. Computers and other digital devices process data for a variety of functions, including graphics, mathematical analysis, word processing, and controlling machines such as car engines or factory robots.*

3 points for a response that correctly answers the question and uses two of the terms accurately

2 points for a response that correctly answers the question and uses one of the terms accurately

1 point for a response that correctly answers the question but does not use the terms accurately

FOCUS

▶ Set Learning Goals

Students will

- Observe how scientists have developed electrical versions of existing devices as well as new electronic technologies
- Examine the various ways electricity is used
- Make a simple capacitor and test its ability to store charge

National Science Education Standards

A.9.a–g Understandings About Scientific Inquiry

E.6.a–c Understandings About Science and Technology

F.5.a–e, F.5.g Science and Technology in Society

G.1.a–b Science as a Human Endeavor

G.2.a Nature of Science

G.3.a–c History of Science

INSTRUCT

The top half of the timeline presents some of the major events in the scientific study of electronics that were historically recorded. The bottom half of the timeline shows advances in technology and the practical applications of electronics. The gap between 600 B.C. and A.D. 1740 represents a period of time that has been omitted on this timeline.

Application

LEYDEN JAR The first actual capacitor—a device for storing electric charge—was the Leyden jar. The basic capacitor design created in 1745 has not changed much even today. A capacitor consists of two parallel conducting plates separated by an insulating layer. While capacitors may vary in size and shape, they work in about the same way, storing and releasing charges. Capacitors are used in various electronic devices from portable radios to the electronic flash on a camera to home computers.

E 72 Unit: **Electricity and Magnetism**

THE STORY OF ELECTRONICS

Inventions such as the battery, the dynamo, and the motor created a revolution in the production and use of electrical energy. Think of how many tools and appliances that people depend on every day run on electric current. Try to imagine not using electricity in any form for an entire day.

The use of electricity as an energy source only begins the story of how electricity has changed our lives. Parts of the story are shown on this timeline. Research in electronics has given us not only electrical versions of machines that already existed but also entirely new technologies.

These technologies include computers. Electricity is used as a signal inside computers to code and transmit information. Electricity can even mimic some of the processes of logical reasoning and decision making, giving computers the power to solve problems.

600 B.C.

Thales Studies Static Electricity
Greek philosopher-scientist Thales of Miletus discovers that when he rubs amber with wool or fur, the amber attracts feathers and straw. The Greek word for amber, *elektron*, is the origin of the word *electricity*.

EVENTS

| 640 B.C. | 620 B.C. | 600 B.C. | A.D. 1740 |

APPLICATIONS AND TECHNOLOGY

APPLICATION

Leyden Jar
In 1745 German inventor Ewald Georg von Kleist invented a device that would store a static charge. The device, called a Leyden jar, was a glass container filled with water. A wire ran from the outside of the jar through the cork into the water. The Leyden jar was the first capacitor, an electronic component that stores and releases charges. Capacitors have been key to the development of computers.

E 72 Unit: Electricity and Magnetism

DIFFERENTIATE INSTRUCTION

Below Level To help students better understand the concept of a static electric charge, have students rub a piece of wool on a blown-up balloon. Then have them touch the balloon with their finger. Discuss with students what happens between the balloon and their finger.

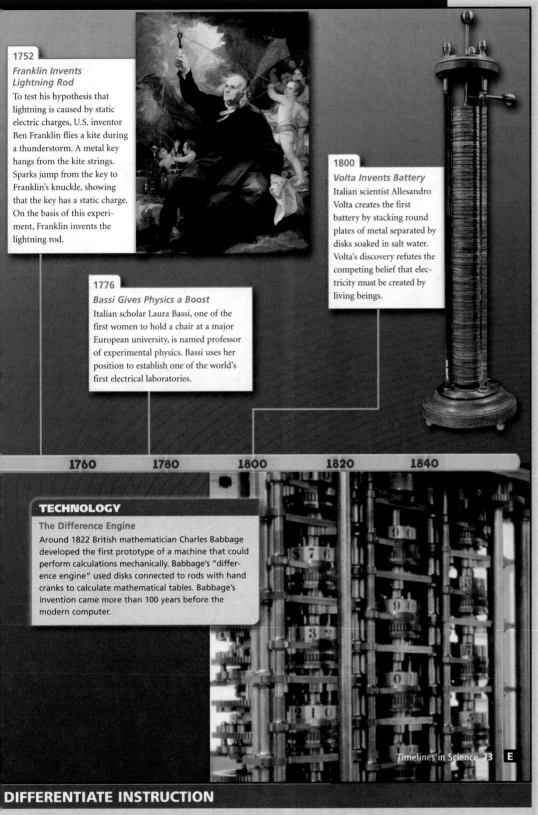

1752
Franklin Invents Lightning Rod
To test his hypothesis that lightning is caused by static electric charges, U.S. inventor Ben Franklin flies a kite during a thunderstorm. A metal key hangs from the kite strings. Sparks jump from the key to Franklin's knuckle, showing that the key has a static charge. On the basis of this experiment, Franklin invents the lightning rod.

1776
Bassi Gives Physics a Boost
Italian scholar Laura Bassi, one of the first women to hold a chair at a major European university, is named professor of experimental physics. Bassi uses her position to establish one of the world's first electrical laboratories.

1800
Volta Invents Battery
Italian scientist Allesandro Volta creates the first battery by stacking round plates of metal separated by disks soaked in salt water. Volta's discovery refutes the competing belief that electricity must be created by living beings.

1760   1780   1800   1820   1840

TECHNOLOGY

The Difference Engine
Around 1822 British mathematician Charles Babbage developed the first prototype of a machine that could perform calculations mechanically. Babbage's "difference engine" used disks connected to rods with hand cranks to calculate mathematical tables. Babbage's invention came more than 100 years before the modern computer.

Scientific Process

When scientists experiment with new technologies, they may find that old theories no longer hold true. In 1799 Alessandro Volta invented the first battery, proving that metallic objects were a source of current electricity. Ask: What old theory did Volta's battery refute? *Electricity must be created by living things.*

Social Studies Connection

1752 Benjamin Franklin is one of the most widely known and recognized Americans in history. He is best known for his invention of the lightning rod but he also invented bifocals, the rocking chair, and the street lamp. Franklin's skills did not stop at just experimentation and invention. He was also a writer and a politician. He founded the first public library, wrote and published *Poor Richard's Almanak*, was elected to Congress, and signed the Declaration of Independence.

Technology

THE DIFFERENCE ENGINE The calculating engines of Charles Babbage had a significant impact on modern computers. His Difference Engine No.1 was the first successful automatic calculator. Discuss with students ways in which computers are used today.

DIFFERENTIATE INSTRUCTION

Advanced Explain to students that batteries produce a continuous flow of charge, or current. Have them compare the simple battery, above, to the Leyden jar capacitor on p. 72. How are they similar? How do they differ? Encourage students to demonstrate their findings through either diagrams or a chart.

Social Studies Connection

1879 It's difficult to go through an ordinary day without using one of Thomas Edison's important discoveries. Edison invented the phonograph, the kinetoscope (a motion-picture device), and the light bulb. He also made improvements to existing technologies, including the electric generator, the telegraph, the typewriter, and the telephone. Having received over 1000 patents, Edison is considered one of the greatest inventors in history.

Scientific Process

1904 While the vacuum tube is an important invention, it produces excess heat and can burn out. By examining the problems of one technology, another technology can be created. Scientists at Bell Labs developed the transistor to replace the vacuum tube in electronic circuitry. Refer students to "Scientists Shrink Circuits to Atomic Level" on page 75, and ask: What is one of the key steps of the scientific process that the IBM researchers might have used to improve on the transistor? *Sample answer: collecting data*

Application

DIGITAL COMPUTER ENIAC was considered to be the world's first digital computer. It weighed 30 tons, used miles of wiring, had thousands of vacuum tubes and manual switches, and was 100 feet long and 10 feet high. Computer technology has come a long way since ENIAC. Have students compare and contrast ENIAC with today's computers. Discuss with students why they think the first digital computer filled a large room while today's computers are small enough to be held in one hand.

1879
Edison Improves Dynamo
To help bring electric lights to the streets of New York City, U.S. inventor Thomas Edison develops an improved dynamo, or generator. Edison's dynamo, known as a long-legged Mary Ann, operates at about twice the efficiency of previous models.

1904
Vacuum Tube Makes Debut
British inventor Ambrose Fleming modifies a light bulb to create an electronic vacuum tube. Fleming's tube, which he calls a valve, allows current to flow in one direction but not the other and can be used to detect weak radio signals.

1947
Transistor Invented
A transistor—a tiny electronic switch made out of a solid material called a semiconductor—is introduced to regulate the flow of electricity. Transistors, which do not produce excess heat and never burn out, can replace the vacuum tube in electronic circuitry and can be used to make smaller, cheaper, and more powerful computers.

1860   1880   1900   1920   1940

APPLICATION

First Electronic Digital Computer
Electronic Numerical Integrator and Computer (ENIAC) was the first digital computer. It was completed and installed in 1944 at the Moore School of Electrical Engineering at the University of Pennsylvania. Weighing more than 30 tons, ENIAC contained 19,000 vacuum tubes, 70,000 resistors, and 6000 switches, and it used almost 200 kilowatts of electric power. ENIAC could perform 5000 additions per second.

E **74** Unit: Electricity and Magnetism

DIFFERENTIATE INSTRUCTION

English Learners English learners may be confused by the use of present tense above the timeline. They have learned to use past tense when writing about events in the past. Explain that the use of present tense in the timeline imitates a newspaper of the times, as if the inventions or discoveries were just made. It allows the reader to feel connected to that time period.

1958

Chip Inventors Think Small

Jack Kilby, a U.S. electrical engineer, conceives the idea of making an entire circuit out of a single piece of germanium. The integrated circuit, or "computer chip," is born. This invention enables computers and other electronic devices to be made much smaller than before.

2001

Scientists Shrink Circuits to Atomic Level

Researchers succeed in building a logic circuit, a kind of transistor the size of a single molecule. The molecule, a tube of carbon atoms called a carbon nanotube, can be as small as 10 atoms across—500 times smaller than previous transistors. Computer chips, which currently contain over 40 million transistors, could hold hundreds of millions or even billions of nanotube transistors.

RESOURCE CENTER
CLASSZONE.COM

Explore current research in electronics and computers.

1960   1980   2000

TECHNOLOGY

Miniaturization

Miniaturization has led to an explosion of computer technology. As circuits have shrunk, allowing more components in less space, computers have become smaller and more powerful. They have also become easier to integrate with other technologies, such as telecommunications. When not being used for a phone call, this cell phone can be used to connect to the Internet, to access e-mail, and even to play computer games.

INTO THE FUTURE

Electronic computer components have become steadily smaller and more efficient over the years. However, the basic mechanism of a computer— a switch that can be either on or off depending on whether an electric charge is present—has remained the same. These switches represent the 1s and 0s, or the "bits", of binary code.

Quantum computing is based on an entirely new way of representing information. In quantum physics, individual subatomic particles can be described in terms of three states rather than just two. Quantum bits, or "qubits" can carry much more information than the binary bits of ordinary computers. Using qubits, quantum computers could be both smaller and faster than binary computers and perform operations not possible with current technology.

Quantum computing is possible in theory, but the development of hardware that can process qubits is just beginning. Scientists are currently looking for ways to put the theory into practice and to build computers that will make current models look as bulky and as slow as ENIAC.

ACTIVITIES

Reliving History

Make a Leyden jar capacitor. Line the inside of a jar with aluminum foil. Stop the jar with a cork or rubber stopper. Insert a straight section of bare metal coat hanger through the plug so that one end touches the foil and the other sticks out of the jar about 2 centimeters.

Run a comb through your hair several times and then touch it to the exposed end of the coat hanger. Repeat this process three or four more times. Then use a multimeter to test the voltage on the coat hanger.

Writing About Science

Learn more about the current state of electronic circuit miniaturization. Write up the results of your research in the form of a magazine article.

INTO THE FUTURE

To show how a particular technology changes and improves, make two columns on the board and label them *Existing Technology* and *Improved Technology*. With the class, list existing technologies that have recently been replaced or improved upon. For example, videotape might be listed under the first column and DVDs under the second column. Discuss with students what advantages the improved technology provides over the existing technology.

ACTIVITIES

Reliving History

Before students get started, remind them of the following tips:

- Use a pencil to help you press the aluminum foil against the inside of the glass jar.
- Students should watch the multimeter while they are testing voltage because the number will appear only briefly.

Writing About Science

Suggest that student's imitate the style of the timeline—writing as journalists in the present tense, reporting who, what, where, and why, and using a punchy headline for their magazine article.

Technology Resources

Students can visit **ClassZone.com** for news about advances in technology involving electronics.

DIFFERENTIATE INSTRUCTION

Advanced Ask students why they think the voltage difference shown on the multimeter display drops to zero shortly after the multimeter is touched to the wire. *The Leyden jar stores charge. Because the multimeter contains conductive metal wire, the charge is released into the multimeter when the multimeter and the Leyden jar touch.*

Physical Science
UNIFYING PRINCIPLES

PRINCIPLE 1

Matter is made of particles too small to see.

PRINCIPLE 2

Matter changes form and moves from place to place.

PRINCIPLE 3

Energy changes from one form to another, but it cannot be created or destroyed.

PRINCIPLE 4

Physical forces affect the movement of all matter on Earth and throughout the universe.

Unit: Electricity and Magnetism
BIG IDEAS

CHAPTER 1
Electricity

Charged particles transfer electrical energy.

CHAPTER 2
Circuits and Electronics

Circuits control the flow of electrical charge.

CHAPTER 3
Magnetism

Current can produce magnetism, and magnetism can produce current.

CHAPTER 3
KEY CONCEPTS

SECTION 3.1

Magnetism is a force that acts at a distance.

1. Magnets attract and repel other magnets.

2. Some materials are magnetic.

3. Earth is a magnet.

SECTION 3.2

Current can produce magnetism.

1. An electric current produces a magnetic field.

2. Motors use electromagnets.

SECTION 3.3

Magnetism can produce current.

1. Magnets are used to generate an electric current.

2. Magnets are used to control voltage.

SECTION 3.4

Generators supply electrical energy.

1. Generators provide most of the world's electrical energy.

2. Electric power can be measured.

The Big Idea Flow Chart is available on p. T17 in the **UNIT TRANSPARENCY BOOK.**

Previewing Content

SECTION

 3.1 Magnetism is a force that acts at a distance. pp. 79–87

1. Magnets attract and repel other magnets.

The attraction between the north pole of a **magnet** and the south pole of another magnet is based on the **magnetic field** lines that go from the north to the south pole of a magnet.

- If two like poles are placed near each other, the magnetic fields oppose, and the poles repel each other.
- If opposite poles are brought near each other, the magnetic field goes from one magnet to another, and the poles attract.

Attraction and Repulsion of Magnetic Poles

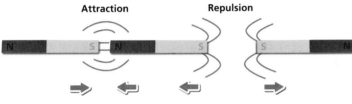

Opposite poles attract.   Like poles repel.

2. Some materials are magnetic.

Of all the common elements on the periodic table, only iron, nickel, cobalt and a few other metals are magnetic. Other materials, such as steel, are magnetic because they contain one or more of these elements. Iron, nickel, and cobalt are next to each other in the periodic table. This location indicates that the properties of these elements are similar because their electron configurations are similar.

Each of these elements contains unpaired electrons that produce a very small but strong magnetic field. Atoms in a magnetic material align so that these small magnetic fields form a **magnetic domain.** When placed within a larger magnetic field, the magnetic domains align, and the material becomes a magnet.

3. Earth is a magnet.

Because the north pole of a suspended magnet always points in a northerly direction on Earth, it can be inferred that Earth itself is a magnet. We commonly call the direction that the north pole of the magnet points to magnetic north. Because it attracts the north pole of a magnet, however, the magnetic north pole of Earth is actually the south pole of the magnet formed by Earth.

SECTION

 3.2 Current can produce magnetism. pp. 88–94

1. An electric current produces a magnetic field.

Both permanent and temporary magnets result from the magnetic field formed from a moving electric charge.

- Permanent magnets result from the spinning of unpaired electrons, which are electrically charged particles.
- Temporary magnets can also be formed in this manner, but the magnetic domains don't remain aligned. Temporary magnets known as **electromagnets** are produced by electric current, which consists of moving charge.

In the electromagnet below, coils of wire around an iron core create a magnet as long as electric charge flows. When current stops, the magnetism stops.

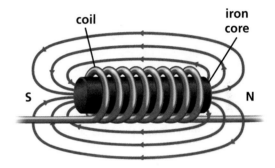

coil   iron core

S   N

2. Motors use electromagnets.

The basic parts of an electric motor are a voltage source, a shaft, a commutator, an electro-magnet, and at least one additional magnet.

- Forces of repulsion between the two magnets turn the shaft.
- The commutator reverses the poles of the electromagnet when it starts to align poles with the permanent magnet.
- This reversal renews the repulsion force and keeps the shaft moving.

 MISCONCEPTION DATABASE
CLASSZONE.COM Background on student misconceptions

Common Misconceptions

MAGNETIC METALS Students may think that all metals are attracted to magnets. Of the common metals, only three—iron, nickel, and cobalt—can become magnetic or be attracted to magnets to a noticeable extent.

 This misconception is addressed on p. 82.

Previewing Content

SECTION 3.3 Magnetism can produce current.
pp. 95–101

1. Magnets are used to generate an electric current.
Essentially, a **generator** is the opposite of an electric motor. The motor uses a moving charge to produce a magnetic field, and a generator uses a moving magnet to produce an electric charge.
- The magnet and the wire through which the current passes must be moving relative to each other. It doesn't matter which one moves, as long as one of them does.
- The current produced can be either **direct current,** which flows in one direction only, or **alternating current,** which changes direction at regular intervals.

2. Magnets are used to control voltage.
A **transformer** either increases or decreases voltage. In a transformer, two coils of wire are wrapped around an iron ring. Current through the first coil causes the ring to become an electromagnet. This electromagnet induces a current in the second coil.
- If the first coil has more loops, the voltage decreases, and the system is a step-down transformer.
- If the first coil has fewer loops than the second coil, the transformer is a step-up transformer, and voltage increases.

Step-Down Transformer

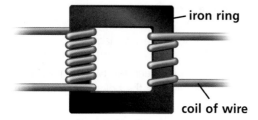

iron ring

coil of wire

SECTION 3.4 Generators supply electrical energy.
pp. 102–107

1. Generators provide most of the world's electrical energy.
Electric power is not the electrical energy produced, but a measure of the rate at which some other form of energy is converted to electrical energy. It is also the rate at which an appliance converts electrical energy into another form of energy, such as light or heat. For example,
- most electrical energy is produced by generators
- the rate at which the chemical energy in fossil fuels or the kinetic energy of falling water is converted to electrical energy is electric power

The current distributed from electrical generating plants is too great to be useful in homes and businesses. The current must pass through step-down transformers before it is useful.

2. Electric power can be measured.
The unit used to measure power is the **watt.** Because the watt is a small unit, **kilowatts** are often used to measure power in a building.

The amount of energy used is the product of the rate at which the energy is used (power) and the time over which it is used. Its unit of measurement is the **kilowatt-hour** (kWh).

Common Misconceptions

ENERGY AND POWER Students might think that electrical energy and power are the same thing because the two terms are often used synonymously. In reality, current supplies electrical energy for use. The rate at which this energy is transformed from or to other forms of energy is electric power.

TE This misconception is addressed on p. 104.

 MISCONCEPTION DATABASE
CLASSZONE.COM Background on student misconceptions

Previewing Labs

Lab Generator CD-ROM
Edit these Pupil Edition labs and generate alternative labs.

EXPLORE (the BIG idea)

Is It Magnetic? p. 77
Students are introduced to magnetism as they test the attraction of various materials to a magnet.

TIME 10 minutes
MATERIALS magnet; various metallic materials with various amounts of iron, such as coins, foil, washers, paper clips, and wire

---

How Can You Make a Chain? p. 77
Students use paperclips to learn about magnetic fields.

TIME 10 minutes
MATERIALS magnet, several metal paper clips

Internet Activity: Electromagnets, p. 77
Students use the Internet to work with a virtual electromagnet.

TIME 20 minutes
MATERIALS computer with Internet access

 SECTION 3.1

EXPLORE Magnetism, p. 79
Students manipulate magnets on a dowel rod to explore attraction and repulsion.

TIME 5 minutes
MATERIALS wooden dowel, spring clothespin, 3 disk magnets, ruler

---

INVESTIGATE Earth's Magnetic Field, p. 85
Students use a magnet, a needle, and aluminum foil to infer the effects of Earth's magnetic field on a compass.

TIME 15 minutes
MATERIALS small square of aluminum foil, bowl, water, strong magnet, sewing needle

---

 SECTION 3.2

EXPLORE Magnetism from Electric Current, p. 88
Students set up a compass and a circuit to observe how an electric current produces a magnetic field.

TIME 10 minutes
MATERIALS electrical tape, 15 cm copper wire, AA battery, compass

---

INVESTIGATE Electromagnets, p. 90
Students make an electromagnet and observe how it uses current to produce magnetism.

TIME 20 minutes
MATERIALS 40 cm insulated wire, large iron nail, 2 D batteries, electrical tape, metal paper clip

---

 SECTION 3.3

EXPLORE Energy Conversion, p. 95
Students use a small motor to produce current.

TIME 10 minutes
MATERIALS small motor with wires, AA battery, light bulb in holder

---

INVESTIGATE Electric Current, p. 98
Students build a current detector using wire and a compass to infer the properties of alternating current.

TIME 15 minutes
MATERIALS 2 m of 22-gauge magnet wire, compass, ruler, 14 in electrical tape, sandpaper, D battery

---

CHAPTER INVESTIGATION
Build a Speaker, pp. 100–101
Students construct a speaker to explore the relationship between the strengths of three magnets and the volume of the speaker.

TIME 40 minutes
MATERIALS 3 magnets of different strengths, metal paper clip, ruler, wire (2 m of 22-gauge), marker, foam cup, masking tape, 2 wire leads with alligator clips, stereo system

---

 SECTION 3.4

INVESTIGATE Power, p. 105
Students model using electrical energy to power appliances.

[R] Power Ratings Chart, p. 176

TIME 30 minutes
MATERIALS graph paper, colored pencils, Power Ratings Chart

[R] **Additional INVESTIGATION,** It's a Plot! A, B, & C, pp. 195–203; Teacher Instructions, pp. 206–207

Previewing Chapter Resources

| | INTEGRATED TECHNOLOGY | LABS AND ACTIVITIES |

Chapter 3
Magnetism

 CLASSZONE.COM
- eEdition Plus
- EasyPlanner Plus
- Misconception Database
- Content Review
- Test Practice
- Simulation
- Visualization
- Resource Centers
- Internet Activity: Electromagnets
- Math Tutorial

 CD-ROMS
- eEdition
- EasyPlanner
- Power Presentations
- Content Review
- Lab Generator
- Test Generator

AUDIO CDS
- Audio Readings
- Audio Readings in Spanish

SCILINKS.ORG

 EXPLORE the Big Idea, p. 77
- Is It Magnetic?
- How Can You Make a Chain?
- Internet Activity: Electromagnets

 UNIT RESOURCE BOOK
Unit Projects, pp. 5–10

 Lab Generator CD-ROM
Generate customized labs.

SECTION
3.1 Magnetism is a force that acts at a distance.
pp. 79–87

Time: 2 periods (1 block)

 Lesson Plan, pp. 134–135

 RESOURCE CENTER, Magnetism

 UNIT TRANSPARENCY BOOK
- Big Idea Flow Chart, p. T17
- Daily Vocabulary Scaffolding, p. T18
- Note-Taking Model, p. T19
- 3-Minute Warm-Up, p. T20
- "How Magnets Differ from Other Materials" Visual, p. T22

 • EXPLORE Magnetism, p. 79
- INVESTIGATE Earth's Magnetic Field, p. 85
- Think Science, p. 87

 UNIT RESOURCE BOOK
- Datasheet, Earth's Magnetic Field, p. 143
- Additional INVESTIGATION, It's a Plot! A, B, & C, pp. 195–203

SECTION
3.2 Current can produce magnetism.
pp. 88–94

Time: 2 periods (1 block)

 Lesson Plan, pp. 145–146

 VISUALIZATION, Motor

 **UNIT TRANSPARENCY BOOK**
- Daily Vocabulary Scaffolding, p. T18
- 3-Minute Warm-Up, p. T20

 • EXPLORE Magnetism from Electric Current, p. 88
- INVESTIGATE Electromagnets, p. 90

UNIT RESOURCE BOOK
Datasheet, Electromagnets, p. 154

SECTION
3.3 Magnetism can produce current.
pp. 95–101

Time: 3 periods (1.5 blocks)

 Lesson Plan, pp. 156–157

UNIT TRANSPARENCY BOOK
- Daily Vocabulary Scaffolding, p. T18
- 3-Minute Warm-Up, p. T21

 • EXPLORE Energy Conversion, p. 95
- INVESTIGATE Electric Current, p. 98
- CHAPTER INVESTIGATION, Build a Speaker, pp. 100–101

 UNIT RESOURCE BOOK
- Datasheet, Electric Current, p. 165
- CHAPTER INVESTIGATION, Build a Speaker, A, B, & C, pp. 186–194

SECTION
3.4 Generators supply electrical energy.
pp. 102–107

Time: 3 periods (1.5 blocks)

 Lesson Plan, pp. 167–168

 • **RESOURCE CENTERS,** Dams, Energy Use
- **MATH TUTORIAL**

UNIT TRANSPARENCY BOOK
- Big Idea Flow Chart, p. T17
- Daily Vocabulary Scaffolding, p. T18
- 3-Minute Warm-Up, p. T21
- Chapter Outline, pp. T23–T24

 • INVESTIGATE Power, p. 105
- Math in Science, p. 107

 UNIT RESOURCE BOOK
- Power Ratings Chart, p. 176
- Datasheet, Power, p. 177
- Math Support, p. 184
- Math Practice, p. 185

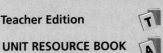

READING AND REINFORCEMENT

ASSESSMENT

STANDARDS

- Description Wheel, B20–21
- Main Idea Web, C38–39
- Daily Vocabulary Scaffolding, H1–8

 UNIT RESOURCE BOOK
- Vocabulary Practice, pp. 181–182
- Decoding Support, p. 183
- Summarizing the Chapter, pp. 204–205

 Audio Readings CD
Listen to Pupil Edition.

 Audio Readings in Spanish CD
Listen to Pupil Edition in Spanish.

 • Chapter Review, pp. 109–110
- Standardized Test Practice, p. 111

 UNIT ASSESSMENT BOOK
- Diagnostic Test, pp. 39–40
- Chapter Test, A, B, & C, pp. 45–56
- Alternative Assessment, pp. 57–58
- Unit Test A, B, C, pp. 59–70

 • Spanish Chapter Test, pp. 309–312
- Spanish Unit Test, pp. 313–316

 Test Generator CD-ROM
Generate customized tests.

 Lab Generator CD-ROM
Rubrics for Labs

National Standards
A.2–8, A.9.a–f, G.1.a–b, G.2.b

See p. 76 for the standards.

 UNIT RESOURCE BOOK
- Reading Study Guide, A & B, pp. 136–139
- Spanish Reading Study Guide, pp. 140–141
- Challenge and Extension, p. 142
- Reinforcing Key Concepts, p. 144

 Ongoing Assessment, pp. 80–84, 86

 Section 3.1 Review, p. 86

 UNIT ASSESSMENT BOOK
Section 3.1 Quiz, p. 41

National Standards
A.2–7, A.9.a–b, A.9.e–f

 UNIT RESOURCE BOOK
- Reading Study Guide, A & B, pp. 147–150
- Spanish Reading Study Guide, pp. 151–152
- Challenge and Extension, p. 153
- Reinforcing Key Concepts, p. 155

 Ongoing Assessment, pp. 89–94

 Section 3.2 Review, p. 94

UNIT ASSESSMENT BOOK
Section 3.2 Quiz, p. 42

National Standards
A.2–7, A.9.a–b, A.9.e–f

 UNIT RESOURCE BOOK
- Reading Study Guide, A & B, pp. 158–161
- Spanish Reading Study Guide, pp. 162–163
- Challenge and Extension, p. 164
- Reinforcing Key Concepts, p. 166
- Challenge Reading, pp. 179–180

 Ongoing Assessment, pp. 95–98

 Section 3.3 Review, p. 99

UNIT ASSESSMENT BOOK
Section 3.3 Quiz, p. 43

National Standards
A.2–7, A.9.a–b, A.9.e–f

 UNIT RESOURCE BOOK
- Reading Study Guide, A & B, pp. 169–172
- Spanish Reading Study Guide, pp. 173–174
- Challenge and Extension, p. 175
- Reinforcing Key Concepts, p. 178

 Ongoing Assessment, pp. 102–104, 106

 Section 3.4 Review, p. 106

 UNIT ASSESSMENT BOOK
Section 3.4 Quiz, p. 44

National Standards
A.2–8, A.9.a–c, A.9.e–f

Previewing Resources for Differentiated Instruction

CHAPTER INVESTIGATION

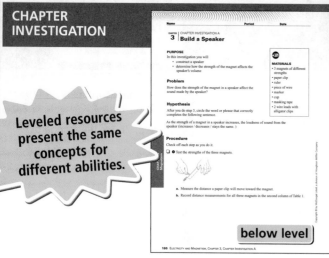

3 Build a Speaker — CHAPTER INVESTIGATION A

below level

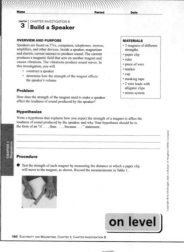

3 Build a Speaker — CHAPTER INVESTIGATION B

on level

3 Build a Speaker — CHAPTER INVESTIGATION C

advanced

Leveled resources present the same concepts for different abilities.

UNIT RESOURCE BOOK, pp. 186–189

pp. 190–193

pp. 190–194

READING STUDY GUIDE

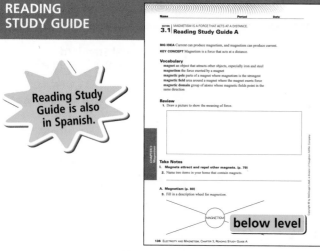

3.1 Reading Study Guide A

below level

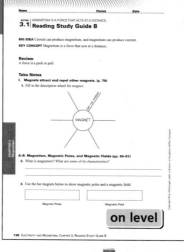

3.1 Reading Study Guide B

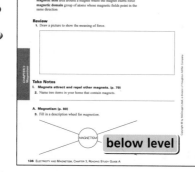

on level

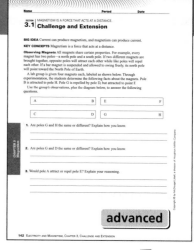

3.1 Challenge and Extension

advanced

Reading Study Guide is also in Spanish.

UNIT RESOURCE BOOK, pp. 136–137

pp. 138–139

p. 142

CHAPTER TEST

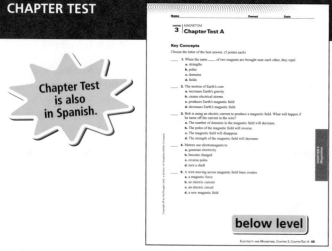

3 Chapter Test A

below level

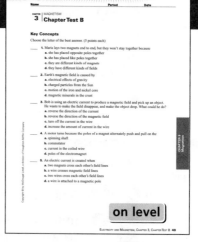

3 Chapter Test B

on level

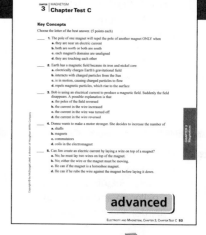

3 Chapter Test C

advanced

Chapter Test is also in Spanish.

**UNIT ASSESSMENT BOOK,** pp. 45–48

pp. 49–52

pp. 53–56

TECHNOLOGY

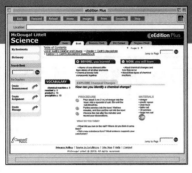

There are three Resource Centers for this chapter.

CLASSZONE.COM

CD/CD-ROMS

CLASSZONE.COM

VISUAL CONTENT

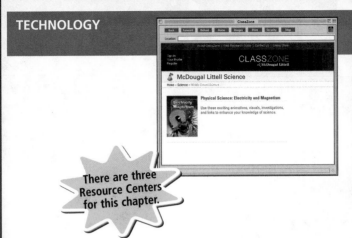

CHAPTER 3 MAGNETISM
Big Idea Flow Chart

CHAPTER 3 MAGNETISM
Note-Taking Model

CHAPTER 3 MAGNETISM
How Magnets Differ from Other Materials

Magnets, and the materials they attract, contain small regions called magnetic domains. In a magnet, the domains are aligned.

Electricity and Magnetism, Chapter 3, Big Idea Flow Chart T17

Electricity and Magnetism, Chapter 3, Note-Taking Model T19

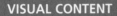
Electricity and Magnetism, Chapter 3, Teaching Visual T22

T **UNIT TRANSPARENCY BOOK,** p. T17

T p. T19

T p. T22

MORE SUPPORT

Reinforcing Key Concepts for each section

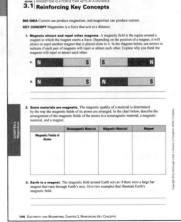

SECTION 3.1 Reinforcing Key Concepts

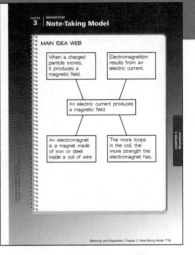

CHAPTER 3 Vocabulary

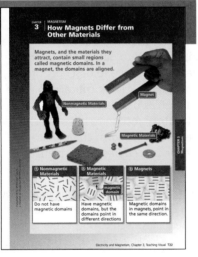

CHAPTER 3 Math Support

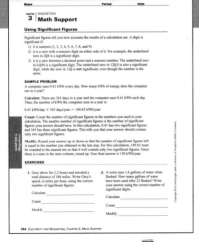

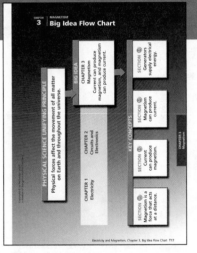

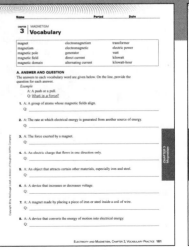

R **UNIT RESOURCE BOOK,** p. 144

R pp. 181–182

R p. 184

INTRODUCE

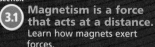

Have students look at the photograph of a hiker using a compass. Discuss how the question in the box links to the Big Idea:

- What makes a compass needle point toward the North Pole?
- What do you think compass needles are made of?

National Science Education Standards

Process

A.2–8 Design and conduct an investigation; use tools to gather and interpret data; use evidence to describe, predict, explain, model; think critically to make relationships between evidence and explanation; recognize different explanations and predictions; communicate scientific procedures and explanations; use mathematics.

A.9.a–f Understand scientific inquiry by using different investigations, methods, mathematics, technology, explanations based on logic, evidence, and skepticism.

G.1.a–b Science as a human endeavor

G.2.b Nature of Science

CHAPTER 3 Magnetism

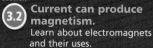

Current can produce magnetism, and magnetism can produce current.

Key Concepts

SECTION 3.1 Magnetism is a force that acts at a distance. Learn how magnets exert forces.

SECTION 3.2 Current can produce magnetism. Learn about electromagnets and their uses.

SECTION 3.3 Magnetism can produce current. Learn how magnetism can produce an electric current.

SECTION 3.4 Generators supply electrical energy. Learn how generators are used in the production of electrical energy.

 Internet Preview

CLASSZONE.COM

Chapter 3 online resources: Content Review, Simulation, Visualization, three Resource Centers, Math Tutorial, Test Practice

E 76 Unit: Electricity and Magnetism

INTERNET PREVIEW

CLASSZONE.COM For student use with the following pages:

Review and Practice
- Content Review, pp. 78, 108
- Math Tutorial: Rounding Decimals, p. 107
- Test Practice, p. 111

Activities and Resources
- Internet Activity: p. 77
- Resource Centers: Magnetism, p. 80; Dams and Electricity, p. 103; Energy Use and Conservation, p. 104
- Visualization: Motor, p. 92

Electromagnetism
Code: MDL067

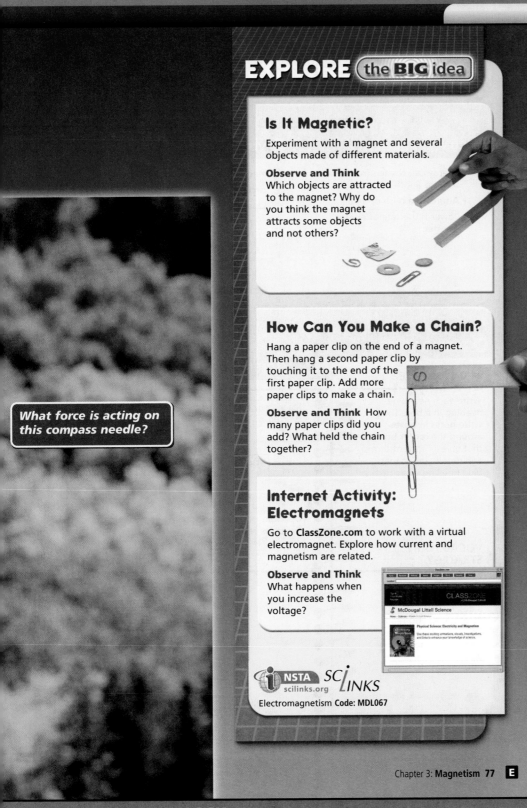

EXPLORE (the BIG idea)

Is It Magnetic?

Experiment with a magnet and several objects made of different materials.

Observe and Think
Which objects are attracted to the magnet? Why do you think the magnet attracts some objects and not others?

What force is acting on this compass needle?

How Can You Make a Chain?

Hang a paper clip on the end of a magnet. Then hang a second paper clip by touching it to the end of the first paper clip. Add more paper clips to make a chain.

Observe and Think How many paper clips did you add? What held the chain together?

Internet Activity: Electromagnets

Go to **ClassZone.com** to work with a virtual electromagnet. Explore how current and magnetism are related.

Observe and Think
What happens when you increase the voltage?

NSTA *SciLINKS*
scilinks.org
Electromagnetism **Code: MDL067**

EXPLORE (the BIG idea)

These inquiry-based activities are appropriate for use at home or as a supplement to classroom instruction.

Is It Magnetic?

PURPOSE To discover that not all metals are magnetic. Students test different materials with a magnet.

TIP *10 min.* Encourage students to predict whether each item is magnetic before testing it.

Answer: Items that contain iron are attracted to the magnet. Some metals are affected by magnetism, and some are not.

REVISIT after p. 83.

How Can You Make a Chain?

PURPOSE To demonstrate the magnetic field that surrounds a magnet. Students observe induced magnetism as evidence of magnetic fields.

TIP *10 min.* Make sure students do not link the paper clips together.

Answer: usually three or four paper clips; the magnetic field

REVISIT after p. 84.

Internet Activity: Electromagnets

PURPOSE To use a simulated electromagnet to explore the relationship between electricity and magnetism.

TIP *20 min.* Have students experiment with the level of voltage supplied to the virtual electromagnet.

Answer: The magnetic field increases.

REVISIT after p. 90.

TEACHING WITH TECHNOLOGY

CBL and Probeware Many activities in this chapter can be made more interesting with a magnetic field sensor. Try using this probeware with activities on pp. 77, 85, 88, and 90.

Ammeter Students could use an AC ammeter instead of a compass in "Investigate Electric Current" on p. 98.

PREPARE

◀ CONCEPT REVIEW

Activate Prior Knowledge

- Turn a flashlight on and ask students how energy is changing. *Electrical energy from the battery is changing to light and heat.*

- Crumple a piece of paper and drop it on the floor.

- Ask students to explain why the paper was pulled to the floor.

- Students should answer that the force of gravity acting on the paper pulled it to the floor.

▶ TAKING NOTES

Main Idea Web

Upon completion of individual main idea webs, have students work in small groups to compare them. Have students discuss any differences. Emphasize that many items are acceptable in the web, but each one must relate to the main idea.

Vocabulary Strategy

Description wheels can contain as much information as students want to add. They can be used as a study guide at the end of the chapter.

Vocabulary and Note-Taking Resources

- Vocabulary Practice, pp. 181–182
- Decoding Support, p. 183

- Daily Vocabulary Scaffolding, p. T18
- Note-Taking Model, p. T19

- Description Wheel, B20–21
- Main Idea Web, C38–39
- Daily Vocabulary Scaffolding, H1–8

◀ CONCEPT REVIEW

- Energy can change from one form to another.
- A force is a push or a pull.
- Power is the rate of energy transfer.

◀ VOCABULARY REVIEW

electric current p. 28

circuit p. 43

kinetic energy *See Glossary.*

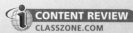

CONTENT REVIEW
CLASSZONE.COM
Review concepts and vocabulary.

▶ TAKING NOTES

MAIN IDEA WEB

Write each new blue heading in a box. Then write notes in boxes around the center box that give important terms and details about that blue heading.

VOCABULARY STRATEGY

Place each vocabulary term at the center of a **description wheel** diagram. As you read about the term, write some words describing it on the spokes.

See the Note-Taking Handbook on pages R45–R51.

SCIENCE NOTEBOOK

| Magnetism is the force exerted by magnets. | All magnets have two poles. |

Magnets attract and repel other magnets.

| Opposite poles attract, and like poles repel. | Magnets have magnetic fields of force around them. |

Magnetic fields of atoms point in the same direction.

In magnets, they line up.

MAGNETIC DOMAINS

Nonmagnetic materials don't have them.

Magnetic materials have them.

CHECK READINESS

Administer the Diagnostic Test to determine students' readiness for new science content and their mastery of requisite math skills.

 Diagnostic Test, pp. 39–40

Technology Resources

Students needing content and math skills should visit **ClassZone.com**.

- CONTENT REVIEW
- MATH TUTORIAL

- CONTENT REVIEW CD-ROM

KEY CONCEPT

3.1 Magnetism is a force that acts at a distance.

BEFORE, you learned
- A force is a push or pull
- Some forces act at a distance
- Atoms contain charged particles

NOW, you will learn
- How magnets attract and repel other magnets
- What makes some materials magnetic
- Why a magnetic field surrounds Earth

VOCABULARY
magnet p. 79
magnetism p. 80
magnetic pole p. 80
magnetic field p. 81
magnetic domain p. 82

EXPLORE Magnetism

How do magnets behave?

PROCEDURE
1. Clamp the clothespin on the dowel so that it makes a stand for the magnets, as shown.
2. Place the three magnets on the dowel. If there is a space between pairs of magnets, measure and record the distance between them.
3. Remove the top magnet, turn it over, and replace it on the dowel. Record your observations. Experiment with different arrangements of the magnets and record your observations.

MATERIALS
- clothespin
- wooden dowel
- 3 disk magnets
- ruler

WHAT DO YOU THINK?
- How did the arrangement of the magnets affect their behavior?
- What evidence indicates that magnets exert a force?

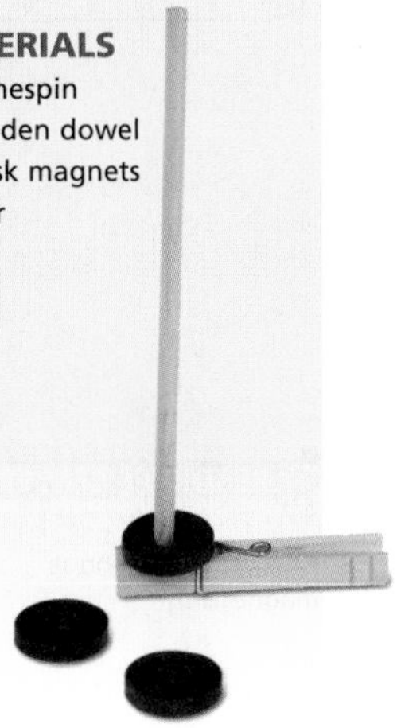

Magnets attract and repel other magnets.

VOCABULARY
Make a description wheel for the term *magnet*.

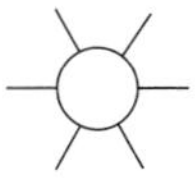

Suppose you get home from school and open the refrigerator to get some milk. As you close the door, it swings freely until it suddenly seems to close by itself. There is a magnet inside the refrigerator door that pulls it shut. A **magnet** is an object that attracts certain other materials, particularly iron and steel.

There may be quite a few magnets in your kitchen. Some are obvious, like the seal of the refrigerator and the magnets that hold notes to its door. Other magnets run the motor in a blender, provide energy in a microwave oven, operate the speakers in a radio on the counter, and make a doorbell ring.

3.1 FOCUS

Set Learning Goals

Students will
- Observe that magnets attract and repel other magnets.
- Discover what makes some materials magnetic.
- Learn why a magnetic field surrounds Earth.
- Infer from an experiment that Earth's magnetic field moves a compass needle.

3-Minute Warm-Up

Display Transparency 20 or copy this exercise on the board:

Decide if these statements are true. If not, correct them.

- A force is a push or a pull. *true*
- All forces act at a distance. *Some forces, such as gravity, act at a distance but others act as a constant force.*
- Atoms contain charged particles. *true*

3-Minute Warm-Up, p. T20

3.1 MOTIVATE

EXPLORE Magnetism

PURPOSE To observe the attraction and repulsion of magnets

TIP *5 min.* If the stand falls over, tape the clothespin to the table.

WHAT DO YOU THINK? *When the magnets are arranged one way, they attract each other; when one magnet is turned over, they push each other apart. Magnets either attract or repel other magnets.*

RESOURCES FOR DIFFERENTIATED INSTRUCTION

Below Level

UNIT RESOURCE BOOK
- Reading Study Guide A, pp. 136–137
- Decoding Support, p. 183

AUDIO CDS

ADDITIONAL INVESTIGATION,
It's a Plot! A, B, & C, pp. 195–203; Teacher Instructions, pp. 206–207

Advanced

UNIT RESOURCE BOOK
Challenge and Extension, p. 142

English Learners

UNIT RESOURCE BOOK
Spanish Reading Study Guide, pp. 140–141

AUDIO CDS
- Audio Readings in Spanish
- Audio Readings (English)

Real World Example

Transportation by maglev was first proposed more than a century ago. The first commercial maglev train started operation in China in 2002, using a train developed in Germany. Similar trains are being tested in Germany and Japan. Developers in the United States have considered similar prototypes, but the cost is prohibitive. It is estimated that it would cost over $25 million a mile to build such a train. The main benefit of maglev trains is that they are faster and smoother than many conventional trains because friction is dramatically reduced.

Ongoing Assessment

CHECK YOUR READING

Answer: Like poles of magnets repel each other, raising the train off the track. Other magnets pull the train forward.

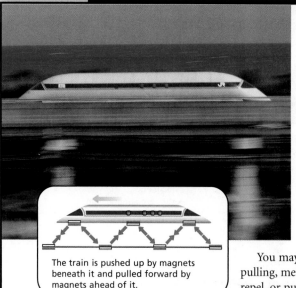

The train is pushed up by magnets beneath it and pulled forward by magnets ahead of it.

RESOURCE CENTER
CLASSZONE.COM

Find out more about magnetism.

Magnetism

The force exerted by a magnet is called **magnetism.** The push or pull of magnetism can act at a distance, which means that the magnet does not have to touch an object to exert a force on it. When you close the refrigerator, you feel the pull before the magnet actually touches the metal frame. There are other forces that act at a distance, including gravity and static electricity. Later you will read how the force of magnetism is related to electricity. In fact, magnetism is the result of a moving electric charge.

You may be familiar with magnets attracting, or pulling, metal objects toward them. Magnets can also repel, or push away, objects. The train in the photograph at the left is called a maglev train. The word *maglev* is short for *magnetic levitation*, or lifting up. As you can see in the diagram, the train does not touch the track. Magnetism pushes the entire train up and pulls it forward. Maglev trains can move as fast as 480 kilometers per hour (300 mi/h).

CHECK YOUR READING How can a train operate without touching the track?

Magnetic Poles

The force of magnetism is not evenly distributed throughout a magnet. **Magnetic poles** are the parts of a magnet where the magnetism is the strongest. Every magnet has two magnetic poles. If a bar magnet is suspended so that it can swing freely, one pole of the magnet always points toward the north. That end of the magnet is known as the north-seeking pole, or north pole. The other end of the magnet is called the south pole. Many magnets are marked with an *N* and an *S* to indicate the poles.

As with electric charges, opposite poles of a magnet attract and like poles—or poles that are the same—repel, or push each other away. Every magnet has both a north pole and a south pole. A horseshoe magnet is like a bar magnet that has been bent into the shape of a *U*. It has a pole at each of its ends. If you break a bar magnet between the two poles, the result is two smaller magnets, each of which has a north pole and a south pole. No matter how many times you break a magnet, the result is smaller magnets.

DIFFERENTIATE INSTRUCTION

? More Reading Support

A What is the name of the force exerted by a magnet? *magnetism*

B What does every magnet have? *a north and south pole*

English Learners The use of dashes in writing may be confusing to English learners. The dash is often used to interject an idea into a sentence. For example, "As with electric charges, opposite poles of a magnet attract and like poles—or poles that are the same—repel, or push each other away." Show students that the sentence would still be complete if the words inside the dashes were not there.

Magnetic Fields

You have read that magnetism is a force that can act at a distance. However magnets cannot exert a force on an object that is too far away. A **magnetic field** is the region around a magnet in which the magnet exerts force. If a piece of iron is within the magnetic field of a magnet, it will be pulled toward the magnet. Many small pieces of iron, called iron filings, are used to show the magnetic field around a magnet. The iron filings form in a pattern of lines called magnetic field lines.

READING TiP
Thin red lines in the illustrations below indicate the magnetic field.

The Magnetic Field Around a Magnet

 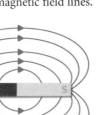 

The arrangement of the magnetic field lines depends on the shape of the magnet, but the lines always extend from one pole to the other pole. The magnetic field lines are always shown as starting from the north pole and ending at the south pole. In the illustrations above, you can see that the lines are closest together near the magnets' poles. That is where the force is strongest. The force is weaker farther away from the magnet.

CHECK YOUR READING Where is the magnetic field of a magnet the strongest?

What happens to the magnetic fields of two magnets when the magnets are brought together? As you can see below, each magnet has an effect on the field of the other magnet. If the magnets are held so that the north pole of one magnet is close to the south pole of the other, the magnetic field lines extend from one magnet to the other. The magnets pull together. On the other hand, if both north poles or both south poles of two magnets are brought near one another, the magnets repel. It is very difficult to push like poles of strong magnets together because magnetic repulsion pushes them apart.

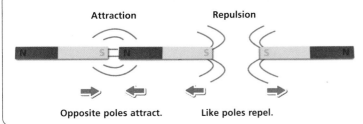

| Attraction | Repulsion |
| Opposite poles attract. | Like poles repel. |

Integrate the Sciences

Magnetic resonance imaging (MRI) uses magnetic fields to create detailed images of the inside of the human body, showing irregularities such as tumors. The patient lies in a stable magnetic field produced by a strong, tubular magnet. As the patient moves through the machine, three other magnets produce varying magnetic fields. Because of these magnetic fields and additional radio waves, hydrogen atoms in the body absorb and release energy. The released energy is detected and used to create images.

History of Science

In the late 1500s, William Gilbert, an English physician, discovered that materials that contained iron could be magnetized by stroking them with lodestone. His other contributions include showing that magnetism gets weaker as a magnet gets hotter, and discovering Earth's magnetic field.

Ongoing Assessment

Observe that magnets attract and repel other magnets.

Ask: What happens when two unlike poles of magnets are brought near each other? *They attract each other.*

CHECK YOUR READING *Answer: at the poles*

---

② More Reading Support

C What is a magnetic field? *the region where a magnet exerts force*

D What happens when you bring the north poles of two magnets together? *They repel.*

Below Level Keep a supply of magnets, magnetic materials, and wire available for students to model chapter concepts.

Inclusion Encourage students with visual impairments to hold two large magnets, one in each hand. What happens if students feel resistance when they bring their hands together? *like poles are facing each other and repelling* What if they feel an attraction pulling their hands together? *opposite poles are attracting*

History of Science

People have used lodestone as a magnet for over 2000 years. When the first European explorers went to China, the Chinese people were using lodestone bowls as compass needles for navigation. A lodestone bowl about the size of the curved part of a spoon was placed on a flat surface. The handle of the bowl always pointed in the same direction and could be used for navigation.

Address Misconceptions

IDENTIFY Ask: Will a magnet pick up an aluminum can, or will there be no magnetic attraction between the two objects? If students answer that the magnet will pick up the can, they may hold the misconception that all metals are attracted to magnets.

CORRECT Provide students with various metals that might be collected for recycling, such as aluminum and steel cans, wire, and copper pipe. Have students use a magnet to determine which objects could be magnetically separated from other items to be recycled. Point out that what is commonly referred to as a tin can is actually steel, which contains iron, covered by a protective layer of another metal.

REASSESS Ask: In general, what metals could be separated from other metals because they are attracted to magnets? *metals that contain iron*

Technology Resources

Visit **ClassZone.com** for background on common student misconceptions.

MISCONCEPTION DATABASE

Ongoing Assessment

 Answer: In magnets, the magnetic fields of atoms align; in nonmagnetic materials, magnetic fields do not align.

Some materials are magnetic.

Some magnets occur naturally. Lodestone is a type of rock that is a natural magnet and formed the earliest magnets that people used. The term *magnet* comes from the name *Magnesia*, a region of Greece where lodestone was discovered. Magnets can also be made from materials that contain certain metallic elements, such as iron.

If you have ever tried picking up different types of objects with a magnet, you have seen that some materials are affected by the magnet and other materials are not. Iron, nickel, cobalt, and a few other metals have properties that make them magnetic. Other materials, such as wood, cannot be made into magnets and are not affected by magnets. Whether a material is magnetic or not depends on its atoms—the particles that make up all matter.

You read in chapter 1 that the protons and electrons of an atom have electric fields. Every atom also has a weak magnetic field, produced by the electron's motion around a nucleus. In addition, each electron spins around its axis, an imaginary line through its center. The spinning motion of the electrons in magnetic materials increases the strength of the magnetic field around each atom. The magnetic effect of one electron is usually cancelled by another electron that spins in the opposite direction.

READING TiP
The red arrows in the illustration on the facing page are tiny magnetic fields.

Inside Magnetic Materials

The illustration on page 83 shows how magnets and the materials they affect differ from other materials.

1. In a material that is not magnetic, such as wood, the magnetic fields of the atoms are weak and point in different directions. The magnetic fields cancel each other out. As a result, the overall material is not magnetic and could not be made into a magnet.

2. In a material that is magnetic, such as iron, the magnetic fields of a group of atoms align, or point in the same direction. A **magnetic domain** is a group of atoms whose magnetic fields are aligned. The domains of a magnetic material are not themselves aligned, so their fields cancel one another out. Magnetic materials are pulled by magnets and can be made into magnets.

3. A magnet is a material in which the magnetic domains are all aligned. The material is said to be magnetized.

CHECK YOUR READING How do magnets differ from materials that are not magnetic?

DIFFERENTIATE INSTRUCTION

 More Reading Support

E What are some magnetic elements? *iron, cobalt, nickel*

F What are a group of atoms whose magnetic fields are aligned? *a magnetic domain*

Alternative Assessment Have groups of students create diagrams that show how magnetic tags are used as antitheft devices in stores. To help them, tell them that some of the detectors people walk through as they leave the store exert a magnetic field. The activated tags on merchandise contain demagnetized magnetic materials. As they pass through the field, the domains align, a signal is sent, and the alarm goes off.

How Magnets Differ from Other Materials

Magnets, and the materials they attract, contain small regions called magnetic domains. In a magnet, the domains are aligned.

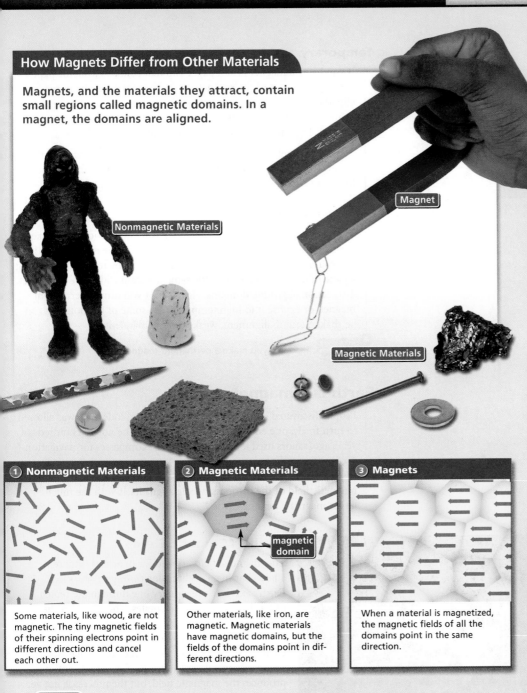

Nonmagnetic Materials

Magnet

Magnetic Materials

① Nonmagnetic Materials

Some materials, like wood, are not magnetic. The tiny magnetic fields of their spinning electrons point in different directions and cancel each other out.

② Magnetic Materials

magnetic domain

Other materials, like iron, are magnetic. Magnetic materials have magnetic domains, but the fields of the domains point in different directions.

③ Magnets

When a material is magnetized, the magnetic fields of all the domains point in the same direction.

READING VISUALS Do the paper clips in this photograph contain magnetic domains? Why or why not?

DIFFERENTIATE INSTRUCTION

Below Level Have students classify each item in the visual according to the boxes at the bottom of the page. Have them work in small groups to compare and discuss their results.

Inclusion Gather a toy, a pencil, a paper clip, a marble, a sponge, a cork, a nail, a washer and if possible a magnetic rock. Pass them around to tactile learners and students with visual impairments. Make sure students know what they have. Then make a chart of magnetic and nonmagnetic materials on the board, and ask students to categorize the items.

Teach from Visuals

Have students examine the visual of magnets. Ask:

• Which box describes electrons in a pencil? *Box 1*

• Do you think that all parts of a pencil are nonmagnetic? If not, what part might be magnetic? *No; the metal below the eraser might be magnetic.*

• How could you test this hypothesis? *Check to see whether the metal is attracted to a magnet.*

• If the metal below the eraser is magnetic, how would these atoms compare to the atoms in the rest of the pencil? *The metal atoms would form magnetic domains, unlike the atoms in the rest of the pencil.*

T This visual is also available as T22 in the Unit Transparency Book.

Develop Critical Thinking

SYNTHESIZE Have students use what they know regarding the arrangement of aligned atoms in a magnet and what happens to particles when they are heated. Ask: Why does heating a magnet tend to weaken it? *The kinetic energy of the aligned atoms increases, causing them to move out of alignment.*

EXPLORE (the **BIG** idea)

Revisit "Is It Magnetic?" on p. 77. Have students explain their results.

Ongoing Assessment

Discover what makes some materials magnetic.

Ask: Why can a magnet pick up an iron nail but not copper wire? *Copper's magnetic fields point in many directions and cancel each other out, so it doesn't have a strong magnetic field like that of iron.*

READING VISUALS *Answer: Yes; they are attracted by a magnet, so they are magnetic.*

Integrate the Sciences

A few strains of bacteria, such as *Magnetospirillum magnetotacticum*, have been observed orienting themselves along the lines of Earth's magnetic field. The bacteria contain small amounts of magnetite, Fe_3O_4, which contains iron and so responds to a magnetic field. The bacteria sense the magnetic force and move in response to it.

EXPLORE (the **BIG** idea)

Revisit "How Can You Make a Chain?" on p. 77. Have students suggest ways to make their chains longer.

Develop Critical Thinking

SYNTHESIZE Have students research the magnetic fields of other planets in the solar system. Ask: How does the strength of the magnetic field provide information about the structure of the planet? *Planets with strong magnetic fields probably have a core consisting of materials that can be magnetized. Those with little or no magnetic field probably do not have such a core.*

Ongoing Assessment

Learn that a magnetic field surrounds Earth.

Ask: What relationship exists between charged particles and Earth's magnetic field? *Charged particles flow within Earth's core, causing a magnetic field.*

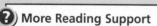 *Answer: Place a magnetic material in very strong magnetic fields.*

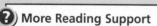 *Answer: the movement of its core, which is mostly iron and nickel*

Temporary and Permanent Magnets

If you bring a magnet near a paper clip that contains iron, the paper clip is pulled toward the magnet. As the magnet nears the paper clip, the domains within the paper clip are attracted to the magnet's nearest pole. As a result, the domains within the paper clip become aligned. The paper clip develops its own magnetic field.

 You can make a chain of paper clips that connect to one another through these magnetic fields. However, if you remove the magnet, the chain falls apart. The paper clips are temporary magnets, and their domains return to a random arrangement when the stronger magnetic field is removed.

Placing magnetic materials in very strong magnetic fields makes permanent magnets, such as the ones you use in the experiments in this chapter. You can make a permanent magnet by repeatedly stroking a piece of magnetic material in the same direction with a strong magnet. This action aligns the domains. However, if you drop a permanent magnet, or expose it to high temperatures, some of the domains can be shaken out of alignment, weakening its magnetism.

CHECK YOUR READING How can you make a permanent magnet?

Earth is a magnet.

People discovered long ago that when a piece of lodestone was allowed to turn freely, one end always pointed toward the north. Hundreds of years ago, sailors used lodestone in the first compasses for navigation. A compass works because Earth itself is a large magnet. A compass is simply a magnet that is suspended so that it can turn freely. The magnetic field of the compass needle aligns itself with the much larger magnetic field of Earth.

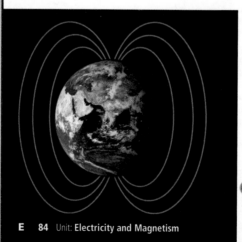

E 84 Unit: Electricity and Magnetism

Earth's Magnetic Field

The magnetic field around Earth acts as if there were a large bar magnet that runs through Earth's axis. Earth's axis is the imaginary line through the center of Earth around which it rotates. The source of the magnetic field that surrounds Earth is the motion of its core, which is composed mostly of iron and nickel. Charged particles flow within the core. Scientists have proposed several explanations of how that motion produces the magnetic field, but the process is not yet completely understood.

CHECK YOUR READING What is the source of Earth's magnetic field?

DIFFERENTIATE INSTRUCTION

 More Reading Support

G What happens when you remove a temporary magnet from a stronger magnetic field? *It stops being a magnet.*

Advanced Ask students to explain why a permanent magnet decreases slightly in strength over time. Refer them back to the visual "How Magnets Differ from Other Materials" on p. 83 if they need help, and point out the alignment of magnetic domains in magnets. *The magnetic domains in a permanent magnet are highly aligned. Over time, disturbances cause some disorder of the domains.*

 Challenge and Extension, p. 142

INVESTIGATE Earth's Magnetic Field

What moves a compass needle?

PROCEDURE

1. Gently place the aluminum foil on the water so that it floats.
2. Rub one pole of the magnet along the needle, from one end of the needle to the other. Lift up the magnet and repeat. Do this about 25 times, rubbing in the same direction each time. Place the magnet far away from your set-up.
3. Gently place the needle on the floating foil to act as a compass.
4. Turn the foil so that the needle points in a different direction. Observe what happens when you release the foil.

WHAT DO YOU THINK?

- What direction did the needle move when you placed it in the bowl?
- What moved the compass's needle?

CHALLENGE How could you use your compass to answer a question of your own about magnetism?

Earth's magnetic field affects all the magnetic materials around you. Even the cans of food in your cupboard are slightly magnetized by this field. Hold a compass close to the bottom of a can and observe what happens. The magnetic domains in the metal can have aligned and produced a weak magnetic field. If you twist the can and check it again several days later, you can observe the effect of the domains changing their alignment.

Sailors learned many centuries ago that the compass does not point exactly toward the North Pole of Earth's axis. Rather, the compass magnet is currently attracted to an area 966 kilometers (600 mi) from the end of the axis of rotation. This area is known as the magnetic north pole. Interestingly, the magnetic poles of Earth can reverse, so that the magnetic north pole becomes the magnetic south pole. This has happened at least 400 times over the last 330 million years. The most recent reversal was about 780,000 years ago.

The evidence that the magnetic north and south poles reverse is found in rocks in which the minerals contain iron. The iron in the minerals lines up with Earth's magnetic field as the rock forms. Once the rock is formed, the domains remain in place. The evidence for the reversing magnetic field is shown in layers of rocks on the ocean floor, where the domains are arranged in opposite directions.

DIFFERENTIATE INSTRUCTION

More Reading Support

H What evidence shows that Earth's magnetic poles have reversed over the years? *rocks on the ocean floor*

Additional Investigation

To reinforce Section 3.1 learning goals, use the following full-period investigation:

Additional INVESTIGATION, It's a Plot! A, B, & C, pp. 195–203, 206–207
(Advanced students should complete Levels B and C.)

Advanced

Have students investigate and report on the geomagnetic reversal of Earth's magnetic field. They should use the terms *epoch* (long-term changes) and *event* (short-term changes) in their reports.

INVESTIGATE Earth's Magnetic Field

PURPOSE To infer that Earth's magnetic field moves a compass needle

TIPS *15 min.*

- Students should use a piece of foil just large enough to float the needle. Make sure the foil floats freely in the bowl.

- Ask students why the compass needle should be kept away from classroom magnets. *The needle will be attracted to the magnet instead of Earth's magnetic field.*

WHAT DO YOU THINK? *The needle pointed north-south. The needle was attracted to Earth's magnetic field.*

CHALLENGE *Sample answer: You could use the compass to determine whether there is a magnetic field around an object.*

Datasheet, Earth's Magnetic Field, p. 143

Technology Resources

Customize this student lab as needed or look for an alternative. Print rubrics to assess student lab reports.

 Lab Generator CD-ROM

Teaching with Technology

If probeware is available, use a magnetic field sensor with "Investigate Earth's Magnetic Field."

Integrate the Sciences

Many organisms that have migratory patterns rely on Earth's magnetic field. Some scientists believe that migratory birds, such as geese and ducks, use it to guide them on seasonal flights that may be hundreds of miles long. Other organisms, such as whales and sea turtles, may use similar methods of navigation. Why would Earth's magnetic field affect these animals and not others? Some scientists hypothesize that the animals might have iron-containing materials in or near their brains that react to Earth's magnetic field, just as a compass does.

CHECK YOUR READING *Answer: These locations are where Earth's magnetic field is strongest.*

Reinforce (the **BIG** idea)

Have students relate the section to the Big Idea.

 R Reinforcing Key Concepts, p. 144

ASSESS & RETEACH

Assess

 A Section 3.1 Quiz, p. 41

Reteach

Have students perform this activity.

- Place a disk magnet on a piece of paper.
- Place a steel ball touching the magnet.
- Place another steel ball on the paper where it touches both the magnet and the other ball.
- Explain what happens. *The magnet makes each ball a temporary magnet, and their like poles repel each other.*

Technology Resources

Have students visit **ClassZone.com** for reteaching of Key Concepts.

 CONTENT REVIEW

CONTENT REVIEW CD-ROM

Magnetism and the Atmosphere

A constant stream of charged particles is released by reactions inside the Sun. These particles could be damaging to living cells if they reached the surface of Earth. One important effect of Earth's magnetic field is that it turns aside, or deflects, the flow of the charged particles.

Observers view a beautiful display of Northern Lights in Alaska.

Many of the particles are deflected toward the magnetic poles, where Earth's magnetic field lines are closest together. As the particles approach Earth, they react with oxygen and nitrogen in Earth's atmosphere. These interactions can be seen at night as vast, moving sheets of color—red, blue, green or violet—that can fill the whole sky. These displays are known as the Northern Lights or the Southern Lights.

CHECK YOUR READING Why do the Northern Lights and the Southern Lights occur near Earth's magnetic poles?

3.1 Review

KEY CONCEPTS

1. What force causes magnets to attract or repel one another?
2. Why are some materials magnetic and not others?
3. Describe three similarities between Earth and a bar magnet.

CRITICAL THINKING

4. **Apply** A needle is picked up by a magnet. What can you say about the needle's atoms?
5. **Infer** The Northern Lights can form into lines in the sky. What do you think causes this effect?

CHALLENGE

6. **Infer** Hundreds of years ago sailors observed that as they traveled farther north, their compass needle tended to point toward the ground as well as toward the north. What can you conclude about the magnet inside Earth from this observation?

ANSWERS

1. magnetism

2. The magnetic fields of the atoms in magnetic materials form magnetic domains. Nonmagnetic materials do not have magnetic domains.

3. They both have magnetic poles, magnetic fields, and can magnetize other objects.

4. The magnetic fields of the atoms are aligned.

5. Charged particles align with the magnetic field lines around Earth.

6. The magnetic poles are within Earth, not on the surface.

Think SCIENCE

Can Magnets Heal People?

Many people believe that a magnetic field can relieve pain and cure injuries or illnesses. They point out that human blood cells contain iron and that magnets attract iron.

▶ Claims

Here are some claims from advertisements and published scientific experiments.

> a. In an advertisement, a person reported back pain that went away overnight when a magnetic pad was taped to his back.
>
> b. In an advertisement, a person used magnets to treat a painful bruise. The pain reportedly stopped soon after the magnet was applied.
>
> c. In a research project, people who had recovered from polio, but still had severe pain, rated the amount of pain they experienced. People who used magnets reported slightly more pain relief than those who used fake magnets that looked like the real magnets.
>
> d. A research project studied people with severe muscle pain. Patients who slept on magnetic pads for six months reported slightly less pain than those who slept on nonmagnetic pads or no pads.
>
> e. A research project studied people with pain in their heels, placing magnets in their shoes. About sixty percent of people with real magnets reported improvements. About sixty percent of people with fake magnets also reported improvements.

▶ Controls

Scientists use control groups to determine whether a change was a result of the experimental variable or some other cause. A control group is the same as an experimental group in every way except for the variable that is tested. For each of the above cases, was a control used? If not, can you think of some other explanation for the result?

▶ Evaluating Conclusions

On Your Own Evaluate each claim or report separately. Based on all the evidence, can you conclude that magnets are useful for relieving pain? What further evidence would help you decide?

As a Group Find advertisements for companies that sell magnets for medical use. Do they provide information about how their tests were conducted and how you can contact the doctors or scientists involved?

CHALLENGE Design an experiment, with controls, that would show whether or not magnets are useful for relieving pain.

Some people believe that pads containing magnets, such as these, can relieve pain.

THINK SCIENCE
Scientific Methods of Thinking

Set Learning Goal

To evaluate claims about whether magnets are of value in healing and pain relief

Present the Science

Magnetic therapy is based on the magnetic fields that result from moving charges in the human body. Unpaired, spinning electrons in atoms and ions create magnetic fields because they are moving electrical charges. Some therapists believe that the human body functions better when these magnetic fields are aligned, and that magnets will cause this alignment.

Guide the Activity

- Remind students that conclusions are human interpretations and thus are not always objective. Tell them that conclusions can be faulty either from inaccurate or unsupported data or from incorrect interpretations of data.

- Help students find information for "As a Group." Because magnet therapy is considered to be alternative medicine, information is more likely to be in health magazines or on the Internet than in medical journals.

COOPERATIVE LEARNING STRATEGY

Have the class decide on a "Challenge" experiment to perform. Divide the class into a control group and a group that uses magnets, perform the experiment, and draw and evaluate conclusions.

Close

In claim e, can you think of an explanation for the results? *The pain might have improved in time, regardless of the magnets.*

ANSWERS

Claims a and b do not show that controls were used, so there is no way to compare results. Only one person is in the "sample." **Claims c and d** show use of controls. The slight differences in pain relief may not be significant. **Claim e** shows no difference between controls and subjects.

ON YOUR OWN Sample answer: There is not enough evidence to decide. Further studies using large sample sizes and controls would provide more evidence.

AS A GROUP Answers will vary; check students' advertisements.

CHALLENGE Experiments might include studies of athletes with similar injuries, half of whom used actual magnets and half of whom used fake magnets.

▶ Set Learning Goals

Students will

- Describe how an electric current can produce a magnetic field.
- Describe some uses of electromagnets.
- Examine how motors use magnets.
- Observe in an experiment how to make an electromagnet.

◀ 3-Minute Warm-Up

Display Transparency 20 or copy this exercise on the board:

Draw a diagram that shows the magnetic fields that result when like poles of two magnets are close to each other and when unlike poles are close. Use lines to show the magnetic fields that result from these situations. *Diagrams should show the repulsion of like poles and the attraction of unlike poles.*

 3-Minute Warm-Up, p. T20

3.2 MOTIVATE

EXPLORE Magnetism from Electric Current

PURPOSE To observe that an electric current produces a magnetic field

TIP *10 min.* The wire and the compass must be close to each other.

WHAT DO YOU THINK? *The compass needle moved when the circuit was completed. When the battery is reversed, the compass needle also reverses.*

Teaching with Technology

If probeware is available, use a magnetic field sensor while performing "Magnetism from Electric Current."

KEY CONCEPT

3.2 Current can produce magnetism.

◀ **BEFORE, you learned**

- Electric current is the flow of charge
- Magnetism is a force exerted by magnets
- Magnets attract or repel other magnets

▶ **NOW, you will learn**

- How an electric current can produce a magnetic field
- How electromagnets are used
- How motors use electromagnets

VOCABULARY

electromagnetism p. 89
electromagnet p. 90

EXPLORE Magnetism from Electric Current

What is the source of magnetism?

PROCEDURE

① Tape one end of the wire to the battery.

② Place the compass on the table. Place the wire so that it is lying beside the compass, parallel to the needle of the compass. Record your observations.

③ Briefly touch the free end of the wire to the other end of the battery. Record your observations.

④ Turn the battery around and tape the other end to the wire. Repeat steps 2 and 3.

MATERIALS

- electrical tape
- copper wire
- AA cell (battery)
- compass

WHAT DO YOU THINK?

- What did you observe?
- What is the relationship between the direction of the battery and the direction of the compass needle?

An electric current produces a magnetic field.

▼ **REMINDER**

Current is the flow of electrons through a conductor.

Like many discoveries, the discovery that electric current is related to magnetism was unexpected. In the 1800s, a Danish physicist named Hans Christian Oersted (UR-stehd) was teaching a physics class. Oersted used a battery and wire to demonstrate some properties of electricity. He noticed that as an electric charge passed through the wire, the needle of a nearby compass moved.

When he turned the current off, the needle returned to its original direction. After more experiments, Oersted confirmed that there is a relationship between magnetism and electricity. He discovered that an electric current produces a magnetic field.

RESOURCES FOR DIFFERENTIATED INSTRUCTION

Below Level
UNIT RESOURCE BOOK
- Reading Study Guide A, pp. 147–148
- Decoding Support, p. 183

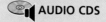 **AUDIO CDS**

Advanced
UNIT RESOURCE BOOK
Challenge and Extension, p. 153

English Learners
UNIT RESOURCE BOOK
Spanish Reading Study Guide, pp. 151–152

AUDIO CDS

- Audio Readings in Spanish
- Audio Readings (English)

Electromagnetism

The relationship between electric current and magnetism plays an important role in many modern technologies. **Electromagnetism** is magnetism that results from an electric current. When a charged particle such as an electron moves, it produces a magnetic field. Because an electric current generally consists of moving electrons, a current in a wire produces a magnetic field. In fact, the wire acts as a magnet. Increasing the amount of current in the wire increases the strength of the magnetic field.

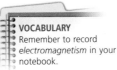

VOCABULARY
Remember to record *electromagnetism* in your notebook.

You have seen how magnetic field lines can be drawn around a magnet. The magnetic field lines around a wire are usually illustrated as a series of circles. The magnetic field of a wire actually forms the shape of a tube around the wire. The direction of the current determines the direction of the magnetic field. If the direction of the electric current is reversed, the magnetic field still exists in circles around the wire, but is reversed.

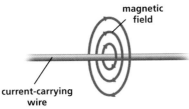

magnetic field

current-carrying wire

If the wire is shaped into a loop, the magnetism becomes concentrated inside the loop. The field is much stronger in the middle of the loop than it is around a straight wire. If you wind the wire into a coil, the magnetic force becomes stronger with each additional turn of wire as the magnetic field becomes more concentrated.

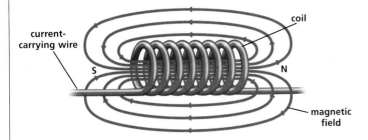

current-carrying wire

coil

S N

magnetic field

A coil of wire with charge flowing through it has a magnetic field that is similar to the magnetic field of a bar magnet. Inside the coil, the field flows in one direction, forming a north pole at one end. The flow outside the coil returns to the south pole. The direction of the electric current in the wire determines which end of the coil becomes the north pole.

CHECK YOUR READING How is a coil of wire that carries a current similar to a bar magnet?

DIFFERENTIATE INSTRUCTION

? More Reading Support

A What is electromagnetism? *magnetism that results from an electric current*

Inclusion Use pipe cleaners for the coil and thin wire for the loops of the magnetic field to model the lower figure on p. 89 for students who have visual impairments.

Teacher Demo

To show a magnetic field, stand up three strong horseshoe magnets in a row about 10 centimeters apart. Place the first and third magnets so that their poles are the same. Reverse the second magnet so that its poles are opposite. Run a strip of extremely thin aluminum foil through the tunnel formed by the poles of the magnets. Use wire to attach the ends of the foil to the terminals of a low-voltage DC power supply that can be varied. Start the current low and increase it. The foil will arch in two places, showing that a force is acting on it.

Teach Difficult Concepts

To help students figure out the direction of the magnetic field for a current-carrying wire, have them use what is known as the right-hand rule.

- Use a pencil to represent the wire, with current flowing from the eraser to the point.
- Have students grasp the pencil with their right hand so that their fingers curl around the pencil and their thumb points to the eraser end of the pencil. In this model, the thumb is pointing in the direction of the flow of the current. The fingers curl around the pencil in the same perpendicular direction as the magnetic field.

Ongoing Assessment

Describe how an electric current can produce a magnetic field.

Ask: Why does an electric current produce a magnetic field? *The current consists of moving charged particles.*

CHECK YOUR READING *Answer: Both have magnetic fields and magnetic poles.*

INVESTIGATE
Electromagnets

PURPOSE To make and observe an electromagnet, which uses current to produce magnetism

TIPS *20 min.*

- Remind students that the batteries must point in the same direction.
- The batteries should make good contact when they are taped together. Make sure tape does not come between the battery terminals.
- Do not leave the batteries connected for more than a few minutes, or the wire and nail will get hot.

WHAT DO YOU THINK? *The nail became magnetized and picked up the paper clip. Yes; the nail was not a magnet before there was current in the wire.*

CHALLENGE *Yes; an aluminum nail would not form a magnet, because aluminum is not a magnetic material.*

 Datasheet, Electromagnets, p. 154

Technology Resources

Customize this student lab as needed or look for an alternative. Print rubrics to assess student lab reports.

 Lab Generator CD-ROM

Teaching with Technology

Use a magnetic field sensor to detect the magnetic field around the nail (after it is connected to the battery).

EXPLORE (the **BIG** idea)

Revisit "Internet Activity: Electromagnets" on p. 77. Have students explain their results.

Ongoing Assessment

CHECK YOUR READING *Answer: Increase the number of coils or the current.*

Making an Electromagnet

? B
? C

Recall that a piece of iron in a strong magnetic field becomes a magnet itself. An **electromagnet** is a magnet made by placing a piece of iron or steel inside a coil of wire. As long as the coil carries a current, the metal acts as a magnet and increases the magnetic field of the coil. But when the current is turned off, the magnetic domains in the metal become random again and the magnetic field disappears.

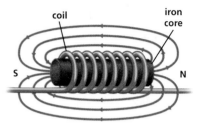
coil   iron core
S   N

By increasing the number of loops in the coil, you can increase the strength of the electromagnet. Electromagnets exert a much more powerful magnetic field than a coil of wire without a metal core. They can also be much stronger than the strongest permanent magnets made of metal alone. You can increase the field strength of an electromagnet by adding more coils or a stronger current. Some of the most powerful magnets in the world are huge electromagnets that are used in scientific instruments.

 CHECK YOUR READING How can you increase the strength of an electromagnet?

INVESTIGATE Electromagnets

How can you make an electromagnet?

PROCEDURE

① Starting about 25 cm from one end of the wire, wrap the wire in tight coils around the nail. The coils should cover the nail from the head almost to the point.

② Tape the two batteries together as shown. Tape one end of the wire to a free battery terminal.

③ Touch the point of the nail to a paper clip and record your observations.

④ Connect the other end of the wire to the other battery terminal. Again touch the point of the nail to a paper clip. Disconnect the wire from the battery. Record your observations.

WHAT DO YOU THINK?

- What did you observe?
- Did you make an electromagnet? How do you know?

CHALLENGE Do you think the result would be different if you used an aluminum nail instead of an iron nail? Why?

SKILL FOCUS
Observing

MATERIALS
- insulated wire
- large iron nail
- 2 D cells
- electrical tape
- paper clip

TIME
20 minutes

DIFFERENTIATE INSTRUCTION

? **More Reading Support**

B What is a piece of iron inside a wire coil? *an electromagnet*

C What happens when you turn off the current? *The magnetic field disappears.*

Below Level Have students relate the parts of the electromagnet that they made during "Investigate Electromagnets" with the diagram of a coil and iron core at the top of the page. *The iron core is the nail, the coil is the wire that they wrapped around the nail, and the magnetic field lines are invisible but caused the nail to attract the paper clip.*

Uses of Electromagnets

Because electromagnets can be turned on and off, they have more uses than permanent magnets. The photograph below shows a powerful electromagnet on a crane. While the electric charge flows through the coils of the magnet, it lifts hundreds of cans at a recycling plant. When the crane operator turns off the current, the magnetic field disappears and the cans drop from the crane.

A permanent magnet would not be nearly as useful for this purpose. Although you could use a large permanent magnet to lift the cans, it would be hard to remove them from the magnet.

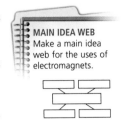

MAIN IDEA WEB
Make a main idea web for the uses of electromagnets.

This powerful electromagnet can be turned on and off to collect and move cans at a recycling plant.

electromagnet

wire supplying electric current

You use an electromagnet every time you store information on a computer. The computer hard drive contains disks that have billions of tiny magnetic domains in them. When you save a file, a tiny electromagnet in the computer is activated. The magnetic field of the electromagnet changes the orientation of the small magnetic domains. The small magnets store your file in a form that can be read later by the computer. A similar system is used to store information on magnetic tape of an audiocassette or videocassette. Sound and pictures are stored on the tape by the arrangement of magnets embedded in the plastic film.

Magnetic information is often stored on credit cards and cash cards. A black strip on the back of the card contains information about the account number and passwords. The cards can be damaged if they are frequently exposed to magnetic fields. For example, cards should not be stored with their strips facing each other, or near a magnetic clasp on a purse or wallet. These magnetic fields can change the arrangement of the tiny magnetic domains on the card and erase the stored information.

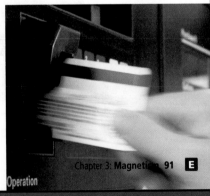

Chapter 3: Magnetism **91** **E**

Operation

Teach from Visuals

Have students examine the photograph of the credit card being used. Ask:

- If the stored information on the credit card can be damaged by a magnetic field, what can you infer about the material that makes up the strip? *It is magnetic.*
- What can you infer about the domains in the material that make up the strip? *They are not aligned in one direction but will align in one direction in a magnetic field.*

Ongoing Assessment

Describe some uses of electromagnets.

Ask: Which would probably have more wire coils, an electromagnet used to separate types of cans or one used to lift cars? *one used to lift cars*

DIFFERENTIATE INSTRUCTION

? More Reading Support

D Why is an electromagnet more useful than a permanent magnet to separate a mixture of metals? *Turning off the magnet is an easy way to release the objects.*

Advanced Have students work in pairs to investigate and explain the operation of a metal detector.

R Challenge and Extension, p. 153

Chapter 3 **91** **E**

History of Science

In 1821, Michael Faraday wrote *Historical Sketch of Electromagnetism,* a paper that summarizes the work of other scientists on electromagnetism. That same year, he made the first simple electrical motor. Many scientists think that one of Faraday's most important contributions to scientific knowledge was his study of electromagnetic induction. He was the first scientist to report that a current is induced by a change in magnetism, not by steady magnetism.

Real World Example

The first operating micromotor was developed in 1988 by a team of electrical engineers headed by Richard Muller at the University of California, Berkeley. Their motor has approximately the same diameter as a human hair. Currently, there are no industrial applications for a motor this small, but similar micromachines are used in airbag deployment systems and heart pacemakers. Even smaller motors have been built, using nanotechnology.

Ongoing Assessment

 Answer: the push and pull of the other magnet in the motor

Motors use electromagnets.

 E

Because magnetism is a force, magnets can be used to move things. Electric motors convert the energy of an electric current into motion by taking advantage of the interaction between current and magnetism.

There are hundreds of devices that contain electric motors. Examples include power tools, electrical kitchen appliances, and the small fans in a computer. Almost anything with moving parts that uses current has an electric motor.

VISUALIZATION
CLASSZONE.COM
See a motor in motion.

Motors

Page 93 shows how a simple motor works. The photograph at the top of the page shows a motor that turns the blades of a fan. The illustration in the middle of the page shows the main parts of a simple motor. Although they may look different from each other, all motors have similar parts and work in a similar way. The main parts of an electrical motor include a voltage source, a shaft, an electromagnet, and at least one additional magnet. The shaft of the motor turns other parts of the device.

Recall that an electromagnet consists of a coil of wire with current flowing through it. Find the electromagnet in the illustration on page 93. The electromagnet is placed between the poles of another magnet.

When current from the voltage source flows through the coil, a magnetic field is produced around the electromagnet. The poles of the magnet interact with the poles of the electromagnet, causing the motor to turn.

1 The poles of the magnet push on the like poles of the electromagnet, causing the electromagnet to turn.

2 As the motor turns, the opposite poles pull on each other.

3 When the poles of the electromagnet line up with the opposite poles of the magnet, a part of the motor called the commutator reverses the polarity of the electromagnet. Now, the poles push on each other again and the motor continues to turn.

The illustration of the motor on page 93 is simplified so that you can see all of the parts. If you saw the inside of an actual motor, it might look like the illustration on the left. Notice that the wire is coiled many times. The electromagnet in a strong motor may coil hundreds of times. The more coils, the stronger the motor.

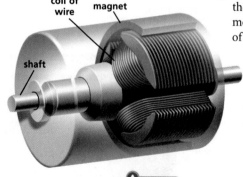

coil of wire   magnet

shaft

CHECK YOUR READING What causes the electromagnet in a motor to turn?

 F

DIFFERENTIATE INSTRUCTION

? More Reading Support

E Motors convert electrical energy into what form of energy? *motion*

F How could you make a motor stronger? *Increase the number of coils.*

Alternative Assessment Have students sketch a concept map showing the operation of a motor. Graphics should include all steps from electrical energy entering the motor to the shaft's turning.

How a Motor Works

Although motors may look different from each other, they all have similar parts and work in a similar way.

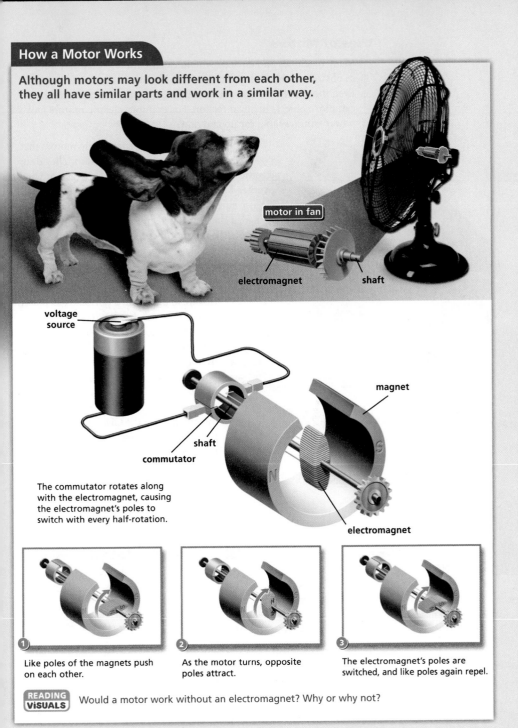

motor in fan

electromagnet

shaft

voltage source

magnet

shaft

commutator

N

S

electromagnet

The commutator rotates along with the electromagnet, causing the electromagnet's poles to switch with every half-rotation.

1. Like poles of the magnets push on each other.

2. As the motor turns, opposite poles attract.

3. The electromagnet's poles are switched, and like poles again repel.

READING ViSUALS Would a motor work without an electromagnet? Why or why not?

Language Arts Connection

Explain that when people go from home to work or school and back, they are said to *commute*. One definition of commute is to travel but it also has other meanings. The word *commute* comes from the Middle English *com-muten*, meaning "to transform" and from the Latin *commutare*, meaning "to change." The *commutative* math property says that in addition or multiplication, you can move the numbers around without changing the problem. Ask students to relate these meanings to the way "commutator" is used in the text.

Integrate the Sciences

Many motors occur in nature on the molecular scale. These motors convert chemical energy into kinetic energy. Molecular motors perform functions such as intercellular transport, cellular organization, and cell movement and growth. Sets of these motors move the flagella of sperm and bacteria. They also contract muscles.

Ongoing Assessment

Examine how motors use electromagnets.

Ask: Which magnet has its polarity reversed by the commutator in a motor? *the electromagnet*

READING ViSUALS *Answer: No; repulsion of the poles of the magnets is the force that moves the parts of the motor.*

DIFFERENTIATE INSTRUCTION

English Learners Within this section are a variety of introductory clauses and phrases. Give students these examples and have them identify the subject of the sentence.

"While the electric charge flows through the coils of the magnet, it lifts hundreds of cans at a recycling plant." (p. 91)

"When you save a file, a tiny electromagnet in the computer is activated." (p. 91)

Encourage students to use introductory clauses and phrases in their own writing.

Ongoing Assessment

 CHECK YOUR READING *One motor spins the CD while another motor moves a laser across the CD to read it.*

Reinforce the **BIG** idea

Have students relate the section to the Big Idea.

 Reinforcing Key Concepts, p. 155

3.2 ASSESS & RETEACH

Assess

 Section 3.2 Quiz, p. 42

Reteach

Have students name as many machines and devices as they can that may contain one or more motors. Remind students that motors can produce different types of movements, not just a rotating movement. Ask students to consider how the motor(s) in each machine create movement. Example: a car's motor converts energy to turn the wheels. Have pairs of students compare their descriptions and discuss any differences.

Technology Resources

Have students visit **ClassZone.com** for reteaching of Key Concepts.

 CONTENT REVIEW

CONTENT REVIEW CD-ROM

Uses of Motors

Many machines and devices contain electric motors that may not be as obvious as the motor that turns the blades of a fan, for example. Even though the motion produced by the motor is circular, motors can move objects in any direction. For example, electric motors move power windows in a car up and down.

Motors can be very large, such as the motors that power an object as large as a subway train. They draw electric current from a third rail on the track or wires overhead that carry electric current. A car uses an electric current to start the engine. When the key is turned, a circuit is closed, producing a current from the battery to the motor. Other motors are very small, like the battery-operated motors that move the hands of a wristwatch.

The illustration on the left shows the two small motors in a portable CD player. Motor A causes the CD to spin. Motor B is connected to a set of gears. The gears convert the rotational motion of the motor into a straight-line motion, or linear motion. As the CD spins, a laser moves straight across the CD from the center outward. The laser reads the information on the CD. The motion from Motor B moves the laser across the CD.

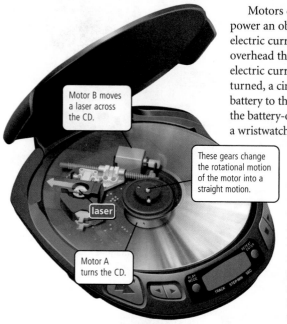

Motor B moves a laser across the CD.

These gears change the rotational motion of the motor into a straight motion.

laser

Motor A turns the CD.

 CHECK YOUR READING Explain the function served by each motor in a CD player.

3.2 Review

KEY CONCEPTS

1. Explain how electric current and magnetism are related.

2. Describe three uses of electromagnets.

3. Explain how electrical energy is converted to motion in a motor.

CRITICAL THINKING

4. **Contrast** How does an electromagnet differ from a permanent magnet?

5. **Apply** Provide examples of two things in your home that use electric motors, and explain why they are easier to use because of the motors.

CHALLENGE

6. **Infer** Why is it necessary to change the direction of the current in the coil of an electric motor as it turns?

ANSWERS

1. Electric current flowing through a wire produces a magnetic field around the wire.

2. to lift iron objects, to store information on a computer, to turn a motor

3. The electric current flowing through the electromagnet produces a magnetic field around the wire. A second magnet causes the electromagnet to turn.

4. The strength of an electromagnet can be altered, and electromagnets can be turned on and off.

5. Sample answer: A blender or mixer makes stirring easier because of its motor. An electrical fan keeps air moving even if the wind dies.

6. After the electromagnet turns half a turn, the poles of the permanent magnet start to repel the electromagnet. Reversing the direction of the current reverses the poles and keeps the motor running.

KEY CONCEPT

Magnetism can produce current.

◀ **BEFORE, you learned**

- Magnetism is a force exerted by magnets
- Electric current can produce a magnetic field
- Electromagnets can make objects move

▶ **NOW, you will learn**

- How a magnetic field can produce an electric current
- How a generator converts energy
- How direct current and alternating current differ

VOCABULARY

generator p. 96
direct current p. 97
alternating current p. 97
transformer p. 99

EXPLORE Energy Conversion

How can a motor produce current?

PROCEDURE

1. Touch the wires on the motor to the battery terminals to see how the motor operates.
2. Connect the wires to the light bulb.
3. Roll the shaft, or the movable part of the motor, between your fingers. Observe the light bulb.
4. Now spin the shaft rapidly. Record your observations.

MATERIALS
- small motor
- AA cell (battery)
- light bulb in holder

WHAT DO YOU THINK?
- How did you produce current?
- What effect did your motion have on the amount of light produced?

Magnets are used to generate an electric current.

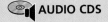

MAIN IDEA WEB
Make a main idea web in your notebook for this heading.

In the 1830s, about ten years after Oersted discovered that an electric current produces magnetism, physicists observed the reverse effect—a moving magnetic field induces an electric current. When a magnet moves inside a coiled wire that is in a circuit, an electric current is generated in the wire.

It is often easier to generate an electric current by moving a wire inside a magnetic field. Whether it is the magnet or the wire that moves, the effect is the same. Current is generated as long as the wire crosses the magnetic field lines.

CHECK YOUR READING What must happen for a magnetic field to produce an electric current?

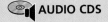
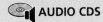

3.3 FOCUS

▶ Set Learning Goals

Students will

- Describe how a magnetic field can produce an electric current.
- Examine how a generator converts energy.
- Explain how direct current and alternating current differ.
- Infer from an experiment how to identify alternating current.

◀ 3-Minute Warm-Up

Display Transparency 21 or copy this exercise on the board:

Decide if these statements are true. If not, correct them.

1. Like charges attract one another. *repel*
2. Electric current can produce a magnetic field. *true*
3. Motors contain magnets. *true*

T 3-Minute Warm-Up, p. T21

3.3 MOTIVATE

EXPLORE Energy Conversion

PURPOSE To observe how a motor uses kinetic energy to produce current

TIP *10 min.* Use small, sensitive motors, such as those used with solar cells.

WHAT DO YOU THINK? *Moving the shaft manually produced current. The faster the spin, the brighter the light.*

Ongoing Assessment

Describe how a magnetic field can produce an electric current.

Ask: To produce a current, which needs to move, the wire or the magnetic field? *either one*

CHECK YOUR READING *Answer: The wire must cross magnetic field lines.*

Teacher Demo

Show the effect of a moving magnetic field on stationary metal. Acquire 1 meter of copper tubing and two neodymium magnets small enough to easily pass through the tubing. Try to pick up the copper with the magnets to show that the copper is not magnetic. Hold the tubing vertically. Drop a small steel ball through the tubing, noticing how long it takes for it to pass through. Then drop the pair of magnets through the tube, again noticing how long it takes. Ask students to explain this difference in time. *The ball dropped as fast as gravity could pull it. The magnets induced a current in the tubing. The current produced a magnetic field that repelled the magnets, making them fall more slowly.*

Teach Difficult Concepts

Compare the relationship between a motor and a generator to other opposite processes. Ask students to explain how the motor-generator relationship is like the relationship between a speaker and a microphone. *In a motor, electrical energy produces a magnetic field, and in a generator, a magnetic field produces electrical energy. In a speaker, electrical signals produce sound, while a microphone transforms sound into electrical signals.*

Ongoing Assessment

Examine how a generator converts energy.

Ask: What energy conversion takes place in a generator? *Kinetic energy is converted to electrical energy.*

CHECK YOUR READING *Answer: the person turning the handle*

READING VISUALS *Answer: It produces a magnetic field that generates current in the wire.*

Generating an Electric Current

 A A **generator** is a device that converts the energy of motion, or kinetic energy, into electrical energy. A generator is similar to a motor in reverse. If you manually turn the shaft of a motor that contains a magnet, you can produce electric current. **B**

The illustration below shows a portable generator that provides electrical energy to charge a cell phone in an emergency. The generator produces current as you turn the handle. Because it does not need to be plugged in, the generator can be used wherever and whenever it is needed to recharge a phone. The energy is supplied by the person turning the handle.

1 As the handle is turned, it rotates a series of gears. The gears turn the shaft of the generator.

2 The rotation of the shaft causes coils of wire to rotate within a magnetic field.

3 As the coils of the wire cross the magnetic field line, electric current is generated. The current recharges the battery of the cell phone.

CHECK YOUR READING What is the source of energy for a cell phone generator?

How a Cell Phone Generator Works

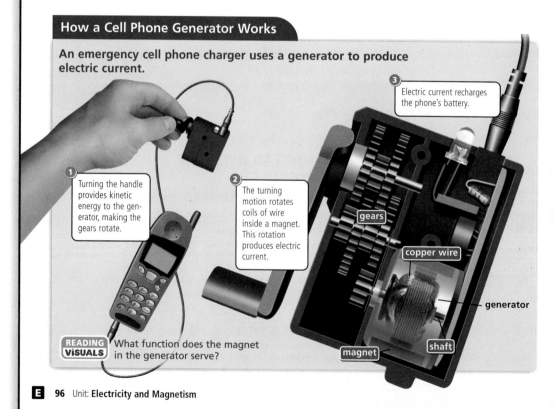

An emergency cell phone charger uses a generator to produce electric current.

1 Turning the handle provides kinetic energy to the generator, making the gears rotate.

2 The turning motion rotates coils of wire inside a magnet. This rotation produces electric current.

3 Electric current recharges the phone's battery.

gears

copper wire

generator

magnet

shaft

READING VISUALS What function does the magnet in the generator serve?

DIFFERENTIATE INSTRUCTION

? More Reading Support

A What is a generator? *a device that converts kinetic energy into electrical energy*

B How are a motor and a generator related? *They are opposites.*

English Learners English learners may be confused by subordinate (dependent) clauses when they appear at the beginning of a sentence. When this happens, tell students to rearrange the clauses into a more familiar pattern. The cause-and-effect relationship in the following sentence, for instance, is the same whether the subordinate clause beginning with "because" starts or completes the sentence. "Because it does not need to be plugged in, the generator can be used wherever and whenever it is needed to recharge a phone."

Direct and Alternating Currents

Think about how current flows in all of the circuits that you have studied so far. Electrons flow from one end of a battery or generator, through the circuit, and eventually back to the battery or generator. Electrons that flow in one direction produce one of two types of current.

- A **direct current** (DC) is electric charge that flows in one direction only. Direct current is produced by batteries and by DC generators such as the cell phone generator.
- An **alternating current** (AC) is a flow of electric charge that reverses direction at regular intervals. The current that enters your home and school is an alternating current.

CHECK YOUR READING What is the difference between direct current and alternating current?

Direct currents and alternating currents are produced by different generators. In an AC generator, the direction in which charge flows depends upon the direction in which the magnet moves in relation to the coil. Because generators use a rotating electromagnet, the poles of the electromagnet alternate between moving toward and moving away from the magnet. The result is a current that reverses with each half-rotation of the coil.

The illustration on the right shows a simple DC generator. DC generators are very similar to AC generators. The main difference is that DC generators have a commutator that causes the current to flow in only one direction.

Many things in your home can work with either direct or alternating currents. In light bulbs, for instance, the resistance to motion of the electrons in the filament makes the filament glow. It doesn't matter in which direction the current is moving.

Some appliances can use only direct current. The black box that is on the plug of some devices is an AC–DC converter. AC–DC converters change the alternating current to direct current. For example, laptop computers use converters like the one shown in the photograph on the right. In a desktop computer, the converter is part of the power supply unit.

commutator

coil of wire

N

S

magnet

DC generator

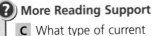

Real World Example

In the United States, the frequency of AC is 60 Hz, which means that it changes direction 60 times a second. It is 50 Hz in most of Europe. Frequencies lower than this cause detectable flickering of lights. European travelers visiting the United States and U.S. travelers visiting Europe have to take adaptors for their appliances, because of the difference in frequencies.

Real World Example

When you turn on an electric car, a controller receives current from a battery. This controller changes the direct current from the battery into alternating current that can be used by the motor. Electric cars are not entirely free of fossil fuel use. Their batteries are recharged by plugging them into a current source. This current source often comes from an electric power plant that burns fossil fuels.

Ongoing Assessment

Explain how direct current and alternating current differ.

Ask: What is the difference between AC and DC generators? *AC generators rapidly alternate direction; DC generators produce current flowing in only one direction.*

CHECK YOUR READING *Answer: In DC, electric charge flows in one direction only. In AC, electric charge reverses direction at regular intervals.*

DIFFERENTIATE INSTRUCTION

More Reading Support

C What type of current is produced by a battery? *direct current*

D What type of generator contains a commutator? *direct current*

Below Level To emphasize how quickly alternating current changes direction, ask groups of students to model it. They might turn their hands from one side to the other as quickly as possible or draw arrows on both sides of a card and flip it back and forth as fast as they can. Have students use a clock with a second hand to reverse direction at regular intervals. Explain that this model is a slower representation of alternating current.

INVESTIGATE Electric Current

PURPOSE To observe characteristics of AC and use inference to identify it

TIPS *15 min.* For best results, use 2 meters of 22-gauge magnet wire and 14 inches of electrical tape. Students must wrap the wire across the face of the compass, not around the circumference. However, they must still be able to see the needle well enough to know in what direction it is pointing.

WHAT DO YOU THINK? *The compass needle moved back-and-forth with AC but not with DC. DC was observed in step 3, and AC in step 4. The type of current could be identified by whether the compass needle moved back-and-forth or not.*

CHALLENGE *Both use magnetism to generate an electrical current that changes direction.*

 Datasheet, Electric Current, p. 165

Technology Resources

Customize this student lab as needed or look for an alternative. Print rubrics to assess student lab reports.

 Lab Generator CD-ROM

Teaching with Technology

Students might want to design and conduct a similar investigation using an AC ammeter instead of a compass.

Metacognitive Strategy

Ask students to write a paragraph about any problems they encountered during the investigation and what they did to solve these problems and complete the activity.

Ongoing Assessment

 Answer: It provides current to the car's electrical devices.

INVESTIGATE Electric Current

How can you identify the current?

PROCEDURE

1. Wrap the wire tightly around the middle of the compass 10–15 times. Leave about 30 cm of wire free at each end. Tape the wire to the back of the compass to keep it in place.

2. Sand the ends of the wire with sandpaper to expose about 2 cm of copper on each end. Arrange the compass on your desk so that the needle is parallel to, or lined up with, the coil. This will serve as your current detector.

3. Tape one end of the wire to one terminal of the battery. Touch the other end of the wire to the other battery terminal. Record your observations.

4. Observe the current detector as you tap the end of the wire to the battery terminal at a steady pace. Speed up or slow down your tapping until the needle of the compass alternates back and forth. Record your observations.

WHAT DO YOU THINK?

• What did you observe?

• What type of current did you detect in step 3? in step 4? How did you identify the type of current?

CHALLENGE How is this setup similar to an AC generator?

SKILL FOCUS
Inferring

MATERIALS
• piece of wire
• compass
• ruler
• tape
• sandpaper
• D cell (battery)

TIME
15 minutes

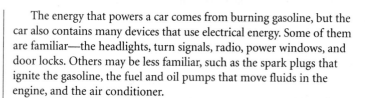

The energy that powers a car comes from burning gasoline, but the car also contains many devices that use electrical energy. Some of them are familiar—the headlights, turn signals, radio, power windows, and door locks. Others may be less familiar, such as the spark plugs that ignite the gasoline, the fuel and oil pumps that move fluids in the engine, and the air conditioner.

A car's engine includes a generator to provide current to its electrical devices. As the engine runs, it converts gasoline to kinetic energy. Some of that energy is transferred to the generator by a belt attached to its shaft. Inside the generator, a complex coil of copper wires turns in a magnetic field, generating a current that operates the electrical devices of the car.

 The generator also recharges the battery, so that power is available when the engine is not running. Because the generator in most cars supplies alternating current, a car generator is usually called an alternator.

 What function does a generator in a car serve?

DIFFERENTIATE INSTRUCTION

 More Reading Support

E If the battery in a car supplies current when the motor is not running, why doesn't it quickly lose current? *The generator recharges the battery.*

Advanced Have students who are interested in how cars operate investigate the way the generator in a car runs the electrical components and what those electrical components are.

 Challenge and Extension, p. 164

Have students who are interested in electricity and circuits read the following article:

 Challenge Reading, pp. 179–180

Magnets are used to control voltage.

A **transformer** is a device that increases or decreases voltage. Transformers use magnetism to control the amount of voltage. A transformer consists of two coils of wire that are wrapped around an iron ring.

An alternating current from the voltage source in the first coil produces a magnetic field. The iron ring becomes an electromagnet. Because the current alternates, the magnetic field is constantly changing. The second coil is therefore within a changing magnetic field. Current is generated in the second coil. If the two coils have the same number of loops, the voltage in the second coil will be the same as the voltage in the first coil.

A change in the voltage is caused when the two coils have different numbers of loops. If the second coil has fewer loops than the first, as in the illustration, the voltage is decreased. This is called a step-down transformer. On the other hand, if the second coil has more loops than the first, the voltage in the second circuit will be higher than the original voltage. This transformer is called a step-up transformer.

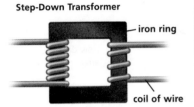

Step-Down Transformer
- iron ring
- coil of wire

Transformers are used in the distribution of current. Current is sent over power lines from power plants at a very high voltage. Step-down transformers on utility poles, such as the one pictured on the right, reduce the voltage available for use in homes. Sending current at high voltages minimizes the amount of energy lost to resistance along the way.

3.3 Review

KEY CONCEPTS

1. What is necessary for a magnetic field to produce an electric current?
2. Explain how electric generators convert kinetic energy into electrical energy.
3. Compare and contrast the ways in which direct current and alternating current are generated.

CRITICAL THINKING

4. **Apply** A radio can be operated either by plugging it into the wall or by using batteries. How can it use either source of current?
5. **Drawing Conclusions** Suppose that all of the electrical devices in a car stop working. Explain what the problem might be.

CHALLENGE

6. **Apply** European power companies deliver current at 220 V. Draw the design for a step-down transformer that would let you operate a CD player made to work at 110 V in France.

3.3 ASSESS & RETEACH

Assess

 Section 3.3 Quiz, p. 43

Reteach

Have students create a graphic organizer that shows how an electrical motor and a generator are related. Tell students to include the terms current and magnetic field in their graphic organizer. *An organizer might have four boxes arranged in a circle, labeled generator, current, motor, and magnetic field. Arrows would then show that electrical current leaves a generator and enters a motor and that a magnetic field is produced by a motor and used by a generator.*

Technology Resources

Have students visit **ClassZone.com** for reteaching of Key Concepts.

 CONTENT REVIEW

CONTENT REVIEW CD-ROM

ANSWERS

1. Magnetic field lines and a wire must cross each other.

2. The kinetic energy turns a magnet, producing an electrical current in a wire.

3. DC is generated without changing the polarity of the magnet that produces it. AC is produced by a magnet that frequently changes polarity.

4. For the radio to use AC, it must contain a converter that changes AC to DC.

5. The most likely cause would be a faulty generator.

6. The transformer should have twice as many coils on the input side as it does on the output side.

Focus

PURPOSE To construct a speaker and determine how the strength of a magnet affects the volume of the speaker

OVERVIEW Students will use various materials to construct a speaker. They will use three magnets of different strengths to determine which one produces the loudest sound, and find that the strongest magnet produces the loudest sound on the speaker.

Lab Preparation

- For best results, use 2 meters of 22-gauge magnet wire. Strip the ends of the wire by scraping them with sandpaper, or use pre-stripped wire.

- Use an old stereo system or radio as it may become damaged by this activity if the volume level is too high. One stereo can be used by the whole class.

- To make the connections to the stereo, cut the wire to the headphone jack, and then separate and strip the two wires within. Students will clip the leads from their speakers to these two wires.

- Prior to the investigation have students read through the investigation and prepare their data tables. Or, copy and distribute datasheets and rubrics.

 UNIT RESOURCE BOOK, pp. 186–194

 SCIENCE TOOLKIT, F14

Lab Management

- Divide students into groups of three or four.

- Each student should do a different task when the group checks a magnet.

SAFETY Warn students not to turn up the sound enough to damage their ears. The volume should be turned down when the investigation is completed.

INCLUSION Ask students with hearing impairments to touch the cup so that they can feel the varying degrees of vibration as a measure of sound intensity. If students have difficulty, have a partner describe the results.

Build a Speaker

OVERVIEW AND PURPOSE Speakers are found on TVs, computers, telephones, stereos, amplifiers, and other devices. Inside a speaker, magnetism and electric current interact to produce sound. The current produces a magnetic field that acts on another magnet and causes vibrations. The vibrations produce sound waves. In this lab, you will

- construct a speaker
- determine how the strength of the magnet affects the speaker's volume

 Problem   Write It Up

How does the strength of the magnet used to make a speaker affect the loudness of sound produced by the speaker?

 Hypothesize   Write It Up

Write a hypothesis that explains how you expect the strength of a magnet to affect the loudness of sound produced by the speaker, and why. Your hypothesis should be in the form of an "if . . . , then . . . , because . . . " statement.

 Procedure

1. Make a data table similar to the one shown on the sample notebook page.

2. Test the strength of each magnet by measuring the distance at which a paper clip will move to the magnet, as shown. Record the measurements in your **Science Notebook**.

step 2

3. Starting about 6 cm from the end of the wire, wrap the wire around the marker 50 times to make a coil.

MATERIALS
- 3 magnets of different strengths
- paper clip
- ruler
- piece of wire
- marker
- cup
- masking tape
- 2 wire leads with alligator clips
- stereo system

INVESTIGATION RESOURCES

 CHAPTER INVESTIGATION, Build a Speaker
- Level A, pp. 186–189
- Level B, pp. 190–193
- Level C, p. 194
Advanced students should complete Levels B & C.

 Writing a Lab Report, D12–13

Technology Resources

Customize this student lab as needed or look for an alternative. Print rubrics to assess student lab reports.

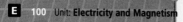 Lab Generator CD-ROM

4. Carefully slide the coil off the marker. Wrap the ends of the wire around the coil to keep it in the shape of a circle, as shown.

step 4

5. Place the cup upside-down on your table. Tape the coil to the bottom of the cup. Clip the leads to the ends of the wire. Tape the alligator clips to the sides of the cup, as shown.

coil

step 5

6. Take turns attaching the alligator clips to the stereo as instructed by your teacher. Place each magnet on the table near the stereo. Test the speaker by holding the cup directly over each magnet and listening. Record your observations.

Observe and Analyze
Write It Up

1. **RECORD OBSERVATIONS** Be sure to record your observations in the data table.

2. **INFER** Why is the coil of wire held near the magnet?

3. **APPLY** The diaphragm on a speaker vibrates to produce sound. What part of your stereo is the diaphragm?

4. **IDENTIFY** What was the independent variable in this experiment? What was the dependent variable?

Conclude
Write It Up

1. **INTERPRET** Which magnet produced the loudest noise when used with your speaker? Answer the question posed in the problem.

2. **ANALYZE** Compare your results with your hypothesis. Did your results support your hypothesis?

3. **IDENTIFY LIMITS** Describe possible limitations or sources of error in the procedure or any places where errors might have occurred.

4. **APPLY** You have built a simple version of a real speaker. Apply what you have learned in this lab to explain how a real speaker might work.

INVESTIGATE Further

CHALLENGE In what ways might you vary the design of the speaker to improve its functioning? Review the procedure to identify variables that might be changed to improve the speaker. Choose one variable and design an experiment to test that variable.

Build a Speaker

Problem How does the strength of the magnet used to make a speaker affect the loudness of sound produced by the speaker?

Hypothesize

Observe and Analyze
Table 1. Strength of Magnet and Loudness of Sound

Magnet	Strength (paper clip distance)	Observations
1		
2		
3		

Conclude

Observe and Analyze
Write It Up

1. Data tables should be complete. Sample data: Magnet 1: 1.5 cm, could not hear any sound; Magnet 2: 2 cm, could hear a faint sound; Magnet 3: 4 cm, could hear the sound clearly

2. The magnetic field produced by the coil of wire should be close enough to affect the magnet on the table.

3. the cup

4. strength of the magnet; loudness of the sound

Conclude
Write It Up

1. The strongest magnet produced the loudest sound. The stronger the magnet used to make a speaker, the louder the sound produced.

2. Answers will vary depending on students' results and original hypotheses.

3. Sample answer: The wires might not be adequately connected, the magnets might not be strong enough, and sounds in the room may have interfered with the observations.

4. Current travels through a coil of wire to produce a magnetic field. The magnetic field of the coil interacts with the magnetic field of the permanent magnet, causing sound vibrations.

INVESTIGATE Further

CHALLENGE Variables might include the number of coils used, the number of permanent magnets used, and the length of the wires attached to the alligator clips.

Post-Lab Discussion

- Ask: Would the experiment have been different if some other object, such as a metal can or a glass jar, had been used instead of the paper cup? Explain. *The ability of the material to vibrate would affect the sound produced. The more the material vibrated, the louder the sound would be.*

- Ask: What questions do you still have upon completing the "Build a Speaker" activity?

○ Set Learning Goals

Students will

- Discover how power plants generate electrical energy.
- Describe how electric power is measured.
- Calculate energy usage.
- Model in an experiment how electrical energy is used.

○ 3-Minute Warm-Up

Display Transparency 21 or copy this exercise on the board:

Match the definitions to the terms.

Definitions

1. the force exerted by magnets *b*
2. a device that uses magnetism to produce current *d*
3. a device that uses current to produce magnetism *e*

Terms

a. electricity   d. generator
b. magnetism   e. electromagnet
c. compass

 3-Minute Warm-Up, p. T21

3.4 MOTIVATE

THINK ABOUT

PURPOSE To think about how the kinetic energy of falling water can be used to generate electricity

DISCUSS Ask students how many of them have been in the ocean or a river and felt the energy of the moving water. Ask them to use this information to answer the question. *The force of the water can be used to turn the moving parts of a generator.*

Ongoing Assessment

 CHECK YOUR READING *Answer: Kinetic energy turns the moving parts of a generator to produce electricity.*

E **102** Unit: **Electricity and Magnetism**

KEY CONCEPT

3.4 Generators supply electrical energy.

◀ **BEFORE, you learned**

- Magnetism is a force exerted by magnets
- A moving magnetic field can generate an electric current in a conductor
- Generators use magnetism to produce current

▶ **NOW, you will learn**

- How power plants generate electrical energy
- How electric power is measured
- How energy usage is calculated

VOCABULARY

electric power p. 102
watt p. 104
kilowatt p. 104
kilowatt-hour p. 105

THINK ABOUT

How can falling water generate electrical energy?

This photograph shows the Hoover Dam on the Nevada/Arizona border, which holds back a large lake, almost 600 feet deep, on the Colorado River. It took thousands of workers nearly five years to build the dam, and it cost millions of dollars. One of the main purposes of the Hoover Dam is the generation of current. Think about what you have read about generators. How could the energy of falling water be used to generate current?

Generators provide most of the world's electrical energy.

The tremendous energy produced by falling water provides the turning motion for large generators at a power plant. The power plant at the Hoover Dam supplies energy to more than a million people.

Other sources of energy at power plants include steam from burning fossil fuels, nuclear reactions, wind, solar heating, and ocean tides. Each source provides the energy of motion to the generators, producing electrical energy. **Electric power** is the rate at which electrical energy is generated from another source of energy.

 **VOCABULARY** Use a description wheel to take notes about *electrical power.*

CHECK YOUR READING What do power plants that use water, steam, and wind all have in common?

E 102 Unit: **Electricity and Magnetism**

RESOURCES FOR DIFFERENTIATED INSTRUCTION

Below Level
UNIT RESOURCE BOOK
- Reading Study Guide A, pp. 169–170
- Decoding Support, p. 183

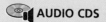 **AUDIO CDS**

Advanced
UNIT RESOURCE BOOK
Challenge and Extension, p. 175

English Learners
UNIT RESOURCE BOOK
Spanish Reading Study Guide, pp. 173–174

AUDIO CDS

- Audio Readings in Spanish
- Audio Readings (English)

Generating Electrical Energy

How does the power plant convert the energy of motion into electrical energy? Very large generators in the plant hold powerful electromagnets surrounded by massive coils of copper wire. The illustration below shows how the energy from water falling from the reservoir to the river far below a dam is converted to electrical energy.

RESOURCE CENTER
CLASSZONE.COM
Find out more about dams that generate current.

❶ As the water falls from the reservoir, its kinetic energy increases and it flows very fast. The falling stream of water turns a fan-like device, called a turbine, which is connected to the generator's shaft.

❷ The rotation of the shaft turns powerful electromagnets that are surrounded by the coil of copper wires. The coil is connected to a step-up transformer that sends high-voltage current to power lines.

❸ Far from the plant, step-down transformers reduce the voltage so that current can be sent through smaller lines to neighborhoods. Another transformer reduces the voltage to the level needed to operate lights and appliances.

How Electrical Power Is Generated

Power plants use generators to convert kinetic energy into electrical energy.

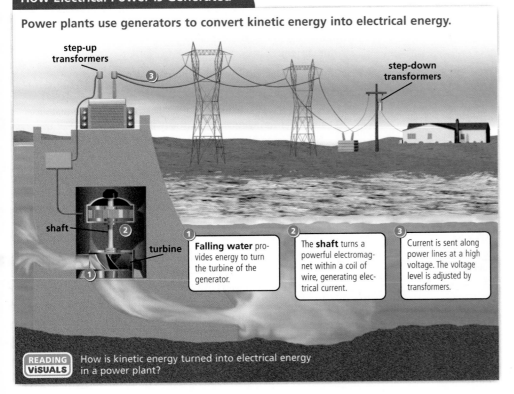

step-up transformers

step-down transformers

shaft

turbine

❶ **Falling water** provides energy to turn the turbine of the generator.

❷ The **shaft** turns a powerful electromagnet within a coil of wire, generating electrical current.

❸ Current is sent along power lines at a high voltage. The voltage level is adjusted by transformers.

READING VISUALS How is kinetic energy turned into electrical energy in a power plant?

Chapter 3: **Magnetism 103** **E**

3.4 INSTRUCT

Teach from Visuals

Have students study the generator schematic. Ask:

- Which has a higher voltage, the current flowing from tower to tower or the current entering the house? *the current flowing from tower to tower*

- Explain your answer. *The current from the tower passes through a step-down transformer before it enters the house.*

- What role does kinetic energy play in the production of electricity? *Kinetic energy turns the turbine in a generator.*

Ongoing Assessment

Discover how power plants generate electrical energy.

Ask: The step-up transformer in a generator sends what type of current to power lines? *high-voltage*

READING VISUALS *Answer: Kinetic energy turns a turbine, which turns a magnetic field inside a coil, producing current.*

DIFFERENTIATE INSTRUCTION

❓ More Reading Support

A What do you call the fan-like device in a generator? *a turbine*

B Which type of transformer does current pass through before it goes into neighborhoods? *step-down*

English Learners The paragraphs above describe in detail how electrical power is generated. Students from homogenous groups may need to practice describing processes or giving directions so an "outsider" can understand. Ask students to pick a place in their school and write out directions from their classroom to that place. Students should write as if the person reading has never been inside their school before—thus, they would not know landmarks such as "Mrs. Lopez's room" or "the broken water fountain."

Address Misconceptions

IDENTIFY Ask: What is the difference between electrical energy and electric power? If students answer "Nothing," they hold the misconception that electric energy and electrical power are the same thing.

CORRECT Point out that power is the rate at which energy is used. Provide students with an example of another type of power. Chemical energy in your body can be used to produce kinetic energy when you move. A person who is running is using energy at a higher rate—and therefore has a higher power— than a person who is walking. Solicit other examples from students of energy being used at different rates.

REASSESS Which measures a rate, electrical energy or power? *power*

Technology Resources

Visit **ClassZone.com** for background on common student misconceptions.

 MISCONCEPTION DATABASE

Teach from Visuals

Have students refer to the power rating table on this page. Ask: If you have a room that is wired for 2000 watts, which of the following groups of appliances could you use in the room? *a, c*

a. refrigerator, stereo system, microwave oven

b. computer, clothes dryer

c. video game system, hair dryer, window fan, television

d. microwave oven, hair dryer

Ongoing Assessment

Describe how electric power is measured.

Ask: What equation shows the relationship among electric power, voltage, and current? *P = VI*

CHECK YOUR READING *Answer: Kilowatts measure a large amount of electric power, such as that used in a building.*

Electric power can be measured.

You have read that electric power is the rate at which electrical energy is generated from another source of energy. Power also refers to the rate at which an appliance converts electrical energy back into another form of energy, such as light, heat, or sound.

In order to provide electrical energy to homes and factories, power companies need to know the rate at which energy is needed. Power can be measured so that companies can determine how much energy is used and where it is used. This information is used to figure out how much to charge customers, and it is used to determine whether more electrical energy needs to be generated. To provide energy to an average home, a power plant needs to burn about four tons of coal each year.

RESOURCE CENTER
CLASSZONE.COM

Learn more about energy use and conservation.

Watts and Kilowatts

? **C** The unit of measurement for power is the **watt** (W). Watts measure the rate at which energy is used by an electrical appliance. For instance, a light bulb converts energy to light and heat. The power rating of the bulb, or of any device that consumes electrical energy, depends on both the voltage and the current. The formula for finding power, in watts, from voltage and current, is shown below. The letter *I* stands for current.

$$\text{Electrical Power} = \text{Voltage} \cdot \text{Current}$$
$$P = V \cdot I$$

You have probably seen the label on a light bulb that gives its power rating in watts—usually in the range of 40 W to 100 W. A brighter bulb converts energy at a higher rate than one with a lower power rating.

The chart at the left shows typical power ratings, in watts, for some appliances that you might have in your home. The exact power rating depends on how each brand of appliance uses energy. You can find the actual power rating for an appliance on its label.

The combined power rating in a building is likely to be a fairly large number. A **kilowatt** (kW) is a unit of power equal to one thousand watts. All of the appliances in a room may have a combined power rating of several kilowatts, but all appliances are not in use all of the time. That is why energy is usually calculated based on how long the appliances are in use.

? **D**

CHECK YOUR READING Explain what kilowatts are used to measure.

Typical Power Ratings

Appliance	Watts
DVD player	20
Radio	20
Video game system	25
Electric blanket	60
Light bulb	75
Stereo system	100
Window fan	100
Television	110
Computer	120
Computer monitor	150
Refrigerator	700
Air conditioner	1000
Microwave oven	1000
Hair dryer	1200
Clothes dryer	3000

DIFFERENTIATE INSTRUCTION

? **More Reading Support**

C What unit measures electric power? *the watt (W)*

D What unit is used for large amounts of electric power? *the kilowatt (kW)*

Below Level Have students look around the room and identify all of the appliances in use that are listed on the table on this page, and the wattage of each appliance. Students should then use the table to estimate the total power rating of the room.

INVESTIGATE Power

How would you use your electrical energy?

PROCEDURE

1. On a sheet of graph paper, outline a box that is 10 squares long by 18 squares wide. The box represents a room that is wired to power a total of 1800 W. Each square represents 10 W of power.

2. From the chart on page 104, choose appliances that you want in your room. Using colored pencils, fill in the appropriate number of boxes for each appliance.

3. All of the items that you choose must fit within the total power available, represented by the 180 squares.

WHAT DO YOU THINK?

- How did you decide to use your electrical energy?
- Could you provide enough energy to operate everything you wanted at one time?

CHALLENGE During the summer, power companies sometimes cannot produce enough energy for the demand. Why do you think that happens?

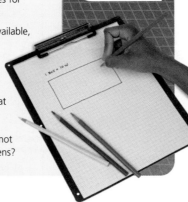

Calculating Energy Use

The electric bill for your energy usage is calculated based on the rate at which energy is used, or the power, and the amount of time it is used at that rate. Total energy used by an appliance is determined by multiplying its power consumption by the amount of time that it is used.

$$\text{Energy used} = \text{Power} \cdot \text{time}$$

$$E = P \cdot t$$

The kilowatt-hour is the unit of measurement for energy usage. A **kilowatt-hour** (kWh) is equal to one kilowatt of power for a one-hour period. Buildings usually have meters that measure how many kilowatt-hours of energy have been used. The meters display four or five small dials in a row, as shown in the photograph on the right. Each dial represents a different place value—ones, tens, hundreds, or thousands. For example, the meter in the photograph shows that the customer has used close to 9,000 kWh of energy—8,933 kWh, to be exact. To find how much energy was used in one month, the last month's reading is subtracted from this total.

INVESTIGATE Power

PURPOSE To make models of energy use

TIPS *30 min.*

- Use preprinted graph paper rather than having students draw their own grid, so that they see that each square represents the identical amount of energy.

- Review how to determine how much of a square to shade for a fraction of 10 watts.

WHAT DO YOU THINK? *Answers will vary, depending on what appliances students choose. The amount of power must total no more than 1800 W.*

CHALLENGE *Appliances used to cool buildings, such as air conditioners and fans, use more electricity than most methods for heating buildings.*

R
- Power Ratings Chart, p. 176
- Datasheet, Power, p. 177

Technology Resources

Customize this student lab as needed or look for an alternative. Print rubrics to assess student lab reports.

Lab Generator CD-ROM

Real World Example

How much electricity a home appliance uses depends on how much time the appliance is actually used and what setting of power is used. If you used a personal computer and monitor for four hours a day, 365 days a year, that would be equal to 394 kWh. At 8.5 cents per kWh that would be $33.51 of electricity costs per year. The wattage of most appliances is stamped on the bottom or back of an appliance. The wattage is usually listed as the maximum power of the appliance. For example, a clock radio uses 10 watts while a dishwasher uses 1200 to 2400 watts.

DIFFERENTIATE INSTRUCTION

? More Reading Support

E What unit represents energy use? *the kilowatt-hour (kWh)*

Advanced Have students sketch the dials on an electric meter if the reading is 6534 kWh. Make sure their sketches show that the first and third dials read counterclockwise, and the second and fourth dials read clockwise.

R Challenge and Extension, p. 175

Ongoing Assessment

Calculate energy usage.

Ask: If you use your computer and monitor for 4 hours a day, how much energy do you use? Use the table on p. 104.

120 + 150 = 270 watts = 0.27kW

E = 0.27kW · 4h = 1.08 kWh

 Practice the Math

Answers:

1. E = 3.3 kW · 6 h
   E = 19.8 kWh

2. E = 1.2 kW · 0.2 h
   E = 0.24 kWh

Reinforce (the **BIG** idea)

Have students relate the section to the Big Idea.

 Reinforcing Key Concepts, p. 178

3.4 ASSESS & RETEACH

Assess

A Section 3.4 Quiz, p. 44

Reteach

Have students use concepts from this section to explain how wind can be used to generate electricity. Have them accompany their explanations with diagrams.

Technology Resources

Have students visit **ClassZone.com** for reteaching of Key Concepts.

 CONTENT REVIEW

 CONTENT REVIEW CD-ROM

To determine the number of kilowatt-hours of energy used by an appliance, find its wattage on the chart on the page 104 or from the label. Then, substitute it into the formula along with the number of hours it was in use. Solve the sample problems below.

Finding Energy Used

▶ Sample Problem

How much energy is used to dry clothes in a 3 kW dryer for 30 minutes?

What do you know?	Power = 3.0 kW, time = 0.5 hr
What do you want to find out?	Energy used
Write the formula:	$E = P \cdot t$
Substitute into the formula:	$E = 3.0 \text{ kW} \cdot 0.5 \text{ hr}$
Calculate and simplify:	$E = 1.5 \text{ kWh}$
Check that your units agree:	Unit is kWh. Unit for energy used is kWh. Units agree.
Answer:	1.5 kWh

▶ Practice the Math

1. All of the appliances in a computer lab are in use for 6 hours every day and together use 3.3 kW. How much energy has been used in 1 day?
2. How much energy is used when a 1.2 kW hair dyer is in use for 0.2 hr?

Energy prices vary, but you can estimate the cost of using an electrical appliance by using a value of about 8 cents/kWh. You can calculate how much energy you can save by turning off the lights or television when you are not using them. Although the number may seem small, try multiplying your savings over the course of a month or year.

3.4 Review

KEY CONCEPTS

1. How do power plants generate electrical energy from kinetic energy?
2. Explain what watts measure.
3. How is energy use determined?

CRITICAL THINKING

4. **Apply** Think about reducing energy usage in your home. What changes would make the largest difference in the amount of energy used?
5. **Calculate** How much energy is used if a 3000 W clothes dryer is used for 4 hours?

⬤ CHALLENGE

6. **Calculate** An electric bill for an apartment shows 396 kWh of energy used over one month. The appliances in the apartment have a total power rating of 2.2 kW. How many hours were the appliances in use?

ANSWERS

1. Kinetic energy turns a turbine in a generator, which produces electric current.

2. Watts measure electric power, or the rate of energy conversion.

3. Multiply the power by the number of hours of use.

4. Reduce usage of high-wattage appliances, such as air conditioners and hair dryers.

5. E = 3 kW × 4 h = 12 kWh

6. t = 396 kWh/2.2 kW
   t = 180 h

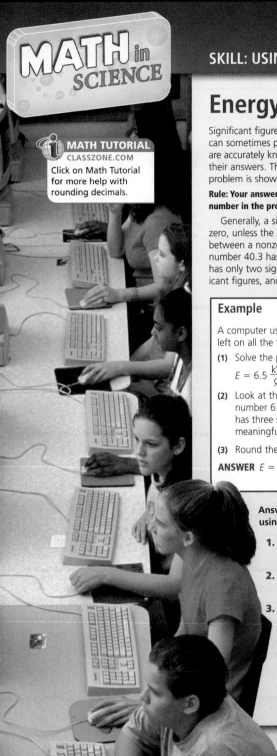

SKILL: USING SIGNIFICANT FIGURES

MATH TUTORIAL
CLASSZONE.COM
Click on Math Tutorial for more help with rounding decimals.

Energy Calculations

Significant figures are meaningful digits in a number. Calculations can sometimes produce answers with more significant figures than are accurately known. Scientists use rules to determine how to round their answers. The rule for writing an answer to a multiplication problem is shown below.

Rule: Your answer may only show as many significant figures as the number in the problem with the fewest significant figures.

Generally, a significant figure is any digit shown except for a zero, unless the zero is contained between two nonzero digits or between a nonzero digit and a decimal point. For example, the number 40.3 has three significant figures, but the number 5.90 has only two significant figures. The number 0.034 has three significant figures, and the number 0.8 has only one significant figure.

Example

A computer uses 6.5 kWh of energy per day. If the computer is left on all the time, how much energy does it use in a year?

(1) Solve the problem.

$$E = 6.5 \; \frac{\text{kWh}}{\text{day}} \cdot 365 \; \frac{\text{days}}{\text{year}} = 2{,}372.5 \; \frac{\text{kWh}}{\text{year}}$$

(2) Look at the number with the fewest significant figures. The number 6.5 has two significant figures and the number 365 has three significant figures. Therefore, the answer is only meaningful to two significant figures.

(3) Round the answer to two significant figures.

ANSWER $E = 2{,}400 \; \frac{\text{kWh}}{\text{year}}$

Answer the following questions. Write your answers using the significant figure rule for multiplication.

1. How much energy is used in a year by a computer that uses 1.7 kWh/day?

2. An energy-efficient computer uses 0.72 kWh/day. How much energy does it use in a week?

3. How much energy is used in one year if a 0.27 kW computer is on for 3 hours/day? (**Hint:** Use the equation $E = P \cdot t$.)

   CHALLENGE The energy usage of a computer is measured to be 0.058030 kWh. How many significant figures does this measurement have?

Set Learning Goal
To determine the number of significant figures in a number

Present the Science
The number 2400 has two significant figures, not four, because the implied decimal point at the end of the number does not count. If the number were written 2400.0, it would have four significant figures.

Develop Number Sense
• The rules for division and significant figures are the same as those for multiplication. Thus, the quotient of 3.4 kWh divided by 2.67 h has two significant figures.

• The rules for addition and subtraction and significant figures differ from those of multiplication and division. When adding or subtracting, align the decimals, then add or subtract. The answer has the same place value as the least precise number in the problem. For example, the sum of 2.3 m, 4.85 m, 0.024 m, and 899 m will be rounded off to the unit's place, because 899 m is the least precise measurement.

Close
Ask: How many significant figures are in the product of 2.3 kW and 3.94 h? *two*

 • Math Support, p. 184
• Math Practice, p. 185

Technology Resources
Students can visit **ClassZone.com** for practice with significant figures.

 MATH TUTORIAL

ANSWERS

1. E = 1.7 kWh/day · 365 days/yr = 620 kWh/yr

2. E = 0.72 kWh/day · 7 days/week = 5 kWh/week

3. E = 0.27 kW · 3 h/day · 365 days/yr = 300 kWh/yr

***CHALLENGE** five*

BACK TO

the **BIG** idea

Have students look at the illustrations on pp. 93 and 97. Ask them to use the diagrams to summarize what they have learned about producing a magnetic field from electric current and producing an electric current from a magnetic field. *A generator produces an electric current from a magnetic field. A motor produces a magnetic field from an electric current.*

◉ KEY CONCEPTS SUMMARY

SECTION 3.1

Ask: How would the magnetic field lines change if the magnet on the left were reversed? *They would show repulsion, not attraction.*

SECTION 3.2

Ask: In the diagram, what pushes on the electromagnet? *the poles of a magnet*

Ask: Which of the magnets is free to move in a motor? *the electromagnet*

SECTION 3.3

Ask: How does a generator differ from a motor? *It works the opposite way.*

SECTION 3.4

Ask: What is the purpose of the turbine in a generator? *It moves a magnet to generate a current.*

Review Concepts

T • Big Idea Flow Chart, p. T17
• Chapter Outline, pp. T23–T24

 Chapter Review

the **BIG** idea

Current can produce magnetism and magnetism can produce current.

CONTENT REVIEW
CLASSZONE.COM

◀ **KEY CONCEPTS SUMMARY**

3.1 **Magnetism is a force that acts at a distance.**

magnetic poles   magnetic field lines

opposite poles attract

All magnets have a north and south pole. The like poles of two magnets repel each other and the opposite poles attract.

VOCABULARY
magnet p. 79
magnetism p. 80
magnetic pole p. 80
magnetic field p. 81
magnetic domain p. 82

3.2 **Current can produce magnetism.**

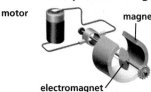

motor   magnet

electromagnet

A magnet that is produced by electric current is called an electromagnet. Motors use electromagnets to convert electrical energy into the energy of motion.

VOCABULARY
electromagnetism p. 89
electromagnet p. 90

3.3 **Magnetism can produce current.**

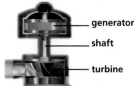

generator

electromagnet   magnet

Magnetism can be used to produce electric current. In a generator the energy of motion is converted into electrical energy.

VOCABULARY
generator p. 96
direct current p. 97
alternating current p. 97
transformer p. 99

3.4 **Generators supply electrical energy.**

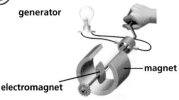

generator
shaft
turbine

Generators at power plants use large magnets to produce electric current, supplying electrical energy to homes and businesses.

VOCABULARY
electric power p. 102
watt p. 104
kilowatt p. 104
kilowatt-hour p. 105

Technology Resources

Have students visit **ClassZone.com** or use the CD-ROM for a cumulative review of concepts.

 CONTENT REVIEW

 CONTENT REVIEW CD-ROM

Engage students in a whole-class interactive review of Key Concepts. Edit content as you wish.

 POWER PRESENTATIONS

Reviewing Vocabulary

Draw a cluster diagram for each of the terms below. Write the vocabulary term in the center circle. In another circle, write the definition of the term in your own words. Add other circles that give examples or characteristics of the term. A sample diagram is completed for you.

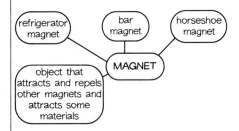

1. magnetism   7. direct current

2. magnetic pole   8. alternating current

3. magnetic field   9. transformer

4. magnetic domain   10. electric power

5. electromagnet   11. watt

6. generator   12. kilowatt-hour

Reviewing Key Concepts

Multiple Choice *Choose the letter of the best answer.*

13. Magnetic field lines flow from a magnet's
    a. north pole to south pole
    b. south pole to north pole
    c. center to the outside
    d. outside to the center

14. Which of the following is characteristic of magnetic materials?
    a. Their atoms are all aligned.
    b. Their atoms are arranged in magnetic domains.
    c. They are all nonmetals.
    d. They are all made of lodestone.

15. The Earth's magnetic field helps to protect living things from
    a. ultraviolet light
    b. meteors
    c. the Northern Lights
    d. charged particles

16. To produce a magnetic field around a copper wire, you have to
    a. place it in Earth's magnetic field
    b. run a current through it
    c. supply kinetic energy to it
    d. place it near a strong magnet

17. An electric current is produced when a wire is
    a. stationary in a magnetic field
    b. moving in a magnetic field
    c. placed between the poles of a magnet
    d. coiled around a magnet

18. In a generator, kinetic energy is converted into
    a. light energy
    b. chemical energy
    c. electrical energy
    d. nuclear energy

19. In an AC circuit, the current moves
    a. back and forth
    b. from one end of a generator to the other
    c. from one end of a battery to the other
    d. in one direction

20. What is the function of the turbine in a power plant?
    a. to increase the voltage
    b. to convert DC to AC
    c. to cool the steam
    d. to turn the coil or magnet

21. The two factors needed to measure usage of electrical energy in a building are
    a. power and time
    b. power and voltage
    c. voltage and time
    d. current and voltage

ASSESSMENT RESOURCES

UNIT ASSESSMENT BOOK
- Chapter Test A, pp. 45–48
- Chapter Test B, pp. 49–52
- Chapter Test C, pp. 53–56
- Alternative Assessment, pp. 57–58
- Unit Test, A, B, C, pp. 59–70

SPANISH ASSESSMENT BOOK
Spanish Chapter Test, pp. 309–312
Spanish Unit Test, pp. 313–316

Technology Resources

Edit test items and answer choices.

 Test Generator CD-ROM

Visit **ClassZone.com** to extend test practice.

 Test Practice

Reviewing Vocabulary

1. magnetism: force exerted by magnets; acts at a distance; affects some metals; push or pull

2. magnetic pole: the part of a magnet where magnetic force is strongest; all magnets have two; north and south; opposite poles attract; like poles repel

3. magnetic field: area of magnetic force around a magnet; force acting at a distance; Earth has a magnetic field

4. magnetic domain: region within magnetic material with a magnetic field; in magnets they are aligned; in magnetic materials they are not aligned

5. electromagnet: a magnet made with electric current; a piece of iron inside a coil of wire; can be turned on and off; can be very strong

6. generator: a device that produces electric current from magnetism; AC; DC; opposite of a motor

7. direct current: current that flows in one direction; DC; used by battery-operated devices

8. alternating current: current that regularly reverses direction; AC; supplied by power plants; can be converted into DC

9. transformer: uses magnetism to adjust voltage level; step-up, more coils on output; step-down, fewer coils on output; used in distribution of electricity

10. electric power: the rate at which energy is generated or used; units of W and kW; appliances have power ratings

11. watt: measure of power of an appliance; light bulbs labeled; small unit

12. kilowatt-hour: measure of energy usage; power times hours of usage; read from electric meters

Reviewing Key Concepts

13. a   18. c

14. b   19. a

15. d   20. d

16. b   21. a

17. b

Thinking Critically

22. *Current will cause a magnetic field to form around the coil, and the iron strip will bend toward it.*

23. *It will switch the poles of the electromagnet.*

24. *The magnet would be attracted to the coil when the current flows one direction and repelled when the current is reversed.*

25. *The motion turns a magnetic field inside a coil, producing current.*

26. *Sample answer: car, computer, hair dryer*

27. *It contains a generator. It contains a battery or some other device that stores energy for later use.*

28. *The moon does not contain a core with the same composition as Earth's core.*

Using Math in Science

29. *approximately 500 kWh*

30. *P = E/t, 400 kWh/2 kW = 200 h*

31. *Assuming usage of 350 kWh, 350 kWh · $0.08 = $28.00*

32. *July, August, September*

the **BIG** idea

33. *Answers should reflect knowledge of the effect of Earth's magnetic field on the compass.*

34. *Magnetism is a force that acts at a distance. Magnets attract and repel other magnets. For example, maglev trains are pushed forward and upward by magnetism. Every magnet has a north and south magnetic pole. A magnetic field is the region around a magnet in which the magnet exerts a force.*

Have students present their projects. Use the appropriate rubrics from the URB to evaluate their work.

R Unit Projects, pp. 5–10

Thinking Critically

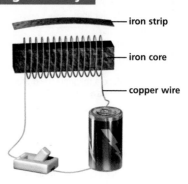
- iron strip
- iron core
- copper wire

Refer to the device in the illustration above to answer the next three questions.

22. **APPLY** What will happen when the switch is closed?

23. **PREDICT** What effect will switching the direction of the current have on the operation of the device?

24. **CONTRAST** If the iron strip is replaced with a thin magnet, how would that affect the answers to the previous two questions?

25. **APPLY** Coal is burned at a power plant to produce steam. The rising steam turns a turbine. Describe how the motion of the turbine produces current at the plant.

26. **CONNECT** List three things that you use in your everyday life that would not exist without the discovery of electromagnetism.

27. **APPLY** A radio for use during power outages works when you crank a handle. How is the radio powered? How can it keep operating even after you stop turning the crank?

28. **HYPOTHESIZE** Use your understanding of magnetic materials and the source of Earth's magnetic field to form a hypothesis about the difference between Earth and the moon that accounts for the fact that the moon does not have a magnetic field.

Using Math in Science

Some electric bills include a bar graph of energy usage similar to the one shown below. Use the information provided in the graph to answer the next four questions.

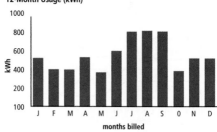
12-Month Usage (kWh)

kWh — vertical axis: 100, 200, 400, 600, 800, 1000
months billed — J F M A M J J A S O N D

29. The first bar in the graph shows energy usage for the month of January. How much energy, in kWh, was used in January?

30. If the appliances in the building have a combined power rating of 2 kW, how many hours were they in use during the month of March? (**Hint:** Use the formula $E = P \cdot t$)

31. The cost of energy was 8 cents per kWh. How much was charged for energy usage in May?

32. The most energy is used when the air conditioner is on. During which three months was the air conditioner on?

the **BIG** idea

33. **ANALYZE** Look back at page 76. Think about the answer you gave to the question about the large photograph. How has your understanding of magnetism changed? Give examples.

34. **SUMMARIZE** Write a paragraph summarizing the first three pages of this chapter. Use the heading at the top of page 79 as your topic sentence. Explain each red and blue heading.

Evaluate all the data, results, and information from your project folder. Prepare to present your project.

MONITOR AND RETEACH

If students have trouble applying the concepts in items 29–32, have them create a table that includes the names of the months written out fully and the approximate number of kWh used each month. Have students round the kWh numbers to the nearest 50. Students may benefit from holding a note card or ruler even with the tops of the bars to help them read the graph.

Students may benefit from summarizing one or more sections of the chapter.

R Summarizing the Chapter, pp. 204–205

Standardized Test Practice

For practice on your state test, go to ...

TEST PRACTICE
CLASSZONE.COM

Analyzing Tables

The table below lists some major advances in the understanding of electromagnetism.

Scientist	Year	Advance
William Gilbert	1600	proposes distinction between magnetism and static electricity
Pieter van Musschenbroek	1745	develops Leyden jar, which stores electric charge
Benjamin Franklin	1752	shows that lightning is a form of electricity
Charles Augustin de Coulomb	1785	proves mathematically that, for electricity and magnetism, force changes with distance
Alessandro Volta	1800	invents battery, first device to generate a continuous current
Hans Christian Oersted	1820	announces he had used electric current to produce magnetic effects
André Marie Ampère	1820	shows that wires carrying current attract and repel each other, just like magnets
Georg Simon Ohm	1827	studies how well different wires conduct electric current
Michael Faraday	1831	produces electricity with a magnet; invents first electric generator

Use the table above to answer the next four questions.

1. Which scientist first produced a device that allowed experimenters to hold an electric charge for later use?

a. Coulomb   **c.** Ohm
b. Franklin   **d.** van Musschenbroek

2. Which scientist developed the first device that could be used to provide a steady source of current to other devices?

a. Ampère   **c.** Volta
b. Faraday   **d.** Gilbert

3. Which scientist had the first experimental evidence that current could produce magnetism?

a. Gilbert   **c.** van Musschenbroek
b. Faraday   **d.** Oersted

4. Why was Coulomb's work important?

a. He showed that electricity and magnetism could be stored.
b. He showed that electricity and magnetism behave similarly.
c. He proved that electricity and magnetism were different.
d. He proved that electricity and magnetism were the same.

Extended Response

Answer the two questions below in detail. Include some of the terms from the word box. Underline each term you use in your answer.

appliance	current	generator
motor	coil	kilowatt-hour

5. How are electromagnets produced? How can the strength of these devices be increased? How can electromagnets be used in ways that permanent magnets cannot?

6. Alix chats online for an average of about an hour a day 6 days a week. Her computer has a power rating of 270 watts. She has a hair dryer with a power rating of 1200 watts. She uses it twice a week for about 15 minutes at a time. Which device is likely to use more power over the course of a year? Why?

Analyzing Tables

1. d   2. c   3. d   4. b

Extended Response

5. RUBRIC

4 points for a response that correctly answers all questions and uses the following terms accurately:

- appliance
- motor
- current
- coil
- generator

Sample: Electromagnets are made by placing a piece of iron or steel inside a <u>coil</u> of wire and running electric <u>current</u> through the wire. Increasing the number of coils or the amount of current increases the strength of the electromagnet. Electromagnets are used in <u>motors</u> of <u>appliances</u> and in <u>generators</u>, because they can be turned off and on.

3 points for a response that correctly answers three questions and uses three terms accurately
2 points for a response that correctly answers two questions and uses two terms accurately
1 point for a response that correctly answers one question or uses one term accurately

6. RUBRIC

4 points for a response that answers both questions and uses the following term accurately: kilowatt-hour

Sample: The computer uses more power over a year because it uses more <u>kilowatt-hours</u>. Computer: 6 days/week · 1 h/day · 52 weeks/yr · 0.270 kW = 84 kWh. Hair dryer: 2 days/week · 0.25 h/day · 52 weeks/yr · 1.2 kW = 31 kWh.

3 points for a response that correctly answers the question and calculates accurately
2 points for a response that uses the formulas correctly but contains errors in the calculations
1 point for a response that attempts to answer the questions but uses the formulas incorrectly

---

METACOGNITIVE ACTIVITY

Have students answer the following questions in their **Science Notebook:**

1. What did you find most challenging to understand about magnets?

2. What questions do you still have about magnetism?

3. Now that you have finished the unit "Electricity and Magnetism," what might you change about your Unit Project?

Student Resource Handbooks

Scientific Thinking Handbook

SCIENTIFIC THINKING HANDBOOK

Making Observations

An **observation** is an act of noting and recording an event, characteristic, behavior, or anything else detected with an instrument or with the senses.

Observations allow you to make informed hypotheses and to gather data for experiments. Careful observations often lead to ideas for new experiments. There are two categories of observations:

- **Quantitative observations** can be expressed in numbers and include records of time, temperature, mass, distance, and volume.

- **Qualitative observations** include descriptions of sights, sounds, smells, and textures.

EXAMPLE

A student dissolved 30 grams of Epsom salts in water, poured the solution into a dish, and let the dish sit out uncovered overnight. The next day, she made the following observations of the Epsom salt crystals that grew in the dish.

To determine the mass, the student found the mass of the dish before and after growing the crystals and then used subtraction to find the difference.

The student measured several crystals and calculated the mean length. (To learn how to calculate the mean of a data set, see page R36.)

Table 1. Observations of Epsom Salt Crystals

Quantitative Observations	Qualitative Observations
• mass = 30 g • mean crystal length = 0.5 cm • longest crystal length = 2 cm	• Crystals are clear. • Crystals are long, thin, and rectangular. • White crust has formed around edge of dish.

Photographs or sketches are useful for recording qualitative observations.

Epsom salt crystals

MORE ABOUT OBSERVING

- Make quantitative observations whenever possible. That way, others will know exactly what you observed and be able to compare their results with yours.

- It is always a good idea to make qualitative observations too. You never know when you might observe something unexpected.

Predicting and Hypothesizing

A **prediction** is an expectation of what will be observed or what will happen. A **hypothesis** is a tentative explanation for an observation or scientific problem that can be tested by further investigation.

EXAMPLE

Suppose you have made two paper airplanes and you wonder why one of them tends to glide farther than the other one.

1. Start by asking a question.

2. Make an educated guess. After examination, you notice that the wings of the airplane that flies farther are slightly larger than the wings of the other airplane.

3. Write a prediction based upon your educated guess, in the form of an "If . . . , then . . ." statement. Write the independent variable after the word *if*, and the dependent variable after the word *then*.

4. To make a hypothesis, explain why you think what you predicted will occur. Write the explanation after the word *because*.

1. Why does one of the paper airplanes glide farther than the other?

2. The size of an airplane's wings may affect how far the airplane will glide.

3. Prediction: If I make a paper airplane with larger wings, then the airplane will glide farther.

> To read about independent and dependent variables, see page R30.

4. Hypothesis: If I make a paper airplane with larger wings, then the airplane will glide farther, because the additional surface area of the wing will produce more lift.

> Notice that the part of the hypothesis after *because* adds an explanation of why the airplane will glide farther.

MORE ABOUT HYPOTHESES

- The results of an experiment cannot prove that a hypothesis is correct. Rather, the results either support or do not support the hypothesis.

- Valuable information is gained even when your hypothesis is not supported by your results. For example, it would be an important discovery to find that wing size is not related to how far an airplane glides.

- In science, a hypothesis is supported only after many scientists have conducted many experiments and produced consistent results.

Inferring

An **inference** is a logical conclusion drawn from the available evidence and prior knowledge. Inferences are often made from observations.

EXAMPLE

A student observing a set of acorns noticed something unexpected about one of them. He noticed a white, soft-bodied insect eating its way out of the acorn.

The student recorded these observations.

Observations

- There is a hole in the acorn, about 0.5 cm in diameter, where the insect crawled out.
- There is a second hole, which is about the size of a pinhole, on the other side of the acorn.
- The inside of the acorn is hollow.

Here are some inferences that can be made on the basis of the observations.

Inferences

- The insect formed from the material inside the acorn, grew to its present size, and ate its way out of the acorn.
- The insect crawled through the smaller hole, ate the inside of the acorn, grew to its present size, and ate its way out of the acorn.
- An egg was laid in the acorn through the smaller hole. The egg hatched into a larva that ate the inside of the acorn, grew to its present size, and ate its way out of the acorn.

When you make inferences, be sure to look at all of the evidence available and combine it with what you already know.

MORE ABOUT INFERENCES

Inferences depend both on observations and on the knowledge of the people making the inferences. Ancient people who did not know that organisms are produced only by similar organisms might have made an inference like the first one. A student today might look at the same observations and make the second inference. A third student might have knowledge about this particular insect and know that it is never small enough to fit through the smaller hole, leading her to the third inference.

Identifying Cause and Effect

In a **cause-and-effect relationship,** one event or characteristic is the result of another. Usually an effect follows its cause in time.

There are many examples of cause-and-effect relationships in everyday life.

Cause	Effect
Turn off a light.	Room gets dark.
Drop a glass.	Glass breaks.
Blow a whistle.	Sound is heard.

Scientists must be careful not to infer a cause-and-effect relationship just because one event happens after another event. When one event occurs after another, you cannot infer a cause-and-effect relationship on the basis of that information alone. You also cannot conclude that one event caused another if there are alternative ways to explain the second event. A scientist must demonstrate through experimentation or continued observation that an event was truly caused by another event.

EXAMPLE

Make an Observation

Suppose you have a few plants growing outside. When the weather starts getting colder, you bring one of the plants indoors. You notice that the plant you brought indoors is growing faster than the others are growing. You cannot conclude from your observation that the change in temperature was the cause of the increased plant growth, because there are alternative explanations for the observation. Some possible explanations are given below.

- The humidity indoors caused the plant to grow faster.

- The level of sunlight indoors caused the plant to grow faster.

- The indoor plant's being noticed more often and watered more often than the outdoor plants caused it to grow faster.

- The plant that was brought indoors was healthier than the other plants to begin with.

To determine which of these factors, if any, caused the indoor plant to grow faster than the outdoor plants, you would need to design and conduct an experiment.

See pages R28–R35 for information about designing experiments.

Recognizing Bias

Television, newspapers, and the Internet are full of experts claiming to have scientific evidence to back up their claims. How do you know whether the claims are really backed up by good science?

Bias is a slanted point of view, or personal prejudice. The goal of scientists is to be as objective as possible and to base their findings on facts instead of opinions. However, bias often affects the conclusions of researchers, and it is important to learn to recognize bias.

When scientific results are reported, you should consider the source of the information as well as the information itself. It is important to critically analyze the information that you see and read.

SOURCES OF BIAS

There are several ways in which a report of scientific information may be biased. Here are some questions that you can ask yourself:

1. **Who is sponsoring the research?**

   Sometimes, the results of an investigation are biased because an organization paying for the research is looking for a specific answer. This type of bias can affect how data are gathered and interpreted.

2. **Is the research sample large enough?**

   Sometimes research does not include enough data. The larger the sample size, the more likely that the results are accurate, assuming a truly random sample.

3. **In a survey, who is answering the questions?**

   The results of a survey or poll can be biased. The people taking part in the survey may have been specifically chosen because of how they would answer. They may have the same ideas or lifestyles. A survey or poll should make use of a random sample of people.

4. **Are the people who take part in a survey biased?**

   People who take part in surveys sometimes try to answer the questions the way they think the researcher wants them to answer. Also, in surveys or polls that ask for personal information, people may be unwilling to answer questions truthfully.

SCIENTIFIC BIAS

It is also important to realize that scientists have their own biases because of the types of research they do and because of their scientific viewpoints. Two scientists may look at the same set of data and come to completely different conclusions because of these biases. However, such disagreements are not necessarily bad. In fact, a critical analysis of disagreements is often responsible for moving science forward.

Identifying Faulty Reasoning

Faulty reasoning is wrong or incorrect thinking. It leads to mistakes and to wrong conclusions. Scientists are careful not to draw unreasonable conclusions from experimental data. Without such caution, the results of scientific investigations may be misleading.

EXAMPLE

Scientists try to make generalizations based on their data to explain as much about nature as possible. If only a small sample of data is looked at, however, a conclusion may be faulty. Suppose a scientist has studied the effects of the El Niño and La Niña weather patterns on flood damage in California from 1989 to 1995. The scientist organized the data in the bar graph below.

The scientist drew the following conclusions:

1. The La Niña weather pattern has no effect on flooding in California.

2. When neither weather pattern occurs, there is almost no flood damage.

3. A weak or moderate El Niño produces a small or moderate amount of flooding.

4. A strong El Niño produces a lot of flooding.

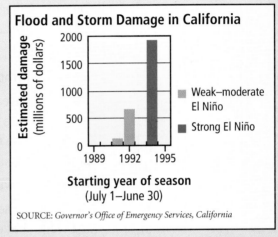

Flood and Storm Damage in California

Estimated damage (millions of dollars)

■ Weak–moderate El Niño
■ Strong El Niño

Starting year of season
(July 1–June 30)

SOURCE: *Governor's Office of Emergency Services, California*

For the six-year period of the scientist's investigation, these conclusions may seem to be reasonable. However, a six-year study of weather patterns may be too small of a sample for the conclusions to be supported. Consider the following graph, which shows information that was gathered from 1949 to 1997.

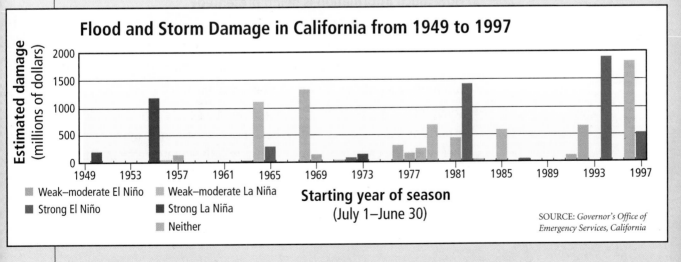

Flood and Storm Damage in California from 1949 to 1997

Estimated damage (millions of dollars)

■ Weak–moderate El Niño   ■ Weak–moderate La Niña
■ Strong El Niño   ■ Strong La Niña
■ Neither

Starting year of season
(July 1–June 30)

SOURCE: *Governor's Office of Emergency Services, California*

The only one of the conclusions that all of this information supports is number 3: a weak or moderate El Niño produces a small or moderate amount of flooding. By collecting more data, scientists can be more certain of their conclusions and can avoid faulty reasoning.

Analyzing Statements

To **analyze** a statement is to examine its parts carefully. Scientific findings are often reported through media such as television or the Internet. A report that is made public often focuses on only a small part of research. As a result, it is important to question the sources of information.

Evaluate Media Claims

To **evaluate** a statement is to judge it on the basis of criteria you've established. Sometimes evaluating means deciding whether a statement is true.

Reports of scientific research and findings in the media may be misleading or incomplete. When you are exposed to this information, you should ask yourself some questions so that you can make informed judgments about the information.

1. **Does the information come from a credible source?**

   Suppose you learn about a new product and it is stated that scientific evidence proves that the product works. A report from a respected news source may be more believable than an advertisement paid for by the product's manufacturer.

2. **How much evidence supports the claim?**

   Often, it may seem that there is new evidence every day of something in the world that either causes or cures an illness. However, information that is the result of several years of work by several different scientists is more credible than an advertisement that does not even cite the subjects of the experiment.

3. **How much information is being presented?**

   Science cannot solve all questions, and scientific experiments often have flaws. A report that discusses problems in a scientific study may be more believable than a report that addresses only positive experimental findings.

4. **Is scientific evidence being presented by a specific source?**

   Sometimes scientific findings are reported by people who are called experts or leaders in a scientific field. But if their names are not given or their scientific credentials are not reported, their statements may be less credible than those of recognized experts.

Differentiate Between Fact and Opinion

Sometimes information is presented as a fact when it may be an opinion. When scientific conclusions are reported, it is important to recognize whether they are based on solid evidence. Again, you may find it helpful to ask yourself some questions.

1. **What is the difference between a fact and an opinion?**

   A **fact** is a piece of information that can be strictly defined and proved true. An **opinion** is a statement that expresses a belief, value, or feeling. An opinion cannot be proved true or false. For example, a person's age is a fact, but if someone is asked how old they feel, it is impossible to prove the person's answer to be true or false.

2. **Can opinions be measured?**

   Yes, opinions can be measured. In fact, surveys often ask for people's opinions on a topic. But there is no way to know whether or not an opinion is the truth.

HOW TO DIFFERENTIATE FACT FROM OPINION

Human Activities and the Environment

Opinions

Notice words or phrases that express beliefs or feelings. The words *unfortunately* and *careless* show that opinions are being expressed.

Opinion

Look for statements that speculate about events. These statements are opinions, because they cannot be proved.

Unfortunately, human use of fossil fuels is one of the most significant developments of the past few centuries. Humans rely on fossil fuels, a non-renewable energy resource, for more than 90 percent of their energy needs.

This careless misuse of our planet's resources has resulted in pollution, global warming, and the destruction of fragile ecosystems. For example, oil pipelines carry more than one million barrels of oil each day across tundra regions. Transporting oil across such areas can only result in oil spills that poison the land for decades.

Facts

Statements that contain statistics tend to be facts. Writers often use facts to support their opinions.

Lab Handbook

Safety Rules

Before you work in the laboratory, read these safety rules twice. Ask your teacher to explain any rules that you do not completely understand. Refer to these rules later on if you have questions about safety in the science classroom.

LAB HANDBOOK

Directions

- Read all directions and make sure that you understand them before starting an investigation or lab activity. If you do not understand how to do a procedure or how to use a piece of equipment, ask your teacher.
- Do not begin any investigation or touch any equipment until your teacher has told you to start.
- Never experiment on your own. If you want to try a procedure that the directions do not call for, ask your teacher for permission first.
- If you are hurt or injured in any way, tell your teacher immediately.

Dress Code

goggles

apron

gloves

- Wear goggles when
  — using glassware, sharp objects, or chemicals
  — heating an object
  — working with anything that can easily fly up into the air and hurt someone's eye
- Tie back long hair or hair that hangs in front of your eyes.
- Remove any article of clothing—such as a loose sweater or a scarf—that hangs down and may touch a flame, chemical, or piece of equipment.
- Observe all safety icons calling for the wearing of eye protection, gloves, and aprons.

Heating and Fire Safety

fire safety

heating safety

- Keep your work area neat, clean, and free of extra materials.
- Never reach over a flame or heat source.
- Point objects being heated away from you and others.
- Never heat a substance or an object in a closed container.
- Never touch an object that has been heated. If you are unsure whether something is hot, treat it as though it is. Use oven mitts, clamps, tongs, or a test-tube holder.
- Know where the fire extinguisher and fire blanket are kept in your classroom.
- Do not throw hot substances into the trash. Wait for them to cool or use the container your teacher puts out for disposal.

Electrical Safety

electrical
safety

- Never use lamps or other electrical equipment with frayed cords.
- Make sure no cord is lying on the floor where someone can trip over it.
- Do not let a cord hang over the side of a counter or table so that the equipment can easily be pulled or knocked to the floor.
- Never let cords hang into sinks or other places where water can be found.
- Never try to fix electrical problems. Inform your teacher of any problems immediately.
- Unplug an electrical cord by pulling on the plug, not the cord.

Chemical Safety

chemical
safety

poison

fumes

- If you spill a chemical or get one on your skin or in your eyes, tell your teacher right away.
- Never touch, taste, or sniff any chemicals in the lab. If you need to determine odor, waft. Wafting consists of holding the chemical in its container 15 centimeters (6 in.) away from your nose, and using your fingers to bring fumes from the container to your nose.
- Keep lids on all chemicals you are not using.
- Never put unused chemicals back into the original containers. Throw away extra chemicals where your teacher tells you to.
- Pour chemicals over a sink or your work area, not over the floor.
- If you get a chemical in your eye, use the eyewash right away.
- Always wash your hands after handling chemicals, plants, or soil.

Wafting

Glassware and Sharp-Object Safety

sharp
objects

- If you break glassware, tell your teacher right away.
- Do not use broken or chipped glassware. Give these to your teacher.
- Use knives and other cutting instruments carefully. Always wear eye protection and cut away from you.

Animal Safety

- Never hurt an animal.
- Touch animals only when necessary. Follow your teacher's instructions for handling animals.
- Always wash your hands after working with animals.

Cleanup

disposal

- Follow your teacher's instructions for throwing away or putting away supplies.
- Clean your work area and pick up anything that has dropped to the floor.
- Wash your hands.

Using Lab Equipment

Different experiments require different types of equipment. But even though experiments differ, the ways in which the equipment is used are the same.

LAB HANDBOOK

Beakers

- Use beakers for holding and pouring liquids.
- Do not use a beaker to measure the volume of a liquid. Use a graduated cylinder instead. (See page R16.)
- Use a beaker that holds about twice as much liquid as you need. For example, if you need 100 milliliters of water, you should use a 200- or 250-milliliter beaker.

Test Tubes

- Use test tubes to hold small amounts of substances.
- Do not use a test tube to measure the volume of a liquid.
- Use a test tube when heating a substance over a flame. Aim the mouth of the tube away from yourself and other people.
- Liquids easily spill or splash from test tubes, so it is important to use only small amounts of liquids.

Test-Tube Holder

- Use a test-tube holder when heating a substance in a test tube.
- Use a test-tube holder if the substance in a test tube is dangerous to touch.
- Make sure the test-tube holder tightly grips the test tube so that the test tube will not slide out of the holder.
- Make sure that the test-tube holder is above the surface of the substance in the test tube so that you can observe the substance.

Test-Tube Rack

- Use a test-tube rack to organize test tubes before, during, and after an experiment.

- Use a test-tube rack to keep test tubes upright so that they do not fall over and spill their contents.

- Use a test-tube rack that is the correct size for the test tubes that you are using. If the rack is too small, a test tube may become stuck. If the rack is too large, a test tube may lean over, and some of its contents may spill or splash.

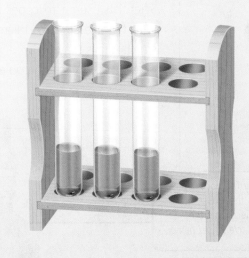

Forceps

- Use forceps when you need to pick up or hold a very small object that should not be touched with your hands.

- Do not use forceps to hold anything over a flame, because forceps are not long enough to keep your hand safely away from the flame. Plastic forceps will melt, and metal forceps will conduct heat and burn your hand.

Hot Plate

- Use a hot plate when a substance needs to be kept warmer than room temperature for a long period of time.

- Use a hot plate instead of a Bunsen burner or a candle when you need to carefully control temperature.

- Do not use a hot plate when a substance needs to be burned in an experiment.

- Always use "hot hands" safety mitts or oven mitts when handling anything that has been heated on a hot plate.

Microscope

Scientists use microscopes to see very small objects that cannot easily be seen with the eye alone. A microscope magnifies the image of an object so that small details may be observed. A microscope that you may use can magnify an object 400 times—the object will appear 400 times larger than its actual size.

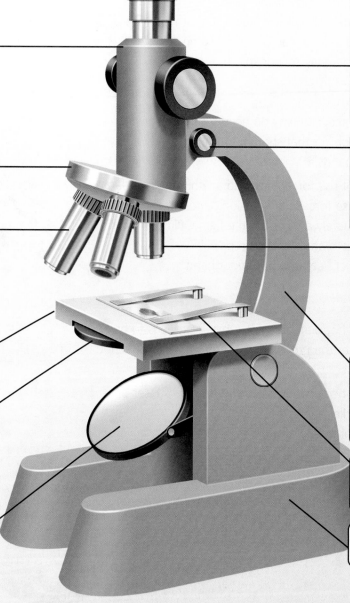

Body The body separates the lens in the eyepiece from the objective lenses below.

Nosepiece The nosepiece holds the objective lenses above the stage and rotates so that all lenses may be used.

High-Power Objective Lens This is the largest lens on the nosepiece. It magnifies an image approximately 40 times.

Stage The stage supports the object being viewed.

Diaphragm The diaphragm is used to adjust the amount of light passing through the slide and into an objective lens.

Mirror or Light Source Some microscopes use light that is reflected through the stage by a mirror. Other microscopes have their own light sources.

Eyepiece Objects are viewed through the eyepiece. The eyepiece contains a lens that commonly magnifies an image 10 times.

Coarse Adjustment This knob is used to focus the image of an object when it is viewed through the low-power lens.

Fine Adjustment This knob is used to focus the image of an object when it is viewed through the high-power lens.

Low-Power Objective Lens This is the smallest lens on the nosepiece. It magnifies an image approximately 10 times.

Arm The arm supports the body above the stage. Always carry a microscope by the arm and base.

Stage Clip The stage clip holds a slide in place on the stage.

Base The base supports the microscope.

VIEWING AN OBJECT

1. Use the coarse adjustment knob to raise the body tube.

2. Adjust the diaphragm so that you can see a bright circle of light through the eyepiece.

3. Place the object or slide on the stage. Be sure that it is centered over the hole in the stage.

4. Turn the nosepiece to click the low-power lens into place.

5. Using the coarse adjustment knob, slowly lower the lens and focus on the specimen being viewed. Be sure not to touch the slide or object with the lens.

6. When switching from the low-power lens to the high-power lens, first raise the body tube with the coarse adjustment knob so that the high-power lens will not hit the slide.

7. Turn the nosepiece to click the high-power lens into place.

8. Use the fine adjustment knob to focus on the specimen being viewed. Again, be sure not to touch the slide or object with the lens.

MAKING A SLIDE, OR WET MOUNT

1 Place the specimen in the center of a clean slide.

2 Place a drop of water on the specimen.

3 Place a cover slip on the slide. Put one edge of the cover slip into the drop of water and slowly lower it over the specimen.

4 Remove any air bubbles from under the cover slip by gently tapping the cover slip.

5 Dry any excess water before placing the slide on the microscope stage for viewing.

Spring Scale (Force Meter)

- Use a spring scale to measure a force pulling on the scale.
- Use a spring scale to measure the force of gravity exerted on an object by Earth.
- To measure a force accurately, a spring scale must be zeroed before it is used. The scale is zeroed when no weight is attached and the indicator is positioned at zero.
- Do not attach a weight that is either too heavy or too light to a spring scale. A weight that is too heavy could break the scale or exert too great a force for the scale to measure. A weight that is too light may not exert enough force to be measured accurately.

Graduated Cylinder

- Use a graduated cylinder to measure the volume of a liquid.
- Be sure that the graduated cylinder is on a flat surface so that your measurement will be accurate.
- When reading the scale on a graduated cylinder, be sure to have your eyes at the level of the surface of the liquid.
- The surface of the liquid will be curved in the graduated cylinder. Read the volume of the liquid at the bottom of the curve, or meniscus (muh-NIHS-kuhs).
- You can use a graduated cylinder to find the volume of a solid object by measuring the increase in a liquid's level after you add the object to the cylinder.

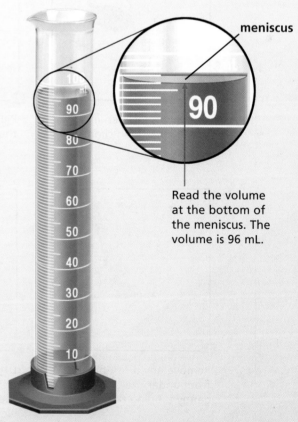

meniscus

Read the volume at the bottom of the meniscus. The volume is 96 mL.

Metric Rulers

- Use metric rulers or meter sticks to measure objects' lengths.

- Do not measure an object from the end of a metric ruler or meter stick, because the end is often imperfect. Instead, measure from the 1-centimeter mark, but remember to subtract a centimeter from the apparent measurement.

- Estimate any lengths that extend between marked units. For example, if a meter stick shows centimeters but not millimeters, you can estimate the length that an object extends between centimeter marks to measure it to the nearest millimeter.

- **Controlling Variables** If you are taking repeated measurements, always measure from the same point each time. For example, if you're measuring how high two different balls bounce when dropped from the same height, measure both bounces at the same point on the balls—either the top or the bottom. Do not measure at the top of one ball and the bottom of the other.

EXAMPLE

How to Measure a Leaf

1. Lay a ruler flat on top of the leaf so that the 1-centimeter mark lines up with one end. Make sure the ruler and the leaf do not move between the time you line them up and the time you take the measurement.

2. Look straight down on the ruler so that you can see exactly how the marks line up with the other end of the leaf.

3. Estimate the length by which the leaf extends beyond a marking. For example, the leaf below extends about halfway between the 4.2-centimeter and 4.3-centimeter marks, so the apparent measurement is about 4.25 centimeters.

4. Remember to subtract 1 centimeter from your apparent measurement, since you started at the 1-centimeter mark on the ruler and not at the end. The leaf is about 3.25 centimeters long (4.25 cm – 1 cm = 3.25 cm).

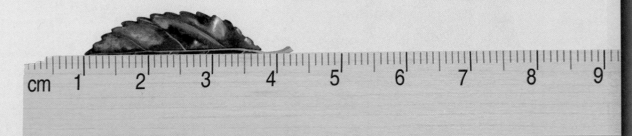

Triple-Beam Balance

This balance has a pan and three beams with sliding masses, called riders. At one end of the beams is a pointer that indicates whether the mass on the pan is equal to the masses shown on the beams.

1. Make sure the balance is zeroed before measuring the mass of an object. The balance is zeroed if the pointer is at zero when nothing is on the pan and the riders are at their zero points. Use the adjustment knob at the base of the balance to zero it.

2. Place the object to be measured on the pan.

3. Move the riders one notch at a time away from the pan. Begin with the largest rider. If moving the largest rider one notch brings the pointer below zero, begin measuring the mass of the object with the next smaller rider.

4. Change the positions of the riders until they balance the mass on the pan and the pointer is at zero. Then add the readings from the three beams to determine the mass of the object.

300 g	position of largest rider
90 g	position of middle rider
+ 3 g	position of smallest rider
393 g	mass of beaker

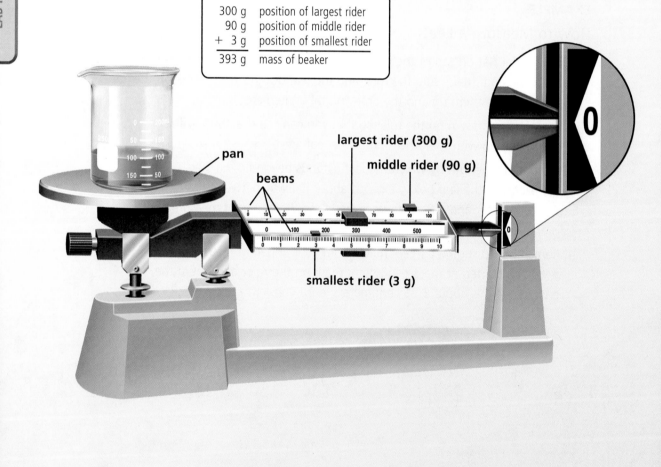

LAB HANDBOOK

Double-Pan Balance

This type of balance has two pans. Between the pans is a pointer that indicates whether the masses on the pans are equal.

1. Make sure the balance is zeroed before measuring the mass of an object. The balance is zeroed if the pointer is at zero when there is nothing on either of the pans. Many double-pan balances have sliding knobs that can be used to zero them.

2. Place the object to be measured on one of the pans.

3. Begin adding standard masses to the other pan. Begin with the largest standard mass. If this adds too much mass to the balance, begin measuring the mass of the object with the next smaller standard mass.

4. Add standard masses until the masses on both pans are balanced and the pointer is at zero. Then add the standard masses together to determine the mass of the object being measured.

LAB HANDBOOK

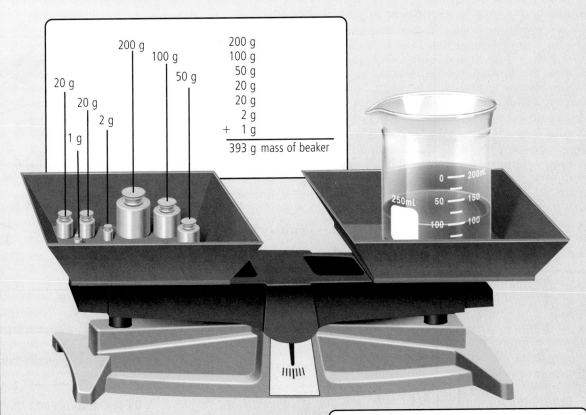

20 g	200 g
20 g	100 g
200 g	50 g
100 g	20 g
50 g	20 g
2 g	2 g
1 g	+ 1 g
	393 g mass of beaker

Never place chemicals or liquids directly on a pan. Instead, use the following procedure:

1 Determine the mass of an empty container, such as a beaker.

2 Pour the substance into the container, and measure the total mass of the substance and the container.

3 Subtract the mass of the empty container from the total mass to find the mass of the substance.

The Metric System and SI Units

Scientists use International System (SI) units for measurements of distance, volume, mass, and temperature. The International System is based on multiples of ten and the metric system of measurement.

Basic SI Units		
Property	**Name**	**Symbol**
length	meter	m
volume	liter	L
mass	kilogram	kg
temperature	kelvin	K

SI Prefixes		
Prefix	**Symbol**	**Multiple of 10**
kilo-	k	1000
hecto-	h	100
deca-	da	10
deci-	d	$0.1 \left(\frac{1}{10}\right)$
centi-	c	$0.01 \left(\frac{1}{100}\right)$
milli-	m	$0.001 \left(\frac{1}{1000}\right)$

Changing Metric Units

You can change from one unit to another in the metric system by multiplying or dividing by a power of 10.

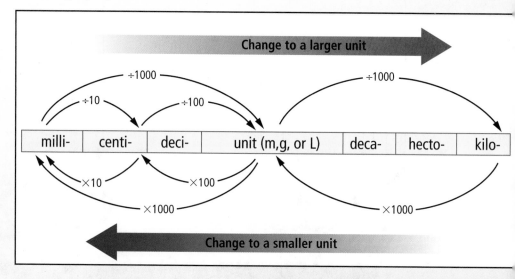

Change to a larger unit

÷1000   ÷10   ÷100   ÷1000

| milli- | centi- | deci- | unit (m,g, or L) | deca- | hecto- | kilo- |

×10   ×100   ×1000   ×1000

Change to a smaller unit

Example

Change 0.64 liters to milliliters.

(1) Decide whether to multiply or divide.

(2) Select the power of 10.

ANSWER 0.64 L = 640 mL

Change to a smaller unit by multiplying.

mL ◄──── × 1000 ──── L

0.64 × 1000 = **640.**

Example

Change 23.6 grams to kilograms.

(1) Decide whether to multiply or divide.

(2) Select the power of 10.

ANSWER 23.6 g = 0.0236 kg

Change to a larger unit by dividing.

g ──── ÷ 1000 ──► kg

23.6 ÷ 1000 = **0.0236**

Temperature Conversions

Even though the kelvin is the SI base unit of temperature, the degree Celsius will be the unit you use most often in your science studies. The formulas below show the relationships between temperatures in degrees Fahrenheit (°F), degrees Celsius (°C), and kelvins (K).

$$°C = \frac{5}{9}(°F - 32)$$

$$°F = \frac{9}{5}°C + 32$$

$$K = °C + 273$$

See page R42 for help with using formulas.

Examples of Temperature Conversions

Condition	Degrees Celsius	Degrees Fahrenheit
Freezing point of water	0	32
Cool day	10	50
Mild day	20	68
Warm day	30	86
Normal body temperature	37	98.6
Very hot day	40	104
Boiling point of water	100	212

Converting Between SI and U.S. Customary Units

Use the chart below when you need to convert between SI units and U.S. customary units.

SI Unit	From SI to U.S. Customary			From U.S. Customary to SI		
Length	**When you know**	**multiply by**	**to find**	**When you know**	**multiply by**	**to find**
kilometer (km) = 1000 m	kilometers	0.62	miles	miles	1.61	kilometers
meter (m) = 100 cm	meters	3.28	feet	feet	0.3048	meters
centimeter (cm) = 10 mm	centimeters	0.39	inches	inches	2.54	centimeters
millimeter (mm) = 0.1 cm	millimeters	0.04	inches	inches	25.4	millimeters
Area	**When you know**	**multiply by**	**to find**	**When you know**	**multiply by**	**to find**
square kilometer (km²)	square kilometers	0.39	square miles	square miles	2.59	square kilometers
square meter (m²)	square meters	1.2	square yards	square yards	0.84	square meters
square centimeter (cm²)	square centimeters	0.155	square inches	square inches	6.45	square centimeters
Volume	**When you know**	**multiply by**	**to find**	**When you know**	**multiply by**	**to find**
liter (L) = 1000 mL	liters	1.06	quarts	quarts	0.95	liters
	liters	0.26	gallons	gallons	3.79	liters
	liters	4.23	cups	cups	0.24	liters
	liters	2.12	pints	pints	0.47	liters
milliliter (mL) = 0.001 L	milliliters	0.20	teaspoons	teaspoons	4.93	milliliters
	milliliters	0.07	tablespoons	tablespoons	14.79	milliliters
	milliliters	0.03	fluid ounces	fluid ounces	29.57	milliliters
Mass	**When you know**	**multiply by**	**to find**	**When you know**	**multiply by**	**to find**
kilogram (kg) = 1000 g	kilograms	2.2	pounds	pounds	0.45	kilograms
gram (g) = 1000 mg	grams	0.035	ounces	ounces	28.35	grams

Precision and Accuracy

When you do an experiment, it is important that your methods, observations, and data be both precise and accurate.

low precision

precision, but not accuracy

precision and accuracy

Precision

In science, **precision** is the exactness and consistency of measurements. For example, measurements made with a ruler that has both centimeter and millimeter markings would be more precise than measurements made with a ruler that has only centimeter markings. Another indicator of precision is the care taken to make sure that methods and observations are as exact and consistent as possible. Every time a particular experiment is done, the same procedure should be used. Precision is necessary because experiments are repeated several times and if the procedure changes, the results will change.

EXAMPLE

Suppose you are measuring temperatures over a two-week period. Your precision will be greater if you measure each temperature at the same place, at the same time of day, and with the same thermometer than if you change any of these factors from one day to the next.

Accuracy

In science, it is possible to be precise but not accurate. **Accuracy** depends on the difference between a measurement and an actual value. The smaller the difference, the more accurate the measurement.

EXAMPLE

Suppose you look at a stream and estimate that it is about 1 meter wide at a particular place. You decide to check your estimate by measuring the stream with a meter stick, and you determine that the stream is 1.32 meters wide. However, because it is hard to measure the width of a stream with a meter stick, it turns out that you didn't do a very good job. The stream is actually 1.14 meters wide. Therefore, even though your estimate was less precise than your measurement, your estimate was actually more accurate.

Making Data Tables and Graphs

Data tables and graphs are useful tools for both recording and communicating scientific data.

Making Data Tables

You can use a **data table** to organize and record the measurements that you make. Some examples of information that might be recorded in data tables are frequencies, times, and amounts.

EXAMPLE

Suppose you are investigating photosynthesis in two elodea plants. One sits in direct sunlight, and the other sits in a dimly lit room. You measure the rate of photosynthesis by counting the number of bubbles in the jar every ten minutes.

1. Title and number your data table.
2. Decide how you will organize the table into columns and rows.
3. Any units, such as seconds or degrees, should be included in column headings, not in the individual cells.

Table 1. Number of Bubbles from Elodea

Time (min)	Sunlight	Dim Light
0	0	0
10	15	5
20	25	8
30	32	7
40	41	10
50	47	9
60	42	9

> Always number and title data tables.

The data in the table above could also be organized in a different way.

Table 1. Number of Bubbles from Elodea

Light Condition	Time (min)						
	0	10	20	30	40	50	60
Sunlight	0	15	25	32	41	47	42
Dim light	0	5	8	7	10	9	9

> Put units in column heading.

Making Line Graphs

You can use a **line graph** to show a relationship between variables. Line graphs are particularly useful for showing changes in variables over time.

EXAMPLE

Suppose you are interested in graphing temperature data that you collected over the course of a day.

Table 1. Outside Temperature During the Day on March 7

	Time of Day						
	7:00 A.M.	9:00 A.M.	11:00 A.M.	1:00 P.M.	3:00 P.M.	5:00 P.M.	7:00 P.M.
Temp (°C)	8	9	11	14	12	10	6

1. Use the vertical axis of your line graph for the variable that you are measuring—temperature.

2. Choose scales for both the horizontal axis and the vertical axis of the graph. You should have two points more than you need on the vertical axis, and the horizontal axis should be long enough for all of the data points to fit.

3. Draw and label each axis.

4. Graph each value. First find the appropriate point on the scale of the horizontal axis. Imagine a line that rises vertically from that place on the scale. Then find the corresponding value on the vertical axis, and imagine a line that moves horizontally from that value. The point where these two imaginary lines intersect is where the value should be plotted.

5. Connect the points with straight lines.

Be sure to add a number and a title to your graph.

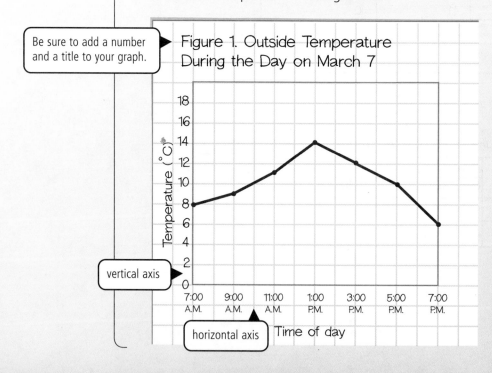

Figure 1. Outside Temperature During the Day on March 7

vertical axis

horizontal axis

Time of day

Making Circle Graphs

You can use a **circle graph,** sometimes called a pie chart, to represent data as parts of a circle. Circle graphs are used only when the data can be expressed as percentages of a whole. The entire circle shown in a circle graph is equal to 100 percent of the data.

EXAMPLE

Suppose you identified the species of each mature tree growing in a small wooded area. You organized your data in a table, but you also want to show the data in a circle graph.

1. To begin, find the total number of mature trees.

    $56 + 34 + 22 + 10 + 28 = 150$

2. To find the degree measure for each sector of the circle, write a fraction comparing the number of each tree species with the total number of trees. Then multiply the fraction by 360°.

    Oak: $\frac{56}{150} \times 360° = 134.4°$

3. Draw a circle. Use a protractor to draw the angle for each sector of the graph.

4. Color and label each sector of the graph.

5. Give the graph a number and title.

Table 1. Tree Species in Wooded Area

Species	Number of Specimens
Oak	56
Maple	34
Birch	22
Willow	10
Pine	28

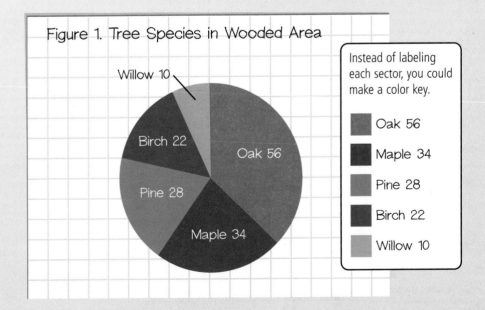

Figure 1. Tree Species in Wooded Area

Instead of labeling each sector, you could make a color key.

- Oak 56
- Maple 34
- Pine 28
- Birch 22
- Willow 10

Bar Graph

A **bar graph** is a type of graph in which the lengths of the bars are used to represent and compare data. A numerical scale is used to determine the lengths of the bars.

EXAMPLE

To determine the effect of water on seed sprouting, three cups were filled with sand, and ten seeds were planted in each. Different amounts of water were added to each cup over a three-day period.

Table 1. Effect of Water on Seed Sprouting

Daily Amount of Water (mL)	Number of Seeds That Sprouted After 3 Days in Sand
0	1
10	4
20	8

1. Choose a numerical scale. The greatest value is 8, so the end of the scale should have a value greater than 8, such as 10. Use equal increments along the scale, such as increments of 2.

2. Draw and label the axes. Mark intervals on the vertical axis according to the scale you chose.

3. Draw a bar for each data value. Use the scale to decide how long to make each bar.

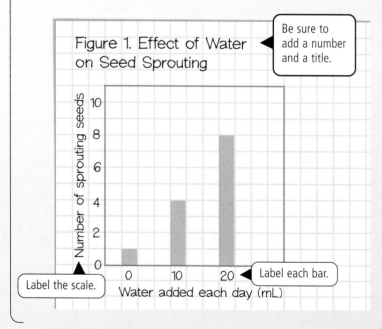

Figure 1. Effect of Water on Seed Sprouting

Be sure to add a number and a title.

Number of sprouting seeds

Water added each day (mL)

Label the scale.

Label each bar.

LAB HANDBOOK

Double Bar Graph

A **double bar graph** is a bar graph that shows two sets of data. The two bars for each measurement are drawn next to each other.

EXAMPLE

The same seed-sprouting experiment was repeated with potting soil. The data for sand and potting soil can be plotted on one graph.

1. Draw one set of bars, using the data for sand, as shown below.

2. Draw bars for the potting-soil data next to the bars for the sand data. Shade them a different color. Add a key.

Table 2. Effect of Water and Soil on Seed Sprouting

Daily Amount of Water (mL)	Number of Seeds That Sprouted After 3 Days in Sand	Number of Seeds That Sprouted After 3 Days in Potting Soil
0	1	2
10	4	5
20	8	9

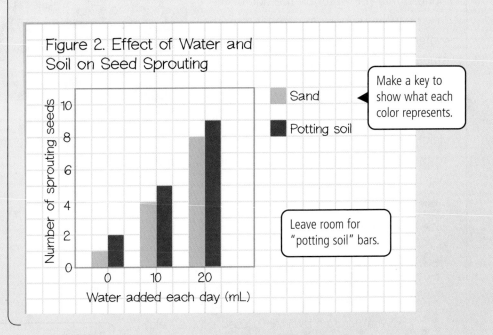

Figure 2. Effect of Water and Soil on Seed Sprouting

Make a key to show what each color represents.

Leave room for "potting soil" bars.

Designing an Experiment

Use this section when designing or conducting an experiment.

Determining a Purpose

Don't forget to learn as much as possible about your topic before you begin.

You can find a purpose for an experiment by doing research, by examining the results of a previous experiment, or by observing the world around you. An **experiment** is an organized procedure to study something under controlled conditions.

1. Write the purpose of your experiment as a question or problem that you want to investigate.

2. Write down research questions and begin searching for information that will help you design an experiment. Consult the library, the Internet, and other people as you conduct your research.

EXAMPLE

Middle school students observed an odor near the lake by their school. They also noticed that the water on the side of the lake near the school was greener than the water on the other side of the lake. The students did some research to learn more about their observations. They discovered that the odor and green color in the lake

came from algae. They also discovered that a new fertilizer was being used on a field nearby. The students inferred that the use of the fertilizer might be related to the presence of the algae and designed a controlled experiment to find out whether they were right.

Problem

How does fertilizer affect the presence of algae in a lake?

Research Questions

- Have other experiments been done on this problem? If so, what did those experiments show?
- What kind of fertilizer is used on the field? How much?
- How do algae grow?
- How do people measure algae?
- Can fertilizer and algae be used safely in a lab? How?

Research
As you research, you may find a topic that is more interesting to you than your original topic, or learn that a procedure you wanted to use is not practical or safe. It is OK to change your purpose as you research.

LAB HANDBOOK

Writing a Hypothesis

A **hypothesis** is a tentative explanation for an observation or scientific problem that can be tested by further investigation. You can write your hypothesis in the form of an "If . . . , then . . . , because . . ." statement.

Hypothesis

If the amount of fertilizer in lake water is increased, then the amount of algae will also increase, because fertilizers provide nutrients that algae need to grow.

Hypotheses
For help with hypotheses, refer to page R3.

Determining Materials

Make a list of all the materials you will need to do your experiment. Be specific, especially if someone else is helping you obtain the materials. Try to think of everything you will need.

Materials

- 1 large jar or container
- 4 identical smaller containers
- rubber gloves that also cover the arms
- sample of fertilizer-and-water solution
- eyedropper
- clear plastic wrap
- scissors
- masking tape
- marker
- ruler

Determining Variables and Constants

EXPERIMENTAL GROUP AND CONTROL GROUP

An experiment to determine how two factors are related always has two groups—a control group and an experimental group.

1. Design an experimental group. Include as many trials as possible in the experimental group in order to obtain reliable results.

2. Design a control group that is the same as the experimental group in every way possible, except for the factor you wish to test.

Experimental Group: two containers of lake water with one drop of fertilizer solution added to each

Control Group: two containers of lake water with no fertilizer solution added

> Go back to your materials list and make sure you have enough items listed to cover both your experimental group and your control group.

VARIABLES AND CONSTANTS

Identify the variables and constants in your experiment. In a controlled experiment, a **variable** is any factor that can change. **Constants** are all of the factors that are the same in both the experimental group and the control group.

1. Read your hypothesis. The **independent variable** is the factor that you wish to test and that is manipulated or changed so that it can be tested. The independent variable is expressed in your hypothesis after the word *if*. Identify the independent variable in your laboratory report.

2. The **dependent variable** is the factor that you measure to gather results. It is expressed in your hypothesis after the word *then*. Identify the dependent variable in your laboratory report.

> **Hypothesis**
> If the amount of fertilizer in lake water is increased, then the amount of algae will also increase, because fertilizers provide nutrients that algae need to grow.

Table 1. Variables and Constants in Algae Experiment

Independent Variable	Dependent Variable	Constants
Amount of fertilizer in lake water	Amount of algae that grow	• Where the lake water is obtained • Type of container used • Light and temperature conditions where water will be stored

> Set up your experiment so that you will test only one variable.

LAB HANDBOOK

MEASURING THE DEPENDENT VARIABLE

Before starting your experiment, you need to define how you will measure the dependent variable. An **operational definition** is a description of the one particular way in which you will measure the dependent variable.

Your operational definition is important for several reasons. First, in any experiment there are several ways in which a dependent variable can be measured. Second, the procedure of the experiment depends on how you decide to measure the dependent variable. Third, your operational definition makes it possible for other people to evaluate and build on your experiment.

EXAMPLE 1

An operational definition of a dependent variable can be qualitative. That is, your measurement of the dependent variable can simply be an observation of whether a change occurs as a result of a change in the independent variable. This type of operational definition can be thought of as a "yes or no" measurement.

Table 2. Qualitative Operational Definition of Algae Growth

Independent Variable	Dependent Variable	Operational Definition
Amount of fertilizer in lake water	Amount of algae that grow	Algae grow in lake water

A qualitative measurement of a dependent variable is often easy to make and record. However, this type of information does not provide a great deal of detail in your experimental results.

EXAMPLE 2

An operational definition of a dependent variable can be quantitative. That is, your measurement of the dependent variable can be a number that shows how much change occurs as a result of a change in the independent variable.

Table 3. Quantitative Operational Definition of Algae Growth

Independent Variable	Dependent Variable	Operational Definition
Amount of fertilizer in lake water	Amount of algae that grow	Diameter of largest algal growth (in mm)

A quantitative measurement of a dependent variable can be more difficult to make and analyze than a qualitative measurement. However, this type of data provides much more information about your experiment and is often more useful.

Writing a Procedure

Write each step of your procedure. Start each step with a verb, or action word, and keep the steps short. Your procedure should be clear enough for someone else to use as instructions for repeating your experiment.

> If necessary, go back to your materials list and add any materials that you left out.

Procedure

1. Put on your gloves. Use the large container to obtain a sample of lake water.

2. Divide the sample of lake water equally among the four smaller containers.

3. Use the eyedropper to add one drop of fertilizer solution to two of the containers.

> **Controlling Variables**
> The same amount of fertilizer solution must be added to two of the four containers.

4. Use the masking tape and the marker to label the containers with your initials, the date, and the identifiers "Jar 1 with Fertilizer," "Jar 2 with Fertilizer," "Jar 1 without Fertilizer," and "Jar 2 without Fertilizer."

5. Cover the containers with clear plastic wrap. Use the scissors to punch ten holes in each of the covers.

6. Place all four containers on a window ledge. Make sure that they all receive the same amount of light.

> **Controlling Variables**
> All four containers must receive the same amount of light.

7. Observe the containers every day for one week.

8. Use the ruler to measure the diameter of the largest clump of algae in each container, and record your measurements daily.

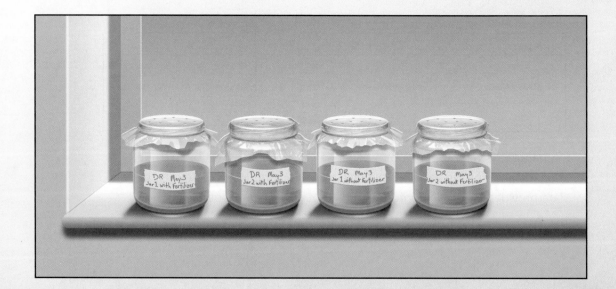

Recording Observations

Once you have obtained all of your materials and your procedure has been approved, you can begin making experimental observations. Gather both quantitative and qualitative data. If something goes wrong during your procedure, make sure you record that too.

> **Observations**
> For help with making qualitative and quantitative observations, refer to page R2.

Table 4. Fertilizer and Algae Growth

Date and Time	Experimental Group		Control Group		
	Jar 1 with Fertilizer (diameter of algae in mm)	Jar 2 with Fertilizer (diameter of algae in mm)	Jar 1 without Fertilizer (diameter of algae in mm)	Jar 2 without Fertilizer (diameter of algae in mm)	Observations
5/3 4:00 P.M.	0	0	0	0	condensation in all containers
5/4 4:00 P.M.	0	3	0	0	tiny green blobs in jar 2 with fertilizer
5/5 4:15 P.M.	4	5	0	3	green blobs in jars 1 and 2 with fertilizer and jar 2 without fertilizer
5/6 4:00 P.M.	5	6	0	4	water light green in jar 2 with fertilizer
5/7 4:00 P.M.	8	10	0	6	water light green in jars 1 and 2 with fertilizer and in jar 2 without fertilizer
5/8 3:30 P.M.	10	18	0	6	cover off jar 2 with fertilizer
5/9 3:30 P.M.	14	23	0	8	drew sketches of each container

> For more examples of data tables, see page R23.

> Notice that on the sixth day, the observer found that the cover was off one of the containers. It is important to record observations of unintended factors because they might affect the results of the experiment.

> Use technology, such as a microscope, to help you make observations when possible.

Drawings of Samples Viewed Under Microscope on 5/9 at 100x

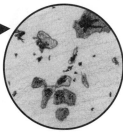

Jar 1 with Fertilizer

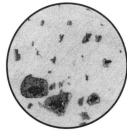

Jar 2 with Fertilizer

Jar 1 without Fertilizer

Jar 2 without Fertilizer

Summarizing Results

To summarize your data, look at all of your observations together. Look for meaningful ways to present your observations. For example, you might average your data or make a graph to look for patterns. When possible, use spreadsheet software to help you analyze and present your data. The two graphs below show the same data.

EXAMPLE 1

Figure 1. Fertilizer and Algae Growth
Diameter of algae (mm)
25
20
15
10
5
0
5/3
5/4
5/5
5/6
5/7
5/8
5/9
Date
Jar 1 with fertilizer
Jar 1 without fertilizer
Jar 2 with fertilizer
Jar 2 without fertilizer

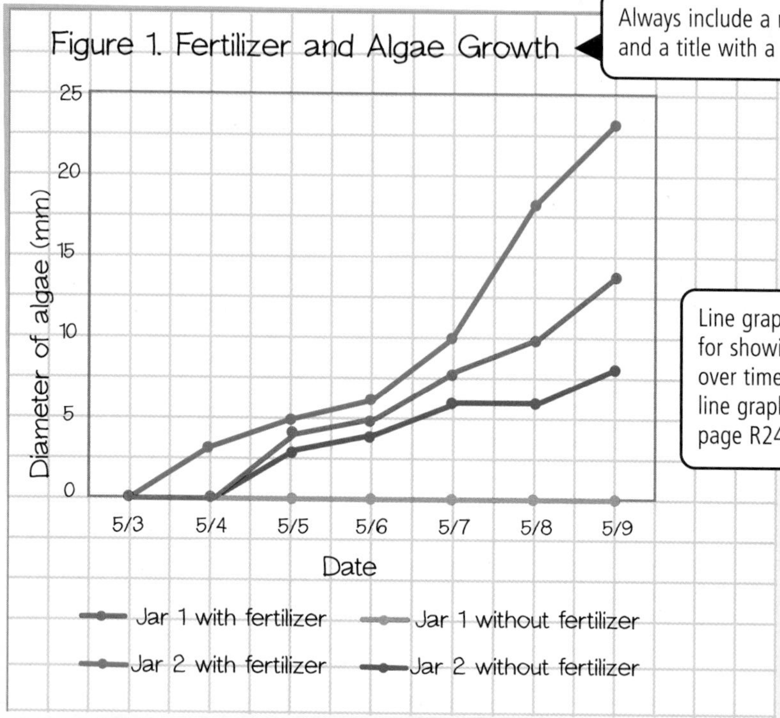

Always include a number and a title with a graph.

Line graphs are useful for showing changes over time. For help with line graphs, refer to page R24.

EXAMPLE 2

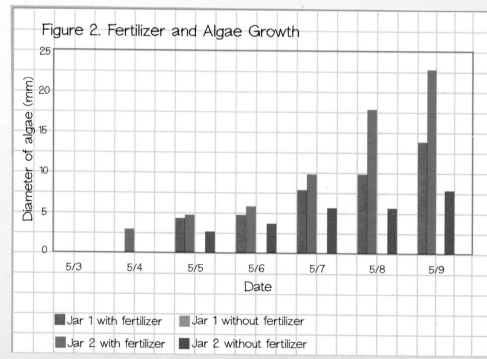

Figure 2. Fertilizer and Algae Growth
Diameter of algae (mm)
25
20
15
10
5
0
5/3
5/4
5/5
5/6
5/7
5/8
5/9
Date
Jar 1 with fertilizer
Jar 1 without fertilizer
Jar 2 with fertilizer
Jar 2 without fertilizer

Bar graphs are useful for comparing different data sets. This bar graph has four bars for each day. Another way to present the data would be to calculate averages for the tests and the controls, and to show one test bar and one control bar for each day.

LAB HANDBOOK

Drawing Conclusions

RESULTS AND INFERENCES

To draw conclusions from your experiment, first write your results. Then compare your results with your hypothesis. Do your results support your hypothesis? Be careful not to make inferences about factors that you did not test.

> For help with making inferences, see page R4.

Results and Inferences

The results of my experiment show that more algae grew in lake water to which fertilizer had been added than in lake water to which no fertilizer had been added. My hypothesis was supported. I infer that it is possible that the growth of algae in the lake was caused by the fertilizer used on the field.

> Notice that you cannot conclude from this experiment that the presence of algae in the lake was due only to the fertilizer.

QUESTIONS FOR FURTHER RESEARCH

Write a list of questions for further research and investigation. Your ideas may lead you to new experiments and discoveries.

Questions for Further Research

- What is the connection between the amount of fertilizer and algae growth?
- How do different brands of fertilizer affect algae growth?
- How would algae growth in the lake be affected if no fertilizer were used on the field?
- How do algae affect the lake and the other life in and around it?
- How does fertilizer affect the lake and the life in and around it?
- If fertilizer is getting into the lake, how is it getting there?

Math Handbook

Describing a Set of Data

Means, medians, modes, and ranges are important math tools for describing data sets such as the following widths of fossilized clamshells.

13 mm   25 mm   14 mm   21 mm   16 mm   23 mm   14 mm

Mean

The **mean** of a data set is the sum of the values divided by the number of values.

Example

To find the mean of the clamshell data, add the values and then divide the sum by the number of values.

$$\frac{13 \text{ mm} + 25 \text{ mm} + 14 \text{ mm} + 21 \text{ mm} + 16 \text{ mm} + 23 \text{ mm} + 14 \text{ mm}}{7} = \frac{126 \text{ mm}}{7} = 18 \text{ mm}$$

ANSWER The mean is 18 mm.

Median

The **median** of a data set is the middle value when the values are written in numerical order. If a data set has an even number of values, the median is the mean of the two middle values.

Example

To find the median of the clamshell data, arrange the values in order from least to greatest. The median is the middle value.

13 mm   14 mm   14 mm   16 mm   21 mm   23 mm   25 mm

ANSWER The median is 16 mm.

Mode

The **mode** of a data set is the value that occurs most often.

> ### Example
>
> To find the mode of the clamshell data, arrange the values in order from least to greatest and determine the value that occurs most often.
>
> 13 mm   14 mm   14 mm   16 mm   21 mm   23 mm   25 mm
>
> **ANSWER** The mode is 14 mm.

A data set can have more than one mode or no mode. For example, the following data set has modes of 2 mm and 4 mm:

2 mm   2 mm   3 mm   4 mm   4 mm

The data set below has no mode, because no value occurs more often than any other.

2 mm   3 mm   4 mm   5 mm

Range

The **range** of a data set is the difference between the greatest value and the least value.

> ### Example
>
> To find the range of the clamshell data, arrange the values in order from least to greatest.
>
> 13 mm   14 mm   14 mm   16 mm   21 mm   23 mm   25 mm
>
> Subtract the least value from the greatest value.
>
> 13 mm is the least value.
> 25 mm is the greatest value.
>
> 25 mm − 13 mm = 12 mm
>
> **ANSWER** The range is 12 mm.

Using Ratios, Rates, and Proportions

You can use ratios and rates to compare values in data sets. You can use proportions to find unknown values.

Ratios

A **ratio** uses division to compare two values. The ratio of a value a to a nonzero value b can be written as $\frac{a}{b}$.

Example

The height of one plant is 8 centimeters. The height of another plant is 6 centimeters. To find the ratio of the height of the first plant to the height of the second plant, write a fraction and simplify it.

$$\frac{8 \text{ cm}}{6 \text{ cm}} = \frac{4 \times \overset{1}{\cancel{2}}}{3 \times \underset{1}{\cancel{2}}} = \frac{4}{3}$$

ANSWER The ratio of the plant heights is $\frac{4}{3}$.

You can also write the ratio $\frac{a}{b}$ as "a to b" or as $a:b$. For example, you can write the ratio of the plant heights as "4 to 3" or as $4:3$.

Rates

A **rate** is a ratio of two values expressed in different units. A unit rate is a rate with a denominator of 1 unit.

Example

A plant grew 6 centimeters in 2 days. The plant's rate of growth was $\frac{6 \text{ cm}}{2 \text{ days}}$. To describe the plant's growth in centimeters per day, write a unit rate.

Divide numerator and denominator by 2: $\quad \dfrac{6 \text{ cm}}{2 \text{ days}} = \dfrac{6 \text{ cm} \div 2}{2 \text{ days} \div 2}$

> You divide 2 days by 2 to get 1 day, so divide 6 cm by 2 also.

Simplify: $\quad = \dfrac{3 \text{ cm}}{1 \text{ day}}$

ANSWER The plant's rate of growth is 3 centimeters per day.

Proportions

A **proportion** is an equation stating that two ratios are equivalent. To solve for an unknown value in a proportion, you can use cross products.

Example

If a plant grew 6 centimeters in 2 days, how many centimeters would it grow in 3 days (if its rate of growth is constant)?

$$\textit{Write a proportion:} \qquad \frac{6 \text{ cm}}{2 \text{ days}} = \frac{x \text{ cm}}{3 \text{ days}}$$

$$\textit{Set cross products:} \qquad 6 \cdot 3 = 2x$$

$$\textit{Multiply 6 and 3:} \qquad 18 = 2x$$

$$\textit{Divide each side by 2:} \qquad \frac{18}{2} = \frac{2x}{2}$$

$$\textit{Simplify:} \qquad 9 = x$$

ANSWER The plant would grow 9 centimeters in 3 days.

Using Decimals, Fractions, and Percents

Decimals, fractions, and percentages are all ways of recording and representing data.

Decimals

A **decimal** is a number that is written in the base-ten place value system, in which a decimal point separates the ones and tenths digits. The values of each place is ten times that of the place to its right.

Example

A caterpillar traveled from point A to point C along the path shown.

A ———— 36.9 cm ———— B ———— 52.4 cm ———— C

ADDING DECIMALS To find the total distance traveled by the caterpillar, add the distance from A to B and the distance from B to C. Begin by lining up the decimal points. Then add the figures as you would whole numbers and bring down the decimal point.

```
  36.9 cm
+ 52.4 cm
─────────
  89.3 cm
```

ANSWER The caterpillar traveled a total distance of 89.3 centimeters.

Example _continued_

SUBTRACTING DECIMALS To find how much farther the caterpillar traveled on the second leg of the journey, subtract the distance from _A_ to _B_ from the distance from _B_ to _C_.

$$\begin{array}{r} 52.4 \text{ cm} \\ -\ 36.9 \text{ cm} \\ \hline 15.5 \text{ cm} \end{array}$$

ANSWER The caterpillar traveled 15.5 centimeters farther on the second leg of the journey.

Example

A caterpillar is traveling from point _D_ to point _F_ along the path shown. The caterpillar travels at a speed of 9.6 centimeters per minute.

D   E   **33.6 cm**   F

MULTIPLYING DECIMALS You can multiply decimals as you would whole numbers. The number of decimal places in the product is equal to the sum of the number of decimal places in the factors.

For instance, suppose it takes the caterpillar 1.5 minutes to go from _D_ to _E_. To find the distance from _D_ to _E_, multiply the caterpillar's speed by the time it took.

$$\begin{array}{r} 9.6 \\ \times\ 1.5 \\ \hline 480 \\ 96 \\ \hline 14.40 \end{array}$$

1   decimal place
+ 1   decimal place

2   decimal places

> Align as shown.

ANSWER The distance from _D_ to _E_ is 14.4 centimeters.

DIVIDING DECIMALS When you divide by a decimal, move the decimal points the same number of places in the divisor and the dividend to make the divisor a whole number.

For instance, to find the time it will take the caterpillar to travel from _E_ to _F_, divide the distance from _E_ to _F_ by the caterpillar's speed.

$$9.6\,\overline{)33.6}$$

> Move each decimal point one place to the right.

$$\begin{array}{r} 3.5 \\ 96\,\overline{)336.} \\ \underline{288} \\ 480 \\ \underline{480} \\ 0 \end{array}$$

> Line up decimal points.

ANSWER The caterpillar will travel from _E_ to _F_ in 3.5 minutes.

Fractions

A **fraction** is a number in the form $\frac{a}{b}$, where b is not equal to 0. A fraction is in **simplest form** if its numerator and denominator have a greatest common factor (GCF) of 1. To simplify a fraction, divide its numerator and denominator by their GCF.

Example

A caterpillar is 40 millimeters long. The head of the caterpillar is 6 millimeters long. To compare the length of the caterpillar's head with the caterpillar's total length, you can write and simplify a fraction that expresses the ratio of the two lengths.

Write the ratio of the two lengths: $\quad \dfrac{\text{Length of head}}{\text{Total length}} = \dfrac{6 \text{ mm}}{40 \text{ mm}}$

Write numerator and denominator as products of numbers and the GCF: $\quad = \dfrac{3 \times 2}{20 \times 2}$

Divide numerator and denominator by the GCF: $\quad = \dfrac{3 \times \cancel{2}^{1}}{20 \times \cancel{2}_{1}}$

Simplify: $\quad = \dfrac{3}{20}$

ANSWER In simplest form, the ratio of the lengths is $\frac{3}{20}$.

Percents

A **percent** is a ratio that compares a number to 100. The word *percent* means "per hundred" or "out of 100." The symbol for *percent* is %.

For instance, suppose 43 out of 100 caterpillars are female. You can represent this ratio as a percent, a decimal, or a fraction.

Percent	Decimal	Fraction
43%	0.43	$\frac{43}{100}$

Example

In the preceding example, the ratio of the length of the caterpillar's head to the caterpillar's total length is $\frac{3}{20}$. To write this ratio as a percent, write an equivalent fraction that has a denominator of 100.

Multiply numerator and denominator by 5: $\quad \dfrac{3}{20} = \dfrac{3 \times 5}{20 \times 5}$

$\quad = \dfrac{15}{100}$

Write as a percent: $\quad = 15\%$

ANSWER The caterpillar's head represents 15 percent of its total length.

Using Formulas

A mathematical **formula** is a statement of a fact, rule, or principle. It is usually expressed as an equation.

The term *variable* is also used in science to refer to a factor that can change during an experiment.

In science, a formula often has a word form and a symbolic form. The formula below expresses Ohm's law.

Word Form

$$\text{Current} = \frac{\text{voltage}}{\text{resistance}}$$

Symbolic Form

$$I = \frac{V}{R}$$

In this formula, I, V, and R are variables. A mathematical **variable** is a symbol or letter that is used to represent one or more numbers.

Example

Suppose that you measure a voltage of 1.5 volts and a resistance of 15 ohms. You can use the formula for Ohm's law to find the current in amperes.

Write the formula for Ohm's law: $\quad I = \dfrac{V}{R}$

Substitute 1.5 volts for V and 15 ohms for R: $\quad I = \dfrac{1.5 \text{ volts}}{15 \text{ ohms}}$

Simplify: $\quad I = 0.1 \text{ amp}$

ANSWER The current is 0.1 ampere.

If you know the values of all variables but one in a formula, you can solve for the value of the unknown variable. For instance, Ohm's law can be used to find a voltage if you know the current and the resistance.

Example

Suppose that you know that a current is 0.2 amperes and the resistance is 18 ohms. Use the formula for Ohm's law to find the voltage in volts.

Write the formula for Ohm's law: $\quad I = \dfrac{V}{R}$

Substitute 0.2 amp for I and 18 ohms for R: $\quad 0.2 \text{ amp} = \dfrac{V}{18 \text{ ohms}}$

Multiply both sides by 18 ohms: $\quad 0.2 \text{ amp} \cdot 18 \text{ ohms} = V$

Simplify: $\quad 3.6 \text{ volts} = V$

ANSWER The voltage is 3.6 volts.

Finding Areas

The area of a figure is the amount of surface the figure covers.

Area is measured in square units, such as square meters (m^2) or square centimeters (cm^2). Formulas for the areas of three common geometric figures are shown below.

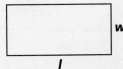

Area = (side length)^2
$A = s^2$

Area = length × width
$A = lw$

Area = $\frac{1}{2}$ × base × height
$A = \frac{1}{2} bh$

Example

Each face of a halite crystal is a square like the one shown. You can find the area of the square by using the steps below.

3 mm

3 mm

Write the formula for the area of a square: $A = s^2$

Substitute 3 mm for s: $= (3 \text{ mm})^2$

Simplify: $= 9 \text{ mm}^2$

ANSWER The area of the square is 9 square millimeters.

Finding Volumes

The volume of a solid is the amount of space contained by the solid.

Volume is measured in cubic units, such as cubic meters (m^3) or cubic centimeters (cm^3). The volume of a rectangular prism is given by the formula shown below.

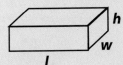

Volume = length × width × height
$V = lwh$

Example

A topaz crystal is a rectangular prism like the one shown. You can find the volume of the prism by using the steps below.

10 mm

12 mm

20 mm

Write the formula for the volume of a rectangular prism: $V = lwh$

Substitute dimensions: $= 20 \text{ mm} \times 12 \text{ mm} \times 10 \text{ mm}$

Simplify: $= 2400 \text{ mm}^3$

ANSWER The volume of the rectangular prism is 2400 cubic millimeters.

Using Significant Figures

The **significant figures** in a decimal are the digits that are warranted by the accuracy of a measuring device.

When you perform a calculation with measurements, the number of significant figures to include in the result depends in part on the number of significant figures in the measurements. When you multiply or divide measurements, your answer should have only as many significant figures as the measurement with the fewest significant figures.

Example

Using a balance and a graduated cylinder filled with water, you determined that a marble has a mass of 8.0 grams and a volume of 3.5 cubic centimeters. To calculate the density of the marble, divide the mass by the volume.

Write the formula for density: $\text{Density} = \dfrac{\text{mass}}{\text{Volume}}$

Substitute measurements: $= \dfrac{8.0 \text{ g}}{3.5 \text{ cm}^3}$

Use a calculator to divide: $\approx 2.285714286 \text{ g/cm}^3$

ANSWER Because the mass and the volume have two significant figures each, give the density to two significant figures. The marble has a density of 2.3 grams per cubic centimeter.

Using Scientific Notation

Scientific notation is a shorthand way to write very large or very small numbers. For example, 73,500,000,000,000,000,000,000 kg is the mass of the Moon. In scientific notation, it is 7.35×10^{22} kg.

Example

You can convert from standard form to scientific notation.

Standard Form	Scientific Notation
720,000	7.2×10^5
5 decimal places left	Exponent is 5.
0.000291	2.91×10^{-4}
4 decimal places right	Exponent is −4.

You can convert from scientific notation to standard form.

Scientific Notation	Standard Form
4.63×10^7	46,300,000
Exponent is 7.	7 decimal places right
1.08×10^{-6}	0.00000108
Exponent is −6.	6 decimal places left

Note-Taking Handbook

Note-Taking Strategies

Taking notes as you read helps you understand the information. The notes you take can also be used as a study guide for later review. This handbook presents several ways to organize your notes.

Content Frame

1. Make a chart in which each column represents a category.
2. Give each column a heading.
3. Write details under the headings.

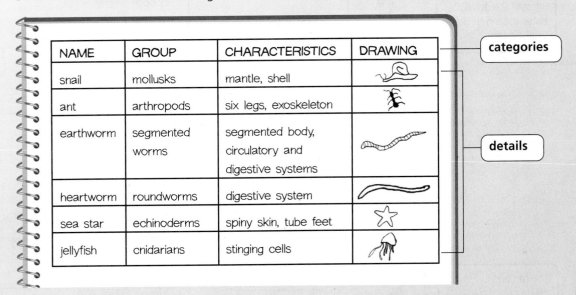

NAME	GROUP	CHARACTERISTICS	DRAWING
snail	mollusks	mantle, shell	
ant	arthropods	six legs, exoskeleton	
earthworm	segmented worms	segmented body, circulatory and digestive systems	
heartworm	roundworms	digestive system	
sea star	echinoderms	spiny skin, tube feet	
jellyfish	cnidarians	stinging cells	

categories

details

Combination Notes

1. For each new idea or concept, write an informal outline of the information.
2. Make a sketch to illustrate the concept, and label it.

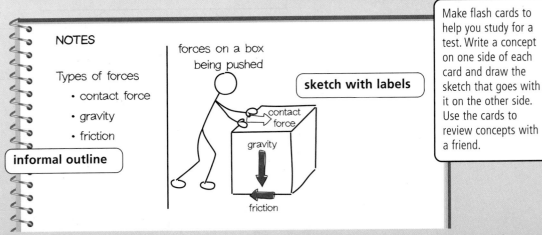

NOTES

Types of forces
- contact force
- gravity
- friction

informal outline

forces on a box being pushed

sketch with labels

contact force

gravity

friction

Make flash cards to help you study for a test. Write a concept on one side of each card and draw the sketch that goes with it on the other side. Use the cards to review concepts with a friend.

Main Idea and Detail Notes

1. In the left-hand column of a two-column chart, list main ideas. The blue headings express main ideas throughout this textbook.

2. In the right-hand column, write details that expand on each main idea.

You can shorten the headings in your chart. Be sure to use the most important words.

When studying for tests, cover up the detail notes column with a sheet of paper. Then use each main idea to form a question—such as "How does latitude affect climate?" Answer the question, and then uncover the detail notes column to check your answer.

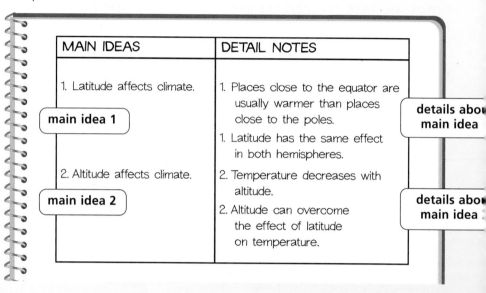

MAIN IDEAS	DETAIL NOTES
1. Latitude affects climate. [main idea 1]	1. Places close to the equator are usually warmer than places close to the poles. [details about main idea] 1. Latitude has the same effect in both hemispheres.
2. Altitude affects climate. [main idea 2]	2. Temperature decreases with altitude. 2. Altitude can overcome the effect of latitude on temperature. [details about main idea]

Main Idea Web

1. Write a main idea in a box.

2. Add boxes around it with related vocabulary terms and important details.

You can find definitions near highlighted terms.

definition of *work*
Work is the use of force to move an object.

formula
Work = force · distance

main idea
Force is necessary to do work.

The joule is the unit used to measure work.
definition of *joule*

Work depends on the size of a force.
important detail

NOTE-TAKING HANDBOOK

Mind Map

1. Write a main idea in the center.

2. Add details that relate to one another and to the main idea.

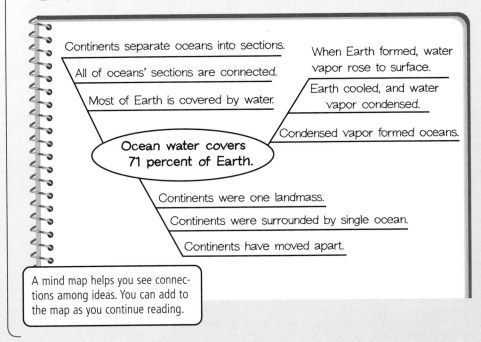

Continents separate oceans into sections.

All of oceans' sections are connected.

Most of Earth is covered by water.

When Earth formed, water vapor rose to surface.

Earth cooled, and water vapor condensed.

Condensed vapor formed oceans.

Ocean water covers 71 percent of Earth.

Continents were one landmass.

Continents were surrounded by single ocean.

Continents have moved apart.

A mind map helps you see connections among ideas. You can add to the map as you continue reading.

Supporting Main Ideas

1. Write a main idea in a box.

2. Add boxes underneath with information—such as reasons, explanations, and examples—that supports the main idea.

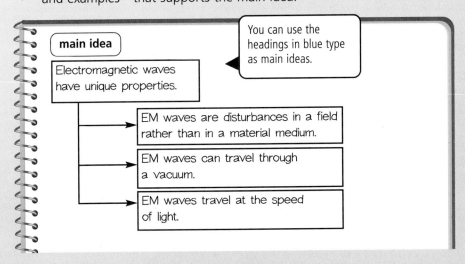

main idea

Electromagnetic waves have unique properties.

You can use the headings in blue type as main ideas.

EM waves are disturbances in a field rather than in a material medium.

EM waves can travel through a vacuum.

EM waves travel at the speed of light.

Outline

1. Copy the chapter title and headings from the book in the form of an outline.

2. Add notes that summarize in your own words what you read.

Cell Processes

1st key idea

I. Cells capture and release energy.

1st subpoint of I

  A. All cells need energy.

2nd subpoint of I

  B. Some cells capture light energy.

1st detail about B

    1. Process of photosynthesis

2nd detail about B

    2. Chloroplasts (site of photosynthesis)

    3. Carbon dioxide and water as raw materials

    4. Glucose and oxygen as products

  C. All cells release energy.

    1. Process of cellular respiration

    2. Fermentation of sugar to carbon dioxide

    3. Bacteria that carry out fermentation

II. Cells transport materials through membranes.

  A. Some materials move by diffusion.

    1. Particle movement from higher to lower concentrations

    2. Movement of water through membrane (osmosis)

  B. Some transport requires energy.

    1. Active transport

    2. Examples of active transport

Correct Outline Form

Include a title.

Arrange key ideas, subpoints, and details as shown.

Indent the divisions of the outline as shown.

Use the same grammatical form for items of the same rank. For example, if A is a sentence, B must also be a sentence.

You must have at least two main ideas or subpoints. That is, every A must be followed by a B, and every 1 must be followed by a 2.

Concept Map

1. Write an important concept in a large oval.

2. Add details related to the concept in smaller ovals.

3. Write linking words on arrows that connect the ovals.

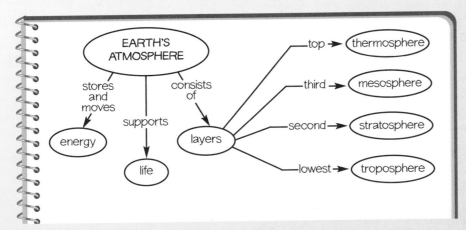

The main ideas or concepts can often be found in the blue headings. An example is "The atmosphere stores and moves energy." Use nouns from these concepts in the ovals, and use the verb or verbs on the lines.

Venn Diagram

1. Draw two overlapping circles, one for each item that you are comparing.

2. In the overlapping section, list the characteristics that are shared by both items.

3. In the outer sections, list the characteristics that are peculiar to each item.

4. Write a summary that describes the information in the Venn diagram.

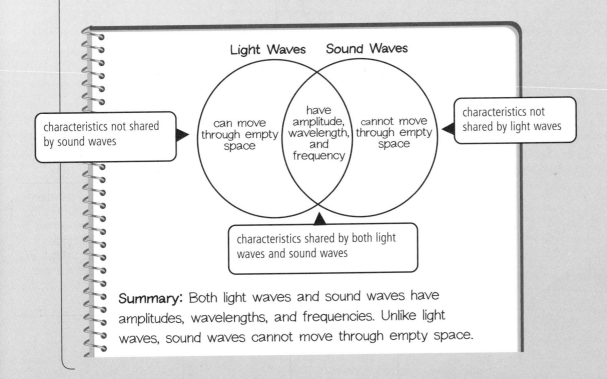

Summary: Both light waves and sound waves have amplitudes, wavelengths, and frequencies. Unlike light waves, sound waves cannot move through empty space.

Vocabulary Strategies

Important terms are highlighted in this book. A definition of each term can be found in the sentence or paragraph where the term appears. You can also find definitions in the Glossary. Taking notes about vocabulary terms helps you understand and remember what you read.

Description Wheel

1. Write a term inside a circle.
2. Write words that describe the term on "spokes" attached to the circle.

When studying for a test with a friend, read the phrases on the spokes one at a time until your friend identifies the correct term.

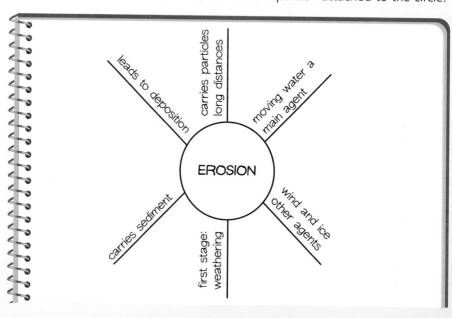

EROSION

- leads to deposition
- carries particles long distances
- moving water a main agent
- wind and ice other agents
- first stage: weathering
- carries sediment

Four Square

1. Write a term in the center.
2. Write details in the four areas around the term.

Definition	Characteristics
any living thing	needs food, water, air; needs energy; grows, develops, reproduces

ORGANISM

Examples	Nonexamples
dogs, cats, birds, insects, flowers, trees	rocks, water, dirt

Include a definition, some characteristics, and examples. You may want to add a formula, a sketch, or examples of things that the term does *not* name.

Frame Game

1. Write a term in the center.
2. Frame the term with details.

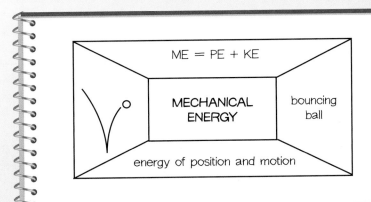

Include examples, descriptions, sketches, or sentences that use the term in context. Change the frame to fit each new term.

Magnet Word

1. Write a term on the magnet.
2. On the lines, add details related to the term.

You can also use phrases or sentences on the lines.

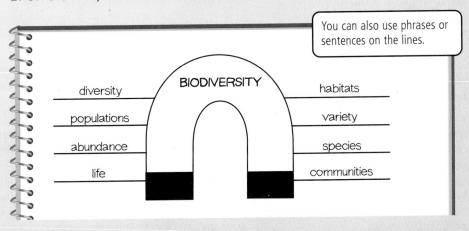

Word Triangle

1. Write a term and its definition in the bottom section.
2. In the middle section, write a sentence in which the term is used correctly.
3. In the top section, draw a small picture to illustrate the term.

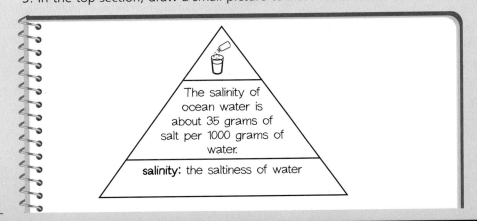

Glossary

A

alternating current AC
Electric current that reverses direction at regular intervals. (p. 97)

    corriente alterna Corriente eléctrica que invierte su dirección a intervalos regulares.

ampere amp
The unit of measurement of electric current, which is equal to one coulomb per second. The number of amps flowing through a circuit equals the circuit's amperage. (p. 29)

    amperio La unidad de medición de la corriente eléctrica, la cual es igual a un culombio por segundo. El número de amperios fluyendo por un circuito es igual al amperaje del circuito.

analog
Represented by a continuous but varying quantity, such as a wave. In electronics, analog information is represented by a continuous but varying electrical signal. (p. 60)

    análogo Que es representado por una cantidad variante pero continua, como una onda. En la electrónica, la información análoga se representa mediante una señal eléctrica continua pero variante.

atom
The smallest particle of an element that has the chemical properties of that element. (p. xv)

    átomo La partícula más pequeña de un elemento que tiene las propiedades químicas de ese elemento.

B

binary code
A coding system in which information is represented by two figures, such as 1 and 0. (p. 58)

    código binario Un sistema de codificación en el cual la información se representa con dos números, como el 1 y el 0.

C

circuit
A closed path through which charge can flow. (p. 43)

    circuito Una trayectoria cerrada por la cual puede fluir una carga.

compound
A substance made up of two or more different types of atoms bonded together.

    compuesto Una sustancia formada por dos o más diferentes tipos de átomos enlazados.

computer
An electronic device that processes digital information. (p. 61)

    computadora Un aparato electrónico que procesa información digital.

conductor
1. A material that transfers electric charge easily (p. 22).
2. A material that transfers energy easily.
conductor 1. Un material que transfiere cargas eléctricas fácilmente. 2. Un material que transfiere energía fácilmente.

cycle
n. A series of events or actions that repeat themselves regularly; a physical and/or chemical process in which one material continually changes locations and/or forms. Examples include the water cycle, the carbon cycle, and the rock cycle.

v. To move through a repeating series of events or actions.

    ciclo *s.* Una serie de eventos o acciones que se repiten regularmente; un proceso físico y/o químico en el cual un material cambia continuamente de lugar y/o forma. Ejemplos: el ciclo del agua, el ciclo del carbono y el ciclo de las rocas.

D

data
Information gathered by observation or experimentation that can be used in calculating or reasoning. *Data* is a plural word; the singular is datum.

    datos Información reunida mediante observación o experimentación y que se puede usar para calcular o para razonar.

density
A property of matter representing the mass per unit volume.

    densidad Una propiedad de la materia que representa la masa por unidad de volumen.

digital
Represented by numbers. In electronics, digital information is represented by the numbers 1 and 0, signaled by a circuit that is either on or off. (p. 58)

    digital Que es representado por números. En la electrónica, la información digital es representada por los números 1 y 0, señalados por un circuito que está encendido o apagado.

direct current DC
Electric current that flows in one direction only. (p. 97)

    corriente directa Corriente eléctrica que fluye en una sola dirección.

E

electric cell
A device that produces electric current using the chemical or physical properties of different materials. A battery consists of two or more cells linked together. (p. 31)

    celda eléctrica Un aparato que produce corriente eléctrica usando las propiedades químicas o físicas de diferentes materiales. Una pila consiste de dos o más celdas conectadas.

electric charge
A property that allows one object to exert a force on another object without touching it. Electric charge can be positive or negative: positive charge is a property of the proton, while negative charge is a property of the electron. (p. 10)

    carga eléctrica Una propiedad que permite a un objeto ejercer una fuerza sobre otro objeto sin tocarlo. La carga eléctrica puede ser positiva o negativa: la carga positiva es una propiedad del protón mientras que la carga negativa es una propiedad del electrón.

electric current
A continuous flow of electric charge, which is measured in amperes. (p. 28)

    corriente eléctrica Un flujo continuo de una carga eléctrica, el cual se mide en amperios.

electric field
An area surrounding a charged object, within which the object can exert a force on another object without touching it. (p. 10)

    campo eléctrico Un área que rodea un objeto con carga, dentro del cual el objeto puede ejercer una fuerza sobre otro objeto sin tocarlo.

electric potential
The amount of potential energy per unit charge that a static charge or electric current has. Electric potential is measured in volts and is often called voltage. (p. 19)

    potencial eléctrico La cantidad de energía potencial por unidad de carga que tiene una carga estática o una corriente eléctrica. El potencial eléctrico se mide en voltios y a menudo se llama voltaje.

electric power
The rate at which electrical energy is generated from, or converted into, another source of energy, such as kinetic energy. (p. 102)

    potencia eléctrica El ritmo al cual se genera energía eléctrica a partir de, o se convierte en, otra fuente de energía, como energía cinética.

electromagnet
A magnet that consists of a piece of iron or steel inside a coil of current-carrying wire. (p. 90)

    electroimán Un imán que consiste de un pedazo de hierro o de acero dentro de una bobina de alambre por la cual fluye una corriente eléctrica.

electromagnetism
Magnetism that results from the flow of electric charge. (p. 89)

    electromagnetismo Magnetismo que resulta del flujo de una carga eléctrica.

electron
A negatively charged particle located outside an atom's nucleus.

    electrón Una partícula con cargada negativamente localizada fuera del núcleo de un átomo.

electronic

adj. Operating by means of an electrical signal. An electronic device is a device that uses electric current to represent coded information. (p. 57)

n. An electronic device or system, such as a computer, calculator, CD player, or game system.

electrónico adj. Que opera por medio de una señal eléctrica. Un aparato electrónico es un aparato que usa corriente eléctrica para representar información codificada.

element

A substance that cannot be broken down into a simpler substance by ordinary chemical changes. An element consists of atoms of only one type. (p. xv)

elemento Una sustancia que no puede descomponerse en otra sustancia más simple por medio de cambios químicos normales. Un elemento consta de átomos de un solo tipo.

energy

The ability to do work or to cause a change. For example, the energy of a moving bowling ball knocks over pins; energy from food allows animals to move and to grow; and energy from the Sun heats Earth's surface and atmosphere, which causes air to move. (p. xix)

energía La capacidad para trabajar o causar un cambio. Por ejemplo, la energía de una bola de boliche en movimiento tumba los pinos; la energía proveniente de su alimento permite a los animales moverse y crecer; la energía del Sol calienta la superficie y la atmósfera de la Tierra, lo que ocasiona que el aire se mueva.

experiment

An organized procedure to study something under controlled conditions. (p. xxiv)

experimento Un procedimiento organizado para estudiar algo bajo condiciones controladas.

F

force

A push or a pull; something that changes the motion of an object. (p. xxi)

fuerza Un empuje o un jalón; algo que cambia el movimiento de un objeto.

friction

A force that resists the motion between two surfaces in contact. (p. xxi)

fricción Una fuerza que resiste el movimiento entre dos superficies en contacto.

G

generator

A device that converts kinetic energy, or the energy of motion, into electrical energy. Generators produce electric current by rotating a magnet within a coil of wire or rotating a coil of wire within a magnetic field. (p. 96)

generador Un aparato que convierte energía cinética, o la energía del movimiento, a energía eléctrica. Los generadores producen corriente eléctrica al girar un imán dentro de una bobina de alambre o haciendo rotar una bobina de alambre dentro de un campo magnético.

gravity

The force that objects exert on each other because of their mass. (p. xxi)

gravedad La fuerza que los objetos ejercen entre sí debido a su masa.

grounding

The creation of a harmless, low-resistance path—a ground—for electricity to follow. Grounding is an important electrical safety procedure. (p. 25)

conexión a tierra La creación de una trayectoria inofensiva, de baja resistencia—una tierra—para que la siga la electricidad. La conexión a tierra es un importante procedimiento de seguridad eléctrica.

H

hypothesis

A tentative explanation for an observation or phenomenon. A hypothesis is used to make testable predictions. (p. xxiv)

hipótesis Una explicación provisional de una observación o de un fenómeno. Una hipótesis se usa para hacer predicciones que se pueden probar.

I

induction

The build-up of a static charge in an object when the object is close to, but not touching, a charged object. (p. 13)

inducción La acumulación de carga estática en un objeto cuando el objeto está cercano a, pero no en contacto con, un objeto con carga.

insulator

1. A material that does not transfer electric charge easily. (p. 22) 2. A material that does not transfer energy easily.

aislante 1. Un material que no transfiere cargas eléctricas fácilmente. 2. Un material que no transfiere energía fácilmente.

J

joule (jool) J

A unit used to measure energy and work. One calorie is equal to 4.18 joules of energy; one joule of work is done when a force of one newton moves an object one meter.

julio Una unidad que se usa para medir la energía y el trabajo. Una caloría es igual a 4.18 julios de energía; se hace un joule de trabajo cuando una fuerza de un newton mueve un objeto un metro.

K

kilowatt kW

A unit of measurement for power equal to 1000 watts. (p. 104)

kilovatio Una unidad de medición para la potencia equivalente a 1000 vatios.

kilowatt-hour kWh

The unit of measurement for electrical energy equal to one kilowatt of power over a one-hour period. (p. 105)

kilovatio-hora La unidad de medición de energía eléctrica igual a un kilovatio de potencia en un período de una hora.

L

law

In science, a rule or principle describing a physical relationship that always works in the same way under the same conditions. The law of conservation of energy is an example.

ley En las ciencias, una regla o un principio que describe una relación física que siempre funciona de la misma manera bajo las mismas condiciones. La ley de la conservación de la energía es un ejemplo.

law of conservation of energy

A law stating that no matter how energy is transferred or transformed, it continues to exist in one form or another. (p. xix)

ley de la conservación de la energía Una ley que establece que no importa cómo se transfiere o transforma la energía, toda la energía sigue presente en alguna forma u otra.

M, N

magnet

An object that attracts certain other materials, particularly iron and steel. (p. 79)

imán Un objeto que atrae a ciertos otros materiales, especialmente al hierro y al acero.

magnetic domain

A group of atoms whose magnetic fields align, or point in the same direction. Magnetic materials have magnetic domains, whereas nonmagnetic materials do not. (p. 83)

dominio magnético Un grupo de átomos cuyos campos magnéticos se alinean, o apuntan en la misma dirección. Los materiales magnéticos tienen dominios magnéticos mientras que los materiales no magnéticos no tienen.

magnetic field

An area surrounding a magnet within which the magnet can exert a force. Magnetic fields are concentrated into a pattern of lines that extend from the magnet's north pole to its south pole. (p. 81)

campo magnético Un área alrededor de un imán dentro del cual el imán puede ejercer una fuerza. Los campos magnéticos se concentran en un patrón de líneas que se extienden del polo norte del imán a su polo sur.

magnetism

The force exerted by a magnet. Opposite poles of two magnets attract, or pull together, whereas like poles of two magnets repel, or push apart. (p. 80)

magnetismo La fuerza que ejerce un imán. Los polos opuestos de dos imanes se atraen, o jalan hacia si, mientras que los polos iguales de dos imanes se repelen, o se empujan para alejarse uno del otro.

mass

A measure of how much matter an object is made of. (p. xv)

masa Una medida de la cantidad de materia de la que está compuesto un objeto.

matter

Anything that has mass and volume. Matter exists ordinarily as a solid, a liquid, or a gas. (p. xv)

materia Todo lo que tiene masa y volumen. Generalmente la materia existe como sólido, líquido o gas.

molecule

A group of atoms that are held together by covalent bonds so that they move as a single unit. (p. xv)

molécula Un grupo de átomos que están unidos mediante enlaces covalentes de tal manera que se mueven como una sola unidad.

O

ohm Ω

The unit of measurement for electrical resistance. (p. 23)

ohmio La unidad de medición para la resistencia eléctrica.

Ohm's law

The mathematical relationship among current, voltage, and resistance, expressed in the formula $I = V/R$ (current = voltage/resistance). (p. 29)

ley de Ohm La relación matemática entre la corriente, el voltaje y la resistencia, expresada en la fórmula $I = V/R$ (corriente = voltaje/resistencia).

P, Q

parallel circuit

A circuit in which current follows more than one path. Each device that is wired in a parallel circuit has its own path to and from the voltage source. (p. 53)

circuito paralelo Un circuito en el cual la corriente sigue más de una trayectoria. Cada aparato que está conectado a un circuito paralelo tiene su propia trayectoria desde y hacia la fuente de voltaje.

proton

A positively charged particle located in an atom's nucleus.

protón Una partícula con carga positiva localizada en el núcleo de un átomo.

R

resistance

The property of a material that determines how easily a charge can move through it. Resistance is measured in ohms. (p. 29)

resistencia La propiedad de un material que determina qué tan fácilmente puede moverse una carga a través de él. La resistencia se mide en ohmios.

resistor

An electrical device that slows the flow of charge in a circuit. (p. 44)

resistencia Un aparato eléctrico que hace más lento el flujo de carga en un circuito.

S

series circuit

A circuit in which current follows a single path. Each device that is wired in a series circuit shares a path to and from the voltage source. (p. 52)

circuito en serie Un circuito en el cual la corriente sigue una sola trayectoria. Cada aparato conectado a un circuito en serie comparte una trayectoria desde y hacia la fuente de voltaje.

short circuit

An unintended and undesired path connecting one part of a circuit with another. (p. 46)

corto circuito Una trayectoria no intencionada y no deseada que conecta una parte de un circuito con otra.

static charge

The buildup of electric charge in an object caused by the uneven distribution of charged particles. (p. 11)

carga estática La acumulación de carga eléctrica en un objeto ocasionada por la desigual distribución de partículas con carga.

system

A group of objects or phenomena that interact. A system can be as simple as a rope, a pulley, and a mass. It also can be as complex as the interaction of energy and matter in the four spheres of the Earth system.

sistema Un grupo de objetos o fenómenos que interactúan. Un sistema puede ser algo tan sencillo como una cuerda, una polea y una masa. También puede ser algo tan complejo como la interacción de la energía y la materia en las cuatro esferas del sistema de la Tierra.

T, U

technology
The use of scientific knowledge to solve problems or engineer new products, tools, or processes.

tecnología El uso de conocimientos científicos para resolver problemas o para diseñar nuevos productos, herramientas o procesos.

theory
In science, a set of widely accepted explanations of observations and phenomena. A theory is a well-tested explanation that is consistent with all available evidence.

teoría En las ciencias, un conjunto de explicaciones de observaciones y fenómenos que es ampliamente aceptado. Una teoría es una explicación bien probada que es consecuente con la evidencia disponible.

transformer
A device that uses electromagnetism to increase or decrease voltage. A transformer is often used in the distribution of current from power plants. (p. 99)

transformador Un aparato que usa electromagnetismo para aumentar o disminuir el voltaje. A menudo se usa un transformador en la distribución de corriente desde las centrales eléctricas.

V

variable
Any factor that can change in a controlled experiment, observation, or model. (p. R30)

variable Cualquier factor que puede cambiar en un experimento controlado, en una observación o en un modelo.

volt V
The unit of measurement for electric potential, which is equal to one joule per coulomb. The number of volts of an electric charge equals the charge's voltage. (p. 19)

voltio La unidad de medición para el potencial eléctrico, el cual es igual a un julio por segundo por culombio. El número de voltios de una carga eléctrica es igual al voltaje de la carga.

volume
An amount of three-dimensional space, often used to describe the space that an object takes up. (p. xv)

volumen Una cantidad de espacio tridimensional; a menudo se usa este término para describir el espacio que ocupa un objeto.

W, X, Y, Z

watt W
The unit of measurement for power, which is equal to one joule of work done or energy transferred in one second. For example, a 75 W light bulb converts electrical energy into heat and light at a rate of 75 joules per second. (p. 104)

vatio La unidad de medición de la potencia, el cual es igual a un julio de trabajo realizado o energía transferida en un segundo. Por ejemplo, una bombilla de 75 W convierte energía eléctrica a calor y luz a un ritmo de 75 julios por segundo.

Index

Page numbers for definitions are printed in **boldface** type.
Page numbers for illustrations, maps, and charts are printed in *italics*.

M

maglev train, 80, *80*
magnetic domain, **82**, 83
magnetic field, **81**, *81*, 89, *89*, 108
    of Earth, *84*, 84–85
magnetic force. *See* magnetism
magnetic north pole, 85
magnetic poles, **80**, *80*, 108
magnetism, xxi, **80**, 108
    atmosphere and, 86
    control of voltage, 99
    electricity and, 88
    electromagnetism, *89*, **89**
    magnetic materials, 82, *82*, 83
    production of electric current, 95, 108
magnets, *xxiii*, **79**, 82, 108
    Earth, *84*, 84–85
    healing properties of, 87
    permanent, 84
    temporary, 84
    vs. other materials, *83*
mass, **xv**
math skills
    area, **R43**
    decimal, **R39**, R40
    describing a set of data, R36–R37
    examples, 35, 56, 107
    formulas, 29, 104, 105, 107, **R42**
    fractions, **R41**
    mean, **R36**
    median, **R36**
    mode, **R37**
    percents, 56, **R41**
    proportions, **R39**
    range, **R37**
    rates, **R38**
    ratios, **R38**
    scientific notation, **R44**
    significant figures, 107, **R44**
    variables, 35
    volume, **R43**
matter, xiv–xvii, **xv**
    conservation of, xvii
    electrical charge property, 9, 36
    forms of, xvi–xvii
    movement of, xvii
    particles and, xiv–xv
    physical forces and, xx
mean, **R36**
median, **R36**
metric system, R20–R21
    changing metric units, R20, *R20*
    converting between US. customary units, R21, *R21*
    temperature conversion, R21, *R21*

microchip, 61
microscope, *R14*, R14–R15
    making a slide or wet mount, R15, *R15*
    scanning tunneling (STM), *xxiv*
    viewing an object, R15
MIDI, 5
mode, **R37**
molecule, **xv**
mosaic, *xiv*
motors, *92*, 92–94, *93*, 108
    electromagnets and, 92
    how they work, 92, *93*
    uses of, 94
multimeter, 30, *30*
music, electronics and, 2–5

N

Neptune, xxiv
net charge, 11
network, computer, 64
north pole, 81, 85
Northern Lights, 86, *86*
note-taking strategies, **R45–R49**
    combination notes, 8, *8*, R45, *R45*
    concept map, R49, *R49*
    content frame, R45, *R45*
    main idea and detail notes, R46, *R46*
    main idea web, 78, *78*, R46, *R46*
    mind map, R47, *R47*
    outline, 42, *42*, R48, *R48*
    supporting main ideas, R47, *R47*
    Venn diagram, R49, *R49*

O

observations, **xxiv, R2,** R5, R33
    qualitative, R2
    quantitative, R2
ohmmeter, 30
ohms, **23**, 29
Ohm's law, **29**, 29–30, 35, 36
open circuit, 45, *45*
operational definition, **R31**
opinion, **R9**
    different from fact, R9
output, 63, *63*
oxygen, fuel cells and, xxvii

V

W, X, Y, Z

Acknowledgments

Photography

Cover © Nick Koudis/Getty Images; **i** © Nick Koudis/Getty Images; **iii** *left (top to bottom)* Photograph of James Trefil by Evan Cantwell; Photograph of Rita Ann Calvo by Joseph Calvo; Photograph of Linda Carnine by Amilcar Cifuentes; Photograph of Sam Miller by Samuel Miller; *right (top to bottom)* Photograph of Kenneth Cutler by Kenneth A. Cutler; Photograph of Donald Steely by Marni Stamm; Photograph of Vicky Vachon by Redfern Photographics; **vi** © 2003 Barbara Ries; **vii** © Philip & Karen Smith/age fotostock america, inc.; **ix** Photographs by Sharon Hoogstraten; **xiv–xv** © Larry Hamill/age fotostock america, inc.; **xvi–xvii** © Fritz Poelking/age fotostock america, inc.; **xviii–xix** © Galen Rowell/Corbis; **xx–xxi** © Jack Affleck/SuperStock; **xxii** AP/Wide World Photos; **xxiii** © David Parker/IMI/University of Birmingham High, TC Consortium/Photo Researchers; **xxiv** *left* AP/Wide World Photos; *right Washington University Record;* **xxv** *top* © Kim Steele/Getty Images; *bottom* Reprinted with permission from S. Zhou et al., *SCIENCE* 291:1944–47. Copyright 2001 AAAS; **xxvi–xxvii** © Mike Fiala/Getty Images; **xxvii** *left* © Derek Trask/Corbis; *right* AP/Wide World Photos; **xxxii** © The Chedd-Angier Production Company; **2–3** © PHISH 2003; **3** © Jacques M. Chenet/Corbis; **4** *top* © John Foxx/ImageState; *bottom* © The Chedd-Angier Production Company; **5** © Stuart Hughes/Corbis; **6–7** AP/Wide World Photos; **7, 9** Photographs by Sharon Hoogstraten; **10** © Roger Ressmeyer/Corbis; **12** © Charles D. Winters/Photo Researchers; **14** Photograph by Sharon Hoogstraten; **16** © Maximilian Stock Ltd./Photo Researchers; **17** *left* © Ann and Rob Simpson; *right* © Patrice Ceisel/Visuals Unlimited; **18** Photograph by Sharon Hoogstraten; **19** © Steve Crise/Corbis; **21** © A & J Verkaik/Corbis; **22** Photograph by Sharon Hoogstraten; **23** *top* © Tim Wright/Corbis; *bottom* © Leland Bobb/Corbis; **24** © James D. Hooker/*Lighting Equipment News (UK);* **26** *top left* © Scott T. Smith; All other photographs by Sharon Hoogstraten; **27, 28, 30, 31** Photographs by Sharon Hoogstraten; **33** © Chip Simons 2003; **34** Photo Courtesy of NASA/Getty Images; **35** © Julian Hirshowitz/Corbis; **36** © James D. Hooker/*Lighting Equipment News (UK);* **40–41** © 2003 Barbara Ries; **41, 43** Photographs by Sharon Hoogstraten; **48** *left* © 1989 Paul Silverman/Fundamental Photographs, NYC; *right* Photograph by Sharon Hoogstraten; **49** © Creative Publishing International, Inc.; **50** *top left* © Gary Rhijnsburger/Masterfile; *center left* © Creative Publishing International, Inc.; **51, 52, 53, 54** Photographs by Sharon Hoogstraten; **56** © Robert Essel NYC/Corbis; **57** Photograph by Sharon Hoogstraten; **59** *top* AP/Wide World Photos; *bottom* Photograph by Sharon Hoogstraten; **61** © Kurt Stier/Corbis; **62, 63** © Gen Nishino/Getty Images; **64** © Donna Cox and Robert Patterson/ National Center for Supercomputing Applications, University of Illinois, Urbana; **65** AP/Wide World Photos; **66** *top* © Sheila Terry/Photo Researchers; *bottom* Photograph by Sharon Hoogstraten; **72** *top* © SPL/Photo Researchers; *bottom* The Granger Collection, New York; **73** *top left* © Philadelphia Museum of Art/Corbis; *top right* © Archivo Iconografico, S.A./Corbis; *bottom* © Adam Hart-Davis/ Photo Researchers; **74** *top* Science Museum/Science & Society Picture Library; *center left* © Bettmann/ Corbis; *center right* © Tony Craddock/Photo Researchers; *bottom* © Bettmann/Corbis; **75** *top* © Alfred Pasieka/Photo Researchers; *bottom* AP/Wide World Photos; **76–77** © Philip & Karen Smith/age foto- stock america, inc.; **77, 79** Photographs by Sharon Hoogstraten; **80** *top* © Michael S. Yamashita/ Corbis; *bottom* Photograph by Sharon Hoogstraten; **81** Photographs by Sharon Hoogstraten; **82** © The Natural History Museum, London; **83** Photograph by Sharon Hoogstraten; **84** NASA; **85** Photograph by Sharon Hoogstraten; **86** © Chris Madeley/Photo Researchers; **87** © Brian Bahr/Getty Images; *inset* Courtesy of Discover Magnetics; **88, 90** Photographs by Sharon Hoogstraten; **91** *top* © George Haling/Photo Researchers; *bottom* © Dick Luria/Photo Researchers; **93** © G. K. & Vikki Hart/Getty Images; **95** Photograph by Sharon Hoogstraten; **97** © Ondrea Barbe/ Corbis; **98** Photograph by Sharon Hoogstraten; **99** © Randy M. Ury/Corbis; **100** *top* © Christopher Gould/Getty Images; *bottom* Photographs by Sharon Hoogstraten; **101** Photographs by Sharon Hoogstraten; **102** Bureau of Reclamation; **104** Courtesy of General Electric; **105** *top* Photograph by Sharon Hoogstraten; *bottom* © Maya Barnes/The Image Works; **107** © Mark Richards/PhotoEdit; **108** *top* Photograph by Sharon Hoogstraten; **R28** © Photodisc/Getty Images.

Illustrations

Ampersand Design Group **50**
Steve Cowden **60, 62–63, 68**
Stephen Durke **11, 12, 13, 21, 25, 31, 36, 39, 44, 46, 53, 68, 103, 108, 110**
Dan Stukenschneider **15, 33, 47, 49, 55, 92, 93, 94, 96, 97, 108, R11–R19, R22, R32**

Content Standards: 5–8

A. Science as Inquiry

As a result of activities in grades 5–8, all students should develop

Abilities Necessary to do Scientific Inquiry

A.1   Identify questions that can be answered through scientific investigations. Students should develop the ability to refine and refocus broad and ill-defined questions. An important aspect of this ability consists of students' ability to clarify questions and inquiries and direct them toward objects and phenomena that can be described, explained, or predicted by scientific investigations. Students should develop the ability to identify their questions with scientific ideas, concepts, and quantitative relationships that guide investigation.

A.2   Design and conduct a scientific investigation. Students should develop general abilities, such as systematic observation, making accurate measurements, and identifying and controlling variables. They should also develop the ability to clarify their ideas that are influencing and guiding the inquiry, and to understand how those ideas compare with current scientific knowledge. Students can learn to formulate questions, design investigations, execute investigations, interpret data, use evidence to generate explanations, propose alternative explanations, and critique explanations and procedures.

A.3   Use appropriate tools and techniques to gather, analyze, and interpret data. The use of tools and techniques, including mathematics, will be guided by the question asked and the investigations students design. The use of computers for the collection, summary, and display of evidence is part of this standard. Students should be able to access, gather, store, retrieve, and organize data, using hardware and software designed for these purposes.

A.4   Develop descriptions, explanations, predictions, and models using evidence. Students should base their explanation on what they observed, and as they develop cognitive skills, they should be able to differentiate explanation from description—providing causes for effects and establishing relationships based on evidence and logical argument. This standard requires a subject matter knowledge base so the students can effectively conduct investigations, because developing explanations establishes connections between the content of science and the contexts within which students develop new knowledge.

A.5   Think critically and logically to make the relationships between evidence and explanations. Thinking critically about evidence includes deciding what evidence should be used and accounting for anomalous data. Specifically, students should be able to review data from a simple experiment, summarize the data, and form a logical argument about the cause-and-effect relationships in the experiment. Students should begin to state some explanations in terms of the relationship between two or more variables.

A.6   Recognize and analyze alternative explanations and predictions. Students should develop the ability to listen to and respect the explanations proposed by other students. They should remain open to and acknowledge different ideas and explanations, be able to accept the skepticism of others, and consider alternative explanations.

A.7 Communicate scientific procedures and explanations. With practice, students should become competent at communicating experimental methods, following instructions, describing observations, summarizing the results of other groups, and telling other students about investigations and explanations.

A.8 Use mathematics in all aspects of scientific inquiry. Mathematics is essential to asking and answering questions about the natural world. Mathematics can be used to ask questions; to gather, organize, and present data; and to structure convincing explanations.

Understandings about Scientific Inquiry

A.9.a Different kinds of questions suggest different kinds of scientific investigations. Some investigations involve observing and describing objects, organisms, or events; some involve collecting specimens; some involve experiments; some involve seeking more information; some involve discovery of new objects and phenomena; and some involve making models.

A.9.b Current scientific knowledge and understanding guide scientific investigations. Different scientific domains employ different methods, core theories, and standards to advance scientific knowledge and understanding.

A.9.c Mathematics is important in all aspects of scientific inquiry.

A.9.d Technology used to gather data enhances accuracy and allows scientists to analyze and quantify results of investigations.

A.9.e Scientific explanations emphasize evidence, have logically consistent arguments, and use scientific principles, models, and theories. The scientific community accepts and uses such explanations until displaced by better scientific ones. When such displacement occurs, science advances.

A.9.f Science advances through legitimate skepticism. Asking questions and querying other scientists' explanations is part of scientific inquiry. Scientists evaluate the explanations proposed by other scientists by examining evidence, comparing evidence, identifying faulty reasoning, pointing out statements that go beyond the evidence, and suggesting alternative explanations for the same observations.

A.9.g Scientific investigations sometimes result in new ideas and phenomena for study, generate new methods or procedures for an investigation, or develop new technologies to improve the collection of data. All of these results can lead to new investigations.

B. Physical Science

As a result of their activities in grades 5–8, all students should develop an understanding of

Properties and Changes of Properties in Matter

B.1.a A substance has characteristic properties, such as density, a boiling point, and solubility, all of which are independent of the amount of the sample. A mixture of substances often can be separated into the original substances using one or more of the characteristic properties.

B.1.b Substances react chemically in characteristic ways with other substances to form new substances (compounds) with different characteristic properties. In chemical reactions, the total mass is conserved. Substances often are placed in categories or groups if they react in similar ways; metals is an example of such a group.

B.1.c Chemical elements do not break down during normal laboratory reactions involving such treatments as heating, exposure to electric current, or reaction with acids. There are more than 100 known elements that combine in a multitude of ways to produce compounds, which account for the living and nonliving substances that we encounter.

Motions and Forces

B.2.a The motion of an object can be described by its position, direction of motion, and speed. That motion can be measured and represented on a graph.

B.2.b An object that is not being subjected to a force will continue to move at a constant speed and in a straight line.

B.2.c If more than one force acts on an object along a straight line, then the forces will reinforce or cancel one another, depending on their direction and magnitude. Unbalanced forces will cause changes in the speed or direction of an object's motion.

Transfer of Energy

B.3.a Energy is a property of many substances and is associated with heat, light, electricity, mechanical motion, sound, nuclei, and the nature of a chemical. Energy is transferred in many ways.

B.3.b Heat moves in predictable ways, flowing from warmer objects to cooler ones, until both reach the same temperature.

B.3.c Light interacts with matter by transmission (including refraction), absorption, or scattering (including reflection). To see an object, light from that object—emitted by or scattered from it—must enter the eye.

B.3.d Electrical circuits provide a means of transferring electrical energy when heat, light, sound, and chemical changes are produced.

B.3.e In most chemical and nuclear reactions, energy is transferred into or out of a system. Heat, light, mechanical motion, or electricity might all be involved in such transfers.

B.3.f The sun is a major source of energy for changes on the earth's surface. The sun loses energy by emitting light. A tiny fraction of that light reaches the earth, transferring energy from the sun to the earth. The sun's energy arrives as light with a range of wavelengths, consisting of visible light, infrared, and ultraviolet radiation.

C. Life Science

As a result of their activities in grades 5–8, all students should develop understanding of

Structure and Function in Living Systems

C.1.a Living systems at all levels of organization demonstrate the complementary nature of structure and function. Important levels of organization for structure and function include cells, organs, tissues, organ systems, whole organisms, and ecosystems.

C.1.b All organisms are composed of cells—the fundamental unit of life. Most organisms are single cells; other organisms, including humans, are multicellular.

C.1.c Cells carry on the many functions needed to sustain life. They grow and divide, thereby producing more cells. This requires that they take in nutrients, which they use to provide energy for the work that cells do and to make the materials that a cell or an organism needs.

C.1.d Specialized cells perform specialized functions in multicellular organisms. Groups of specialized cells cooperate to form a tissue, such as a muscle. Different tissues are in turn grouped together to form larger functional units, called organs. Each type of cell, tissue, and organ has a distinct structure and set of functions that serve the organism as a whole.

C.1.e The human organism has systems for digestion, respiration, reproduction, circulation, excretion, movement, control, and coordination, and for protection from disease. These systems interact with one another.

C.1.f Disease is a breakdown in structures or functions of an organism. Some diseases are the result of intrinsic failures of the system. Others are the result of damage by infection by other organisms.

Reproduction and Heredity

C.2.a Reproduction is a characteristic of all living systems; because no individual organism lives forever, reproduction is essential to the continuation of every species. Some organisms reproduce asexually. Other organisms reproduce sexually.

C.2.b In many species, including humans, females produce eggs and males produce sperm. Plants also reproduce sexually—the egg and sperm are produced in the flowers of flowering plants. An egg and sperm unite to begin development of a new individual. That new individual receives genetic information from its mother (via the egg) and its father (via the sperm). Sexually produced offspring never are identical to either of their parents.

C.2.c Every organism requires a set of instructions for specifying its traits. Heredity is the passage of these instructions from one generation to another.

C.2.d Hereditary information is contained in genes, located in the chromosomes of each cell. Each gene carries a single unit of information. An inherited trait of an individual can be determined by one or by many genes, and a single gene can influence more than one trait. A human cell contains many thousands of different genes.

C.2.e The characteristics of an organism can be described in terms of a combination of traits. Some traits are inherited and others result from interactions with the environment.

Regulation and Behavior

C.3.a All organisms must be able to obtain and use resources, grow, reproduce, and maintain stable internal conditions while living in a constantly changing external environment.

C.3.b Regulation of an organism's internal environment involves sensing the internal environment and changing physiological activities to keep conditions within the range required to survive.

C.3.c Behavior is one kind of response an organism can make to an internal or environmental stimulus. A behavioral response requires coordination and communication at many levels, including cells, organ systems, and whole organisms. Behavioral response is a set of actions determined in part by heredity and in part from experience.

C.3.d An organism's behavior evolves through adaptation to its environment. How a species moves, obtains food, reproduces, and responds to danger are based in the species' evolutionary history.

Populations and Ecosystems

C.4.a A population consists of all individuals of a species that occur together at a given place and time. All populations living together and the physical factors with which they interact compose an ecosystem.

C.4.b Populations of organisms can be categorized by the function they serve in an ecosystem. Plants and some microorganisms are producers—they make their own food. All animals, including humans, are consumers, which obtain food by eating other organisms. Decomposers, primarily bacteria and fungi, are consumers that use waste materials and dead organisms for food. Food webs identify the relationships among producers, consumers, and decomposers in an ecosystem.

C.4.c For ecosystems, the major source of energy is sunlight. Energy entering ecosystems as sunlight is transferred by producers into chemical energy through photosynthesis. That energy then passes from organism to organism in food webs.

C.4.d The number of organisms an ecosystem can support depends on the resources available and abiotic factors, such as quantity of light and water, range of temperatures, and soil composition. Given adequate biotic and abiotic resources and no disease or predators, populations (including humans) increase at rapid rates. Lack of resources and other factors, such as predation and climate, limit the growth of populations in specific niches in the ecosystem.

Diversity and Adaptations of Organisms

C.5.a Millions of species of animals, plants, and microorganisms are alive today. Although different species might look dissimilar, the unity among organisms becomes apparent from an analysis of internal structures, the similarity of their chemical processes, and the evidence of common ancestry.

C.5.b Biological evolution accounts for the diversity of species developed through gradual processes over many generations. Species acquire many of their unique characteristics through biological adaptation, which involves the selection of naturally occurring variations in populations. Biological adaptations include changes in structures, behaviors, or physiology that enhance survival and reproductive success in a particular environment.

C.5.c Extinction of a species occurs when the environment changes and the adaptive characteristics of a species are insufficient to allow its survival. Fossils indicate that many organisms that lived long ago are extinct. Extinction of species is common; most of the species that have lived on the earth no longer exist.

D. Earth and Space Science

As a result of their activities in grades 5–8, all students should develop an understanding of

Structure of the Earth System

D.1.a The solid earth is layered with a lithosphere; hot, convecting mantle; and dense, metallic core.

D.1.b Lithospheric plates on the scales of continents and oceans constantly move at rates of centimeters per year in response to movements in the mantle. Major geological events, such as earthquakes, volcanic eruptions, and mountain building, result from these plate motions.

D.1.c Land forms are the result of a combination of constructive and destructive forces. Constructive forces include crustal deformation, volcanic eruption, and deposition of sediment, while destructive forces include weathering and erosion.

D.1.d Some changes in the solid earth can be described as the "rock cycle." Old rocks at the earth's surface weather, forming sediments that are buried, then compacted, heated, and often recrystallized into new rock. Eventually, those new rocks may be brought to the surface by the forces that drive plate motions, and the rock cycle continues.

D.1.e Soil consists of weathered rocks and decomposed organic material from dead plants, animals, and bacteria. Soils are often found in layers, with each having a different chemical composition and texture.

D.1.f Water, which covers the majority of the earth's surface, circulates through the crust, oceans, and atmosphere in what is known as the "water cycle." Water evaporates from the earth's surface, rises and cools as it moves to higher elevations, condenses as rain or snow, and falls to the surface where it collects in lakes, oceans, soil, and in rocks underground.

D.1.g Water is a solvent. As it passes through the water cycle it dissolves minerals and gases and carries them to the oceans.

D.1.h The atmosphere is a mixture of nitrogen, oxygen, and trace gases that include water vapor. The atmosphere has different properties at different elevations.

D.1.i Clouds, formed by the condensation of water vapor, affect weather and climate.

D.1.j Global patterns of atmospheric movement influence local weather. Oceans have a major effect on climate, because water in the oceans holds a large amount of heat.

D.1.k Living organisms have played many roles in the earth system, including affecting the composition of the atmosphere, producing some types of rocks, and contributing to the weathering of rocks.

Earth's History

D.2.a The earth processes we see today, including erosion, movement of lithospheric plates, and changes in atmospheric composition, are similar to those that occurred in the past. Earth history is also influenced by occasional catastrophes, such as the impact of an asteroid or comet.

D.2.b Fossils provide important evidence of how life and environmental conditions have changed.

Earth in the Solar System

D.3.a The earth is the third planet from the sun in a system that includes the moon, the sun, eight other planets and their moons, and smaller objects, such as asteroids and comets. The sun, an average star, is the central and largest body in the solar system.

D.3.b Most objects in the solar system are in regular and predictable motion. Those motions explain such phenomena as the day, the year, phases of the moon, and eclipses.

D.3.c Gravity is the force that keeps planets in orbit around the sun and governs the rest of the motion in the solar system. Gravity alone holds us to the earth's surface and explains the phenomena of the tides.

D.3.d The sun is the major source of energy for phenomena on the earth's surface, such as growth of plants, winds, ocean currents, and the water cycle. Seasons result from variations in the amount of the sun's energy hitting the surface, due to the tilt of the earth's rotation on its axis and the length of the day.

E. Science and Technology

As a result of activities in grades 5–8, all students should develop

Abilities of Technological Design

E.1 Identify appropriate problems for technological design. Students should develop their abilities by identifying a specified need, considering its various aspects, and talking to different potential users or beneficiaries. They should appreciate that for some needs, the cultural backgrounds and beliefs of different groups can affect the criteria for a suitable product.

E.2 Design a solution or product. Students should make and compare different proposals in the light of the criteria they have selected. They must consider constraints—such as cost, time, trade-offs, and materials needed—and communicate ideas with drawings and simple models.

E.3 Implement a proposed design. Students should organize materials and other resources, plan their work, make good use of group collaboration where appropriate, choose suitable tools and techniques, and work with appropriate measurement methods to ensure adequate accuracy.

E.4 Evaluate completed technological designs or products. Students should use criteria relevant to the original purpose or need, consider a variety of factors that might affect acceptability and suitability for intended users or beneficiaries, and develop measures of quality with respect to such criteria and factors; they should also suggest improvements and, for their own products, try proposed modifications.

E.5 Communicate the process of technological design. Students should review and describe any completed piece of work and identify the stages of problem identification, solution design, implementation, and evaluation.

Understandings about Science and Technology

E.6.a   Scientific inquiry and technological design have similarities and differences. Scientists propose explanations for questions about the natural world, and engineers propose solutions relating to human problems, needs, and aspirations. Technological solutions are temporary; technologies exist within nature and so they cannot contravene physical or biological principles; technological solutions have side effects; and technologies cost, carry risks, and provide benefits.

E.6.b   Many different people in different cultures have made and continue to make contributions to science and technology.

E.6.c   Science and technology are reciprocal. Science helps drive technology, as it addresses questions that demand more sophisticated instruments and provides principles for better instrumentation and technique. Technology is essential to science, because it provides instruments and techniques that enable observations of objects and phenomena that are otherwise unobservable due to factors such as quantity, distance, location, size, and speed. Technology also provides tools for investigations, inquiry, and analysis.

E.6.d   Perfectly designed solutions do not exist. All technological solutions have trade-offs, such as safety, cost, efficiency, and appearance. Engineers often build in back-up systems to provide safety. Risk is part of living in a highly technological world. Reducing risk often results in new technology.

E.6.e   Technological designs have constraints. Some constraints are unavoidable, for example, properties of materials, or effects of weather and friction; other constraints limit choices in the design, for example, environmental protection, human safety, and aesthetics.

E.6.f   Technological solutions have intended benefits and unintended consequences. Some consequences can be predicted, others cannot.

F. Science in Personal and Social Perspectives

As a result of activities in grades 5–8, all students should develop understanding of

Personal Health

F.1.a   Regular exercise is important to the maintenance and improvement of health. The benefits of physical fitness include maintaining healthy weight, having energy and strength for routine activities, good muscle tone, bone strength, strong heart/lung systems, and improved mental health. Personal exercise, especially developing cardiovascular endurance, is the foundation of physical fitness.

F.1.b   The potential for accidents and the existence of hazards imposes the need for injury prevention. Safe living involves the development and use of safety precautions and the recognition of risk in personal decisions. Injury prevention has personal and social dimensions.

F.1.c   The use of tobacco increases the risk of illness. Students should understand the influence of short-term social and psychological factors that lead to tobacco use, and the possible long-term detrimental effects of smoking and chewing tobacco.

F.1.d   Alcohol and other drugs are often abused substances. Such drugs change how the body functions and can lead to addiction.

F.1.e   Food provides energy and nutrients for growth and development. Nutrition requirements vary with body weight, age, sex, activity, and body functioning.

F.1.f   Sex drive is a natural human function that requires understanding. Sex is also a prominent means of transmitting diseases. The diseases can be prevented through a variety of precautions.

F.1.g   Natural environments may contain substances (for example, radon and lead) that are harmful to human beings. Maintaining environmental health involves establishing or monitoring quality standards related to use of soil, water, and air.

Populations, Resources, and Environments

F.2.a   When an area becomes overpopulated, the environment will become degraded due to the increased use of resources.

F.2.b   Causes of environmental degradation and resource depletion vary from region to region and from country to country.

Natural Hazards

F.3.a   Internal and external processes of the earth system cause natural hazards, events that change or destroy human and wildlife habitats, damage property, and harm or kill humans. Natural hazards include earthquakes, landslides, wildfires, volcanic eruptions, floods, storms, and even possible impacts of asteroids.

F.3.b   Human activities also can induce hazards through resource acquisition, urban growth, land-use decisions, and waste disposal. Such activities can accelerate many natural changes.

F.3.c   Natural hazards can present personal and societal challenges because misidentifying the change or incorrectly estimating the rate and scale of change may result in either too little attention and significant human costs or too much cost for unneeded preventive measures.

Risks and Benefits

F.4.a   Risk analysis considers the type of hazard and estimates the number of people that might be exposed and the number likely to suffer consequences. The results are used to determine the options for reducing or eliminating risks.

F.4.b   Students should understand the risks associated with natural hazards (fires, floods, tornadoes, hurricanes, earthquakes, and volcanic eruptions), with chemical hazards (pollutants in air, water, soil, and food), with biological hazards (pollen, viruses, bacterial, and parasites), social hazards (occupational safety and transportation), and with personal hazards (smoking, dieting, and drinking).

F.4.c   Individuals can use a systematic approach to thinking critically about risks and benefits. Examples include applying probability estimates to risks and comparing them to estimated personal and social benefits.

F.4.d   Important personal and social decisions are made based on perceptions of benefits and risks.

Science and Technology in Society

F.5.a Science influences society through its knowledge and world view. Scientific knowledge and the procedures used by scientists influence the way many individuals in society think about themselves, others, and the environment. The effect of science on society is neither entirely beneficial nor entirely detrimental.

F.5.b Societal challenges often inspire questions for scientific research, and social priorities often influence research priorities through the availability of funding for research.

F.5.c Technology influences society through its products and processes. Technology influences the quality of life and the ways people act and interact. Technological changes are often accompanied by social, political, and economic changes that can be beneficial or detrimental to individuals and to society. Social needs, attitudes, and values influence the direction of technological development.

F.5.d Science and technology have advanced through contributions of many different people, in different cultures, at different times in history. Science and technology have contributed enormously to economic growth and productivity among societies and groups within societies.

F.5.e Scientists and engineers work in many different settings, including colleges and universities, businesses and industries, specific research institutes, and government agencies.

F.5.f Scientists and engineers have ethical codes requiring that human subjects involved with research be fully informed about risks and benefits associated with the research before the individuals choose to participate. This ethic extends to potential risks to communities and property. In short, prior knowledge and consent are required for research involving human subjects or potential damage to property.

F.5.g Science cannot answer all questions and technology cannot solve all human problems or meet all human needs. Students should understand the difference between scientific and other questions. They should appreciate what science and technology can reasonably contribute to society and what they cannot do. For example, new technologies often will decrease some risks and increase others.

G. History and Nature of Science

As a result of activities in grades 5–8, all students should develop understanding of

Science as a Human Endeavor

G.1.a Women and men of various social and ethnic backgrounds—and with diverse interests, talents, qualities, and motivations—engage in the activities of science, engineering, and related fields such as the health professions. Some scientists work in teams, and some work alone, but all communicate extensively with others.

G.1.b Science requires different abilities, depending on such factors as the field of study and type of inquiry. Science is very much a human endeavor, and the work of science relies on basic human qualities, such as reasoning, insight, energy, skill, and creativity—as well as on scientific habits of mind, such as intellectual honesty, tolerance of ambiguity, skepticism, and openness to new ideas.

Nature of Science

G.2.a Scientists formulate and test their explanations of nature using observation, experiments, and theoretical and mathematical models. Although all scientific ideas are tentative and subject to change and improvement in principle, for most major ideas in science, there is much experimental and observational confirmation. Those ideas are not likely to change greatly in the future. Scientists do and have changed their ideas about nature when they encounter new experimental evidence that does not match their existing explanations.

G.2.b In areas where active research is being pursued and in which there is not a great deal of experimental or observational evidence and understanding, it is normal for scientists to differ with one another about the interpretation of the evidence or theory being considered. Different scientists might publish conflicting experimental results or might draw different conclusions from the same data. Ideally, scientists acknowledge such conflict and work towards finding evidence that will resolve their disagreement.

G.2.c It is part of scientific inquiry to evaluate the results of scientific investigations, experiments, observations, theoretical models, and the explanations proposed by other scientists. Evaluation includes reviewing the experimental procedures, examining the evidence, identifying faulty reasoning, pointing out statements that go beyond the evidence, and suggesting alternative explanations for the same observations. Although scientists may disagree about explanations of phenomena, about interpretations of data, or about the value of rival theories, they do agree that questioning, response to criticism, and open communication are integral to the process of science. As scientific knowledge evolves, major disagreements are eventually resolved through such interactions between scientists.

History of Science

G.3.a Many individuals have contributed to the traditions of science. Studying some of these individuals provides further understanding of scientific inquiry, science as a human endeavor, the nature of science, and the relationships between science and society.

G.3.b In historical perspective, science has been practiced by different individuals in different cultures. In looking at the history of many peoples, one finds that scientists and engineers of high achievement are considered to be among the most valued contributors to their culture.

G.3.c Tracing the history of science can show how difficult it was for scientific innovators to break through the accepted ideas of their time to reach the conclusions that we currently take for granted.

1. The Nature of Science

By the end of the 8th grade, students should know that

1.A The Scientific World View

1.A.1 When similar investigations give different results, the scientific challenge is to judge whether the differences are trivial or significant, and it often takes further studies to decide. Even with similar results, scientists may wait until an investigation has been repeated many times before accepting the results as correct.

1.A.2 Scientific knowledge is subject to modification as new information challenges prevailing theories and as a new theory leads to looking at old observations in a new way.

1.A.3 Some scientific knowledge is very old and yet is still applicable today.

1.A.4 Some matters cannot be examined usefully in a scientific way. Among them are matters that by their nature cannot be tested objectively and those that are essentially matters of morality. Science can sometimes be used to inform ethical decisions by identifying the likely consequences of particular actions but cannot be used to establish that some action is either moral or immoral.

1.B Scientific Inquiry

1.B.1 Scientists differ greatly in what phenomena they study and how they go about their work. Although there is no fixed set of steps that all scientists follow, scientific investigations usually involve the collection of relevant evidence, the use of logical reasoning, and the application of imagination in devising hypotheses and explanations to make sense of the collected evidence.

1.B.2 If more than one variable changes at the same time in an experiment, the outcome of the experiment may not be clearly attributable to any one of the variables. It may not always be possible to prevent outside variables from influencing the outcome of an investigation (or even to identify all of the variables), but collaboration among investigators can often lead to research designs that are able to deal with such situations.

1.B.3 What people expect to observe often affects what they actually do observe. Strong beliefs about what should happen in particular circumstances can prevent them from detecting other results. Scientists know about this danger to objectivity and take steps to try and avoid it when designing investigations and examining data. One safeguard is to have different investigators conduct independent studies of the same questions.

1.C The Scientific Enterprise

1.C.1 Important contributions to the advancement of science, mathematics, and technology have been made by different kinds of people, in different cultures, at different times.

1.C.2 Until recently, women and racial minorities, because of restrictions on their education and employment opportunities, were essentially left out of much of the formal work of the science establishment; the remarkable few who overcame those obstacles were even then likely to have their work disregarded by the science establishment.

1.C.3 No matter who does science and mathematics or invents things, or when or where they do it, the knowledge and technology that result can eventually become available to everyone in the world.

1.C.4 Scientists are employed by colleges and universities, business and industry, hospitals, and many government agencies. Their places of work include offices, classrooms, laboratories, farms, factories, and natural field settings ranging from space to the ocean floor.

1.C.5 In research involving human subjects, the ethics of science require that potential subjects be fully informed about the risks and benefits associated with the research and of their right to refuse to participate. Science ethics also demand that scientists must not knowingly subject coworkers, students, the neighborhood, or the community to health or property risks without their prior knowledge and consent. Because animals cannot make informed choices, special care must be taken in using them in scientific research.

1.C.6 Computers have become invaluable in science because they speed up and extend people's ability to collect, store, compile, and analyze data, prepare research reports, and share data and ideas with investigators all over the world.

1.C.7 Accurate record-keeping, openness, and replication are essential for maintaining an investigator's credibility with other scientists and society.

3. The Nature of Technology

By the end of the 8th grade, students should know that

3.A Technology and Science

3.A.1 In earlier times, the accumulated information and techniques of each generation of workers were taught on the job directly to the next generation of workers. Today, the knowledge base for technology can be found as well in libraries of print and electronic resources and is often taught in the classroom.

3.A.2 Technology is essential to science for such purposes as access to outer space and other remote locations, sample collection and treatment, measurement, data collection and storage, computation, and communication of information.

3.A.3 Engineers, architects, and others who engage in design and technology use scientific knowledge to solve practical problems. But they usually have to take human values and limitations into account as well.

3.B Design and Systems

3.B.1 Design usually requires taking constraints into account. Some constraints, such as gravity or the properties of the materials to be used, are unavoidable. Other constraints, including economic, political, social, ethical, and aesthetic ones, limit choices.

3.B.2 All technologies have effects other than those intended by the design, some of which may have been predictable and some not. In either case, these side effects may turn out to be unacceptable to some of the population and therefore lead to conflict between groups.

3.B.3 Almost all control systems have inputs, outputs, and feedback. The essence of control is comparing information about what is happening to what people want to happen and then making appropriate adjustments. This procedure requires sensing information, processing it, and making changes. In almost all modern machines, microprocessors serve as centers of performance control.

3.B.4 Systems fail because they have faulty or poorly matched parts, are used in ways that exceed what was intended by the design, or were poorly designed to begin with. The most common ways to prevent failure are pretesting parts and procedures, overdesign, and redundancy.

3.C Issues in Technology

3.C.1 The human ability to shape the future comes from a capacity for generating knowledge and developing new technologies—and for communicating ideas to others.

3.C.2 Technology cannot always provide successful solutions for problems or fulfill every human need.

3.C.3 Throughout history, people have carried out impressive technological feats, some of which would be hard to duplicate today even with modern tools. The purposes served by these achievements have sometimes been practical, sometimes ceremonial.

3.C.4 Technology has strongly influenced the course of history and continues to do so. It is largely responsible for the great revolutions in agriculture, manufacturing, sanitation and medicine, warfare, transportation, information processing, and communications that have radically changed how people live.

3.C.5 New technologies increase some risks and decrease others. Some of the same technologies that have improved the length and quality of life for many people have also brought new risks.

3.C.6 Rarely are technology issues simple and one-sided. Relevant facts alone, even when known and available, usually do not settle matters entirely in favor of one side or another. That is because the contending groups may have different values and priorities. They may stand to gain or lose in different degrees, or may make very different predictions about what the future consequences of the proposed action will be.

3.C.7 Societies influence what aspects of technology are developed and how these are used. People control technology (as well as science) and are responsible for its effects.

4. The Physical Setting

By the end of the 8th grade, students should know that

4.A The Universe

4.A.1 The sun is a medium-sized star located near the edge of a disk-shaped galaxy of stars, part of which can be seen as a glowing band of light that spans the sky on a very clear night. The universe contains many billions of galaxies, and each galaxy contains many billions of stars. To the naked eye, even the closest of these galaxies is no more than a dim, fuzzy spot.

4.A.2 The sun is many thousands of times closer to the earth than any other star. Light from the sun takes a few minutes to reach the earth, but light from the next nearest star takes a few years to arrive. The trip to that star would take the fastest rocket thousands of years. Some distant galaxies are so far away that their light takes several billion years to reach the earth. People on earth, therefore, see them as they were that long ago in the past.

4.A.3 Nine planets of very different size, composition, and surface features move around the sun in nearly circular orbits. Some planets have a great variety of moons and even flat rings of rock and ice particles orbiting around them. Some of these planets and moons show evidence of geologic activity. The earth is orbited by one moon, many artificial satellites, and debris.

4.A.4 Large numbers of chunks of rock orbit the sun. Some of those that the earth meets in its yearly orbit around the sun glow and disintegrate from friction as they plunge through the atmosphere—and sometimes impact the ground. Other chunks of rocks mixed with ice have long, off-center orbits that carry them close to the sun, where the sun's radiation (of light and particles) boils off frozen material from their surfaces and pushes it into a long, illuminated tail.

4.B The Earth

4.B.1 We live on a relatively small planet, the third from the sun in the only system of planets definitely known to exist (although other, similar systems may be discovered in the universe).

4.B.2 The earth is mostly rock. Three-fourths of its surface is covered by a relatively thin layer of water (some of it frozen), and the entire planet is surrounded by a relatively thin blanket of air. It is the only body in the solar system that appears able to support life. The other planets have compositions and conditions very different from the earth's.

4.B.3 Everything on or anywhere near the earth is pulled toward the earth's center by gravitational force.

4.B.4 Because the earth turns daily on an axis that is tilted relative to the plane of the earth's yearly orbit around the sun, sunlight falls more intensely on different parts of the earth during the year. The difference in heating of the earth's surface produces the planet's seasons and weather patterns.

4.B.5 The moon's orbit around the earth once in about 28 days changes what part of the moon is lighted by the sun and how much of that part can be seen from the earth—the phases of the moon.

4.B.6 Climates have sometimes changed abruptly in the past as a result of changes in the earth's crust, such as volcanic eruptions or impacts of huge rocks from space. Even relatively small changes in atmospheric or ocean content can have widespread effects on climate if the change lasts long enough.

4.B.7   The cycling of water in and out of the atmosphere plays an important role in determining climatic patterns. Water evaporates from the surface of the earth, rises and cools, condenses into rain or snow, and falls again to the surface. The water falling on land collects in rivers and lakes, soil, and porous layers of rock, and much of it flows back into the ocean.

4.B.8   Fresh water, limited in supply, is essential for life and also for most industrial processes. Rivers, lakes, and groundwater can be depleted or polluted, becoming unavailable or unsuitable for life.

4.B.9   Heat energy carried by ocean currents has a strong influence on climate around the world.

4.B.10 Some minerals are very rare and some exist in great quantities, but—for practical purposes— the ability to recover them is just as important as their abundance. As minerals are depleted, obtaining them becomes more difficult. Recycling and the development of substitutes can reduce the rate of depletion but may also be costly.

4.B.11 The benefits of the earth's resources—such as fresh water, air, soil, and trees—can be reduced by using them wastefully or by deliberately or inadvertently destroying them. The atmosphere and the oceans have a limited capacity to absorb wastes and recycle materials naturally. Cleaning up polluted air, water, or soil or restoring depleted soil, forests, or fishing grounds can be very difficult and costly.

4.C Processes that Shape the Earth

4.C.1   The interior of the earth is hot. Heat flow and movement of material within the earth cause earthquakes and volcanic eruptions and create mountains and ocean basins. Gas and dust from large volcanoes can change the atmosphere.

4.C.2   Some changes in the earth's surface are abrupt (such as earthquakes and volcanic eruptions) while other changes happen very slowly (such as uplift and wearing down of mountains). The earth's surface is shaped in part by the motion of water and wind over very long times, which act to level mountain ranges.

4.C.3   Sediments of sand and smaller particles (sometimes containing the remains of organisms) are gradually buried and are cemented together by dissolved minerals to form solid rock again.

4.C.4   Sedimentary rock buried deep enough may be reformed by pressure and heat, perhaps melting and recrystallizing into different kinds of rock. These re-formed rock layers may be forced up again to become land surface and even mountains. Subsequently, this new rock too will erode. Rock bears evidence of the minerals, temperatures, and forces that created it.

4.C.5   Thousands of layers of sedimentary rock confirm the long history of the changing surface of the earth and the changing life forms whose remains are found in successive layers. The youngest layers are not always found on top, because of folding, breaking, and uplift of layers.

4.C.6   Although weathered rock is the basic component of soil, the composition and texture of soil and its fertility and resistance to erosion are greatly influenced by plant roots and debris, bacteria, fungi, worms, insects, rodents, and other organisms.

4.C.7   Human activities, such as reducing the amount of forest cover, increasing the amount and variety of chemicals released into the atmosphere, and intensive farming, have changed the earth's land, oceans, and atmosphere. Some of these changes have decreased the capacity of the environment to support some life forms.

4.D Structure of Matter

4.D.1 All matter is made up of atoms, which are far too small to see directly through a microscope. The atoms of any element are alike but are different from atoms of other elements. Atoms may stick together in well-defined molecules or may be packed together in large arrays. Different arrangements of atoms into groups compose all substances.

4.D.2 Equal volumes of different substances usually have different weights.

4.D.3 Atoms and molecules are perpetually in motion. Increased temperature means greater average energy, so most substances expand when heated. In solids, the atoms are closely locked in position and can only vibrate. In liquids, the atoms or molecules have higher energy, are more loosely connected, and can slide past one another; some molecules may get enough energy to escape into a gas. In gases, the atoms or molecules have still more energy and are free of one another except during occasional collisions.

4.D.4 The temperature and acidity of a solution influence reaction rates. Many substances dissolve in water, which may greatly facilitate reactions between them.

4.D.5 Scientific ideas about elements were borrowed from some Greek philosophers of 2,000 years earlier, who believed that everything was made from four basic substances: air, earth, fire, and water. It was the combinations of these "elements" in different proportions that gave other substances their observable properties. The Greeks were wrong about those four, but now over 100 different elements have been identified, some rare and some plentiful, out of which everything is made. Because most elements tend to combine with others, few elements are found in their pure form.

4.D.6 There are groups of elements that have similar properties, including highly reactive metals, less-reactive metals, highly reactive nonmetals (such as chlorine, fluorine, and oxygen), and some almost completely nonreactive gases (such as helium and neon). An especially important kind of reaction between substances involves combination of oxygen with something else—as in burning or rusting. Some elements don't fit into any of the categories; among them are carbon and hydrogen, essential elements of living matter.

4.D.7 No matter how substances within a closed system interact with one another, or how they combine or break apart, the total weight of the system remains the same. The idea of atoms explains the conservation of matter: If the number of atoms stays the same no matter how they are rearranged, then their total mass stays the same.

4.E Energy Transformations

4.E.1 Energy cannot be created or destroyed, but only changed from one form into another.

4.E.2 Most of what goes on in the universe—from exploding stars and biological growth to the operation of machines and the motion of people—involves some form of energy being transformed into another. Energy in the form of heat is almost always one of the products of an energy transformation.

4.E.3 Heat can be transferred through materials by the collisions of atoms or across space by radiation. If the material is fluid, currents will be set up in it that aid the transfer of heat.

4.E.4 Energy appears in different forms. Heat energy is in the disorderly motion of molecules; chemical energy is in the arrangement of atoms; mechanical energy is in moving bodies or in elastically distorted shapes; gravitational energy is in the separation of mutually attracting masses.

4.F Motion

4.F.1   Light from the sun is made up of a mixture of many different colors of light, even though to the eye the light looks almost white. Other things that give off or reflect light have a different mix of colors.

4.F.2   Something can be "seen" when light waves emitted or reflected by it enter the eye—just as something can be "heard" when sound waves from it enter the ear.

4.F.3   An unbalanced force acting on an object changes its speed or direction of motion, or both. If the force acts toward a single center, the object's path may curve into an orbit around the center.

4.F.4   Vibrations in materials set up wavelike disturbances that spread away from the source. Sound and earthquake waves are examples. These and other waves move at different speeds in different materials.

4.F.5   Human eyes respond to only a narrow range of wavelengths of electromagnetic radiation—visible light. Differences of wavelength within that range are perceived as differences in color.

4.G Forces of Nature

4.G.1   Every object exerts gravitational force on every other object. The force depends on how much mass the objects have and on how far apart they are. The force is hard to detect unless at least one of the objects has a lot of mass.

4.G.2   The sun's gravitational pull holds the earth and other planets in their orbits, just as the planets' gravitational pull keeps their moons in orbit around them.

4.G.3   Electric currents and magnets can exert a force on each other.

5. The Living Environment

By the end of the 8th grade, students should know that

5.A Diversity of Life

5.A.1 One of the most general distinctions among organisms is between plants, which use sunlight to make their own food, and animals, which consume energy-rich foods. Some kinds of organisms, many of them microscopic, cannot be neatly classified as either plants or animals.

5.A.2 Animals and plants have a great variety of body plans and internal structures that contribute to their being able to make or find food and reproduce.

5.A.3 Similarities among organisms are found in internal anatomical features, which can be used to infer the degree of relatedness among organisms. In classifying organisms, biologists consider details of internal and external structures to be more important than behavior or general appearance.

5.A.4 For sexually reproducing organisms, a species comprises all organisms that can mate with one another to produce fertile offspring.

5.A.5 All organisms, including the human species, are part of and depend on two main interconnected global food webs. One includes microscopic ocean plants, the animals that feed on them, and finally the animals that feed on those animals. The other web includes land plants, the animals that feed on them, and so forth. The cycles continue indefinitely because organisms decompose after death to return food material to the environment.

5.B Heredity

5.B.1 In some kinds of organisms, all the genes come from a single parent, whereas in organisms that have sexes, typically half of the genes come from each parent.

5.B.2 In sexual reproduction, a single specialized cell from a female merges with a specialized cell from a male. As the fertilized egg, carrying genetic information from each parent, multiplies to form the complete organism with about a trillion cells, the same genetic information is copied in each cell.

5.B.3 New varieties of cultivated plants and domestic animals have resulted from selective breeding for particular traits.

5.C Cells

5.C.1 All living things are composed of cells, from just one to many millions, whose details usually are visible only through a microscope. Different body tissues and organs are made up of different kinds of cells. The cells in similar tissues and organs in other animals are similar to those in human beings but differ somewhat from cells found in plants.

5.C.2 Cells repeatedly divide to make more cells for growth and repair. Various organs and tissues function to serve the needs of cells for food, air, and waste removal.

5.C.3 Within cells, many of the basic functions of organisms—such as extracting energy from food and getting rid of waste—are carried out. The way in which cells function is similar in all living organisms.

5.C.4 About two-thirds of the weight of cells is accounted for by water, which gives cells many of their properties.

5.D Interdependence of Life

5.D.1 In all environments—freshwater, marine, forest, desert, grassland, mountain, and others—organisms with similar needs may compete with one another for resources, including food, space, water, air, and shelter. In any particular environment, the growth and survival of organisms depend on the physical conditions.

5.D.2 Two types of organisms may interact with one another in several ways: They may be in a producer/consumer, predator/prey, or parasite/host relationship. Or one organism may scavenge or decompose another. Relationships may be competitive or mutually beneficial. Some species have become so adapted to each other that neither could survive without the other.

5.E Flow of Matter and Energy

5.E.1 Food provides molecules that serve as fuel and building material for all organisms. Plants use the energy in light to make sugars out of carbon dioxide and water. This food can be used immediately for fuel or materials or it may be stored for later use. Organisms that eat plants break down the plant structures to produce the materials and energy they need to survive. Then they are consumed by other organisms.

5.E.2 Over a long time, matter is transferred from one organism to another repeatedly and between organisms and their physical environment. As in all material systems, the total amount of matter remains constant, even though its form and location change.

5.E.3 Energy can change from one form to another in living things. Animals get energy from oxidizing their food, releasing some of its energy as heat. Almost all food energy comes originally from sunlight.

5.F Evolution of Life

5.F.1 Small differences between parents and offspring can accumulate (through selective breeding) in successive generations so that descendants are very different from their ancestors.

5.F.2 Individual organisms with certain traits are more likely than others to survive and have offspring. Changes in environmental conditions can affect the survival of individual organisms and entire species.

5.F.3 Many thousands of layers of sedimentary rock provide evidence for the long history of the earth and for the long history of changing life forms whose remains are found in the rocks. More recently deposited rock layers are more likely to contain fossils resembling existing species.

6. The Human Organism

By the end of the 8th grade, students should know that

6.A Human Identity

6.A.1   Like other animals, human beings have body systems for obtaining and providing energy, defense, reproduction, and the coordination of body functions.

6.A.2   Human beings have many similarities and differences. The similarities make it possible for human beings to reproduce and to donate blood and organs to one another throughout the world. Their differences enable them to create diverse social and cultural arrangements and to solve problems in a variety of ways.

6.A.3   Fossil evidence is consistent with the idea that human beings evolved from earlier species.

6.A.4   Specialized roles of individuals within other species are genetically programmed, whereas human beings are able to invent and modify a wider range of social behavior.

6.A.5   Human beings use technology to match or excel many of the abilities of other species. Technology has helped people with disabilities survive and live more conventional lives.

6.A.6   Technologies having to do with food production, sanitation, and disease prevention have dramatically changed how people live and work and have resulted in rapid increases in the human population.

6.B Human Development

6.B.1   Fertilization occurs when sperm cells from a male's testes are deposited near an egg cell from the female ovary, and one of the sperm cells enters the egg cell. Most of the time, by chance or design, a sperm never arrives or an egg isn't available.

6.B.2   Contraception measures may incapacitate sperm, block their way to the egg, prevent the release of eggs, or prevent the fertilized egg from implanting successfully.

6.B.3   Following fertilization, cell division produces a small cluster of cells that then differentiate by appearance and function to form the basic tissues of an embryo. During the first three months of pregnancy, organs begin to form. During the second three months, all organs and body features develop. During the last three months, the organs and features mature enough to function well after birth. Patterns of human development are similar to those of other vertebrates.

6.B.4   The developing embryo—and later the newborn infant—encounters many risks from faults in its genes, its mother's inadequate diet, her cigarette smoking or use of alcohol or other drugs, or from infection. Inadequate child care may lead to lower physical and mental ability.

6.B.5   Various body changes occur as adults age. Muscles and joints become less flexible, bones and muscles lose mass, energy levels diminish, and the senses become less acute. Women stop releasing eggs and hence can no longer reproduce. The length and quality of human life are influenced by many factors, including sanitation, diet, medical care, sex, genes, environmental conditions, and personal health behaviors.

6.C Basic Functions

6.C.1 Organs and organ systems are composed of cells and help to provide all cells with basic needs.

6.C.2 For the body to use food for energy and building materials, the food must first be digested into molecules that are absorbed and transported to cells.

6.C.3 To burn food for the release of energy stored in it, oxygen must be supplied to cells, and carbon dioxide removed. Lungs take in oxygen for the combustion of food and they eliminate the carbon dioxide produced. The urinary system disposes of dissolved waste molecules, the intestinal tract removes solid wastes, and the skin and lungs rid the body of heat energy. The circulatory system moves all these substances to or from cells where they are needed or produced, responding to changing demands.

6.C.4 Specialized cells and the molecules they produce identify and destroy microbes that get inside the body.

6.C.5 Hormones are chemicals from glands that affect other body parts. They are involved in helping the body respond to danger and in regulating human growth, development, and reproduction.

6.C.6 Interactions among the senses, nerves, and brain make possible the learning that enables human beings to cope with changes in their environment.

6.D Learning

6.D.1 Some animal species are limited to a repertoire of genetically determined behaviors; others have more complex brains and can learn a wide variety of behaviors. All behavior is affected by both inheritance and experience.

6.D.2 The level of skill a person can reach in any particular activity depends on innate abilities, the amount of practice, and the use of appropriate learning technologies.

6.D.3 Human beings can detect a tremendous range of visual and olfactory stimuli. The strongest stimulus they can tolerate may be more than a trillion times as intense as the weakest they can detect. Still, there are many kinds of signals in the world that people cannot detect directly.

6.D.4 Attending closely to any one input of information usually reduces the ability to attend to others at the same time.

6.D.5 Learning often results from two perceptions or actions occurring at about the same time. The more often the same combination occurs, the stronger the mental connection between them is likely to be. Occasionally a single vivid experience will connect two things permanently in people's minds.

6.D.6 Language and tools enable human beings to learn complicated and varied things from others.

6.E Physical Health

6.E.1   The amount of food energy (calories) a person requires varies with body weight, age, sex, activity level, and natural body efficiency. Regular exercise is important to maintain a healthy heart/lung system, good muscle tone, and bone strength.

6.E.2   Toxic substances, some dietary habits, and personal behavior may be bad for one's health. Some effects show up right away, others may not show up for many years. Avoiding toxic substances, such as tobacco, and changing dietary habits to reduce the intake of such things as animal fat increases the chances of living longer.

6.E.3   Viruses, bacteria, fungi, and parasites may infect the human body and interfere with normal body functions. A person can catch a cold many times because there are many varieties of cold viruses that cause similar symptoms.

6.E.4   White blood cells engulf invaders or produce antibodies that attack them or mark them for killing by other white cells. The antibodies produced will remain and can fight off subsequent invaders of the same kind.

6.E.5   The environment may contain dangerous levels of substances that are harmful to human beings. Therefore, the good health of individuals requires monitoring the soil, air, and water and taking steps to keep them safe.

6.F Mental Health

6.F.1   Individuals differ greatly in their ability to cope with stressful situations. Both external and internal conditions (chemistry, personal history, values) influence how people behave.

6.F.2   Often people react to mental distress by denying that they have any problem. Sometimes they don't know why they feel the way they do, but with help they can sometimes uncover the reasons.

8. The Designed World

By the end of the 8th grade, students should know that

8.A Agriculture

8.A.1   Early in human history, there was an agricultural revolution in which people changed from hunting and gathering to farming. This allowed changes in the division of labor between men and women and between children and adults, and the development of new patterns of government.

8.A.2   People control the characteristics of plants and animals they raise by selective breeding and by preserving varieties of seeds (old and new) to use if growing conditions change.

8.A.3   In agriculture, as in all technologies, there are always trade-offs to be made. Getting food from many different places makes people less dependent on weather in any one place, yet more dependent on transportation and communication among far-flung markets. Specializing in one crop may risk disaster if changes in weather or increases in pest populations wipe out that crop. Also, the soil may be exhausted of some nutrients, which can be replenished by rotating the right crops.

8.A.4   Many people work to bring food, fiber, and fuel to U.S. markets. With improved technology, only a small fraction of workers in the United States actually plant and harvest the products that people use. Most workers are engaged in processing, packaging, transporting, and selling what is produced.

8.B Materials and Manufacturing

8.B.1 The choice of materials for a job depends on their properties and on how they interact with other materials. Similarly, the usefulness of some manufactured parts of an object depends on how well they fit together with the other parts.

8.B.2 Manufacturing usually involves a series of steps, such as designing a product, obtaining and preparing raw materials, processing the materials mechanically or chemically, and assembling, testing, inspecting, and packaging. The sequence of these steps is also often important.

8.B.3 Modern technology reduces manufacturing costs, produces more uniform products, and creates new synthetic materials that can help reduce the depletion of some natural resources.

8.B.4 Automation, including the use of robots, has changed the nature of work in most fields, including manufacturing. As a result, high-skill, high-knowledge jobs in engineering, computer programming, quality control, supervision, and maintenance are replacing many routine, manual-labor jobs. Workers therefore need better learning skills and flexibility to take on new and rapidly changing jobs.

8.C Energy Sources and Use

8.C.1 Energy can change from one form to another, although in the process some energy is always converted to heat. Some systems transform energy with less loss of heat than others.

8.C.2 Different ways of obtaining, transforming, and distributing energy have different environmental consequences.

8.C.3 In many instances, manufacturing and other technological activities are performed at a site close to an energy source. Some forms of energy are transported easily, others are not.

8.C.4 Electrical energy can be produced from a variety of energy sources and can be transformed into almost any other form of energy. Moreover, electricity is used to distribute energy quickly and conveniently to distant locations.

8.C.5 Energy from the sun (and the wind and water energy derived from it) is available indefinitely. Because the flow of energy is weak and variable, very large collection systems are needed. Other sources don't renew or renew only slowly.

8.C.6 Different parts of the world have different amounts and kinds of energy resources to use and use them for different purposes.

8.D Communication

8.D.1 Errors can occur in coding, transmitting, or decoding information, and some means of checking for accuracy is needed. Repeating the message is a frequently used method.

8.D.2 Information can be carried by many media, including sound, light, and objects. In this century, the ability to code information as electric currents in wires, electromagnetic waves in space, and light in glass fibers has made communication millions of times faster than is possible by mail or sound.

8.E Information Processing

8.E.1   Most computers use digital codes containing only two symbols, 0 and 1, to perform all operations. Continuous signals (analog) must be transformed into digital codes before they can be processed by a computer.

8.E.2   What use can be made of a large collection of information depends upon how it is organized. One of the values of computers is that they are able, on command, to reorganize information in a variety of ways, thereby enabling people to make more and better uses of the collection.

8.E.3   Computer control of mechanical systems can be much quicker than human control. In situations where events happen faster than people can react, there is little choice but to rely on computers. Most complex systems still require human oversight, however, to make certain kinds of judgments about the readiness of the parts of the system (including the computers) and the system as a whole to operate properly, to react to unexpected failures, and to evaluate how well the system is serving its intended purposes.

8.E.4   An increasing number of people work at jobs that involve processing or distributing information. Because computers can do these tasks faster and more reliably, they have become standard tools both in the workplace and at home.

8.F Health Technology

8.F.1   Sanitation measures such as the use of sewers, landfills, quarantines, and safe food handling are important in controlling the spread of organisms that cause disease. Improving sanitation to prevent disease has contributed more to saving human life than any advance in medical treatment.

8.F.2   The ability to measure the level of substances in body fluids has made it possible for physicians to make comparisons with normal levels, make very sophisticated diagnoses, and monitor the effects of the treatments they prescribe.

8.F.3   It is becoming increasingly possible to manufacture chemical substances such as insulin and hormones that are normally found in the body. They can be used by individuals whose own bodies cannot produce the amounts required for good health.

9. The Mathematical World

By the end of the 8th grade, students should know that

9.A Numbers

9.A.1 There have been systems for writing numbers other than the Arabic system of place values based on tens. The very old Roman numerals are now used only for dates, clock faces, or ordering chapters in a book. Numbers based on 60 are still used for describing time and angles.

9.A.2 A number line can be extended on the other side of zero to represent negative numbers. Negative numbers allow subtraction of a bigger number from a smaller number to make sense, and are often used when something can be measured on either side of some reference point (time, ground level, temperature, budget).

9.A.3 Numbers can be written in different forms, depending on how they are being used. How fractions or decimals based on measured quantities should be written depends on how precise the measurements are and how precise an answer is needed.

9.A.4 The operations + and – are inverses of each other—one undoes what the other does; likewise x and ÷ .

9.A.5 The expression *a/b* can mean different things: *a* parts of size *1/b* each, *a* divided by *b*, or *a* compared to *b*.

9.A.6 Numbers can be represented by using sequences of only two symbols (such as 1 and 0, on and off); computers work this way.

9.A.7 Computations (as on calculators) can give more digits than make sense or are useful.

9.B Symbolic Relationships

9.B.1 An equation containing a variable may be true for just one value of the variable.

9.B.2 Mathematical statements can be used to describe how one quantity changes when another changes. Rates of change can be computed from differences in magnitudes and vice versa.

9.B.3 Graphs can show a variety of possible relationships between two variables. As one variable increases uniformly, the other may do one of the following: increase or decrease steadily, increase or decrease faster and faster, get closer and closer to some limiting value, reach some intermediate maximum or minimum, alternately increase and decrease indefinitely, increase or decrease in steps, or do something different from any of these.

9.C Shapes

9.C.1 Some shapes have special properties: triangular shapes tend to make structures rigid, and round shapes give the least possible boundary for a given amount of interior area. Shapes can match exactly or have the same shape in different sizes.

9.C.2 Lines can be parallel, perpendicular, or oblique.

9.C.3 Shapes on a sphere like the earth cannot be depicted on a flat surface without some distortion.

9.C.4 The graphic display of numbers may help to show patterns such as trends, varying rates of change, gaps, or clusters. Such patterns sometimes can be used to make predictions about the phenomena being graphed.

9.C.5 It takes two numbers to locate a point on a map or any other flat surface. The numbers may be two perpendicular distances from a point, or an angle and a distance from a point.

9.C.6 The scale chosen for a graph or drawing makes a big difference in how useful it is.

9.D Uncertainty

9.D.1 How probability is estimated depends on what is known about the situation. Estimates can be based on data from similar conditions in the past or on the assumption that all the possibilities are known.

9.D.2 Probabilities are ratios and can be expressed as fractions, percentages, or odds.

9.D.3 The mean, median, and mode tell different things about the middle of a data set.

9.D.4 Comparison of data from two groups should involve comparing both their middles and the spreads around them.

9.D.5 The larger a well-chosen sample is, the more accurately it is likely to represent the whole. But there are many ways of choosing a sample that can make it unrepresentative of the whole.

9.D.6 Events can be described in terms of being more or less likely, impossible, or certain.

9.E Reasoning

9.E.1 Some aspects of reasoning have fairly rigid rules for what makes sense; other aspects don't. If people have rules that always hold, and good information about a particular situation, then logic can help them to figure out what is true about it. This kind of reasoning requires care in the use of key words such as if, and, not, or, all, and some. Reasoning by similarities can suggest ideas but can't prove them one way or the other.

9.E.2 Practical reasoning, such as diagnosing or troubleshooting almost anything, may require many-step, branching logic. Because computers can keep track of complicated logic, as well as a lot of information, they are useful in a lot of problem-solving situations.

9.E.3 Sometimes people invent a general rule to explain how something works by summarizing observations. But people tend to overgeneralize, imagining general rules on the basis of only a few observations.

9.E.4 People are using incorrect logic when they make a statement such as "If *A* is true, then *B* is true; but *A* isn't true, therefore *B* isn't true either."

9.E.5 A single example can never prove that something is always true, but sometimes a single example can prove that something is not always true.

9.E.6 An analogy has some likenesses to but also some differences from the real thing.

10. Historical Perspectives

By the end of the 8th grade, students should know that

10.A Displacing the Earth from the Center of the Universe

10.A.1 The motion of an object is always judged with respect to some other object or point and so the idea of absolute motion or rest is misleading.

10.A.2 Telescopes reveal that there are many more stars in the night sky than are evident to the unaided eye, the surface of the moon has many craters and mountains, the sun has dark spots, and Jupiter and some other planets have their own moons.

10.F Understanding Fire

10.F.1 From the earliest times until now, people have believed that even though millions of different kinds of material seem to exist in the world, most things must be made up of combinations of just a few basic kinds of things. There has not always been agreement, however, on what those basic kinds of things are. One theory long ago was that the basic substances were earth, water, air, and fire. Scientists now know that these are not the basic substances. But the old theory seemed to explain many observations about the world.

10.F.2 Today, scientists are still working out the details of what the basic kinds of matter are and of how they combine, or can be made to combine, to make other substances.

10.F.3 Experimental and theoretical work done by French scientist Antoine Lavoisier in the decade between the American and French revolutions led to the modern science of chemistry.

10.F.4 Lavoisier's work was based on the idea that when materials react with each other many changes can take place but that in every case the total amount of matter afterward is the same as before. He successfully tested the concept of conservation of matter by conducting a series of experiments in which he carefully measured all the substances involved in burning, including the gases used and those given off.

10.F.5 Alchemy was chiefly an effort to change base metals like lead into gold and to produce an elixir that would enable people to live forever. It failed to do that or to create much knowledge of how substances react with each other. The more scientific study of chemistry that began in Lavoisier's time has gone far beyond alchemy in understanding reactions and producing new materials.

10.G Splitting the Atom

10.G.1 The accidental discovery that minerals containing uranium darken photographic film, as light does, led to the idea of radioactivity.

10.G.2 In their laboratory in France, Marie Curie and her husband, Pierre Curie, isolated two new elements that caused most of the radioactivity of the uranium mineral. They named one radium because it gave off powerful, invisible rays, and the other polonium in honor of Madame Curie's country of birth. Marie Curie was the first scientist ever to win the Nobel prize in two different fields—in physics, shared with her husband, and later in chemistry.

10.I Discovering Germs

10.I.1 Throughout history, people have created explanations for disease. Some have held that disease has spiritual causes, but the most persistent biological theory over the centuries was that illness resulted from an imbalance in the body fluids. The introduction of germ theory by Louis Pasteur and others in the 19th century led to the modern belief that many diseases are caused by microorganisms—bacteria, viruses, yeasts, and parasites.

10.I.2 Pasteur wanted to find out what causes milk and wine to spoil. He demonstrated that spoilage and fermentation occur when microorganisms enter from the air, multiply rapidly, and produce waste products. After showing that spoilage could be avoided by keeping germs out or by destroying them with heat, he investigated animal diseases and showed that microorganisms were involved. Other investigators later showed that specific kinds of germs caused specific diseases.

10.I.3 Pasteur found that infection by disease organisms—germs—caused the body to build up an immunity against subsequent infection by the same organisms. He then demonstrated that it was possible to produce vaccines that would induce the body to build immunity to a disease without actually causing the disease itself.

10.I.4 Changes in health practices have resulted from the acceptance of the germ theory of disease. Before germ theory, illness was treated by appeals to supernatural powers or by trying to adjust body fluids through induced vomiting, bleeding, or purging. The modern approach emphasizes sanitation, the safe handling of food and water, the pasteurization of milk, quarantine, and aseptic surgical techniques to keep germs out of the body; vaccinations to strengthen the body's immune system against subsequent infection by the same kind of microorganisms; and antibiotics and other chemicals and processes to destroy microorganisms.

10.I.5 In medicine, as in other fields of science, discoveries are sometimes made unexpectedly, even by accident. But knowledge and creative insight are usually required to recognize the meaning of the unexpected.

10.J Harnessing Power

10.J.1 Until the 1800s, most manufacturing was done in homes, using small, handmade machines that were powered by muscle, wind, or running water. New machinery and steam engines to drive them made it possible to replace craftsmanship with factories, using fuels as a source of energy. In the factory system, workers, materials, and energy could be brought together efficiently.

10.J.2 The invention of the steam engine was at the center of the Industrial Revolution. It converted the chemical energy stored in wood and coal, which were plentiful, into mechanical work. The steam engine was invented to solve the urgent problem of pumping water out of coal mines. As improved by James Watt, it was soon used to move coal, drive manufacturing machinery, and power locomotives, ships, and even the first automobiles.

11. Common Themes

By the end of the 8th grade, students should know that

11.A Systems

11.A.1 A system can include processes as well as things.

11.A.2 Thinking about things as systems means looking for how every part relates to others. The output from one part of a system (which can include material, energy, or information) can become the input to other parts. Such feedback can serve to control what goes on in the system as a whole.

11.A.3 Any system is usually connected to other systems, both internally and externally. Thus a system may be thought of as containing subsystems and as being a subsystem of a larger system.

11.B Models

11.B.1 Models are often used to think about processes that happen too slowly, too quickly, or on too small a scale to observe directly, or that are too vast to be changed deliberately, or that are potentially dangerous.

11.B.2 Mathematical models can be displayed on a computer and then modified to see what happens.

11.B.3 Different models can be used to represent the same thing. What kind of a model to use and how complex it should be depends on its purpose. The usefulness of a model may be limited if it is too simple or if it is needlessly complicated. Choosing a useful model is one of the instances in which intuition and creativity come into play in science, mathematics, and engineering.

11.C Constancy and Change

11.C.1 Physical and biological systems tend to change until they become stable and then remain that way unless their surroundings change.

11.C.2 A system may stay the same because nothing is happening or because things are happening but exactly counterbalance one another.

11.C.3 Many systems contain feedback mechanisms that serve to keep changes within specified limits.

11.C.4 Symbolic equations can be used to summarize how the quantity of something changes over time or in response to other changes.

11.C.5 Symmetry (or the lack of it) may determine properties of many objects, from molecules and crystals to organisms and designed structures.

11.C.6 Cycles, such as the seasons or body temperature, can be described by their cycle length or frequency, what their highest and lowest values are, and when these values occur. Different cycles range from many thousands of years down to less than a billionth of a second.

11.D Scale

11.D.1 Properties of systems that depend on volume, such as capacity and weight, change out of proportion to properties that depend on area, such as strength or surface processes.

11.D.2 As the complexity of any system increases, gaining an understanding of it depends increasingly on summaries, such as averages and ranges, and on descriptions of typical examples of that system.

12. Habits of Mind

By the end of the 8th grade, students should know that

12.A Values and Attitudes

12.A.1 Know why it is important in science to keep honest, clear, and accurate records.

12.A.2 Know that hypotheses are valuable, even if they turn out not to be true, if they lead to fruitful investigations.

12.A.3 Know that often different explanations can be given for the same evidence, and it is not always possible to tell which one is correct.

12.B Computation and Estimation

12.B.1 Find what percentage one number is of another and figure any percentage of any number.

12.B.2 Use, interpret, and compare numbers in several equivalent forms such as integers, fractions, decimals, and percents.

12.B.3 Calculate the circumferences and areas of rectangles, triangles, and circles, and the volumes of rectangular solids.

12.B.4 Find the mean and median of a set of data.

12.B.5 Estimate distances and travel times from maps and the actual size of objects from scale drawings.

12.B.6 Insert instructions into computer spreadsheet cells to program arithmetic calculations.

12.B.7 Determine what unit (such as seconds, square inches, or dollars per tankful) an answer should be expressed in from the units of the inputs to the calculation, and be able to convert compound units (such as yen per dollar into dollar per yen, or miles per hour into feet per second).

12.B.8 Decide what degree of precision is adequate and round off the result of calculator operations to enough significant figures to reasonably reflect those of the inputs.

12.B.9 Express numbers like 100, 1,000, and 1,000,000 as powers of 10.

12.B.10 Estimate probabilities of outcomes in familiar situations, on the basis of history or the number of possible outcomes.

12.C Manipulation and Observation

12.C.1 Use calculators to compare amounts proportionally.

12.C.2 Use computers to store and retrieve information in topical, alphabetical, numerical, and key-word files, and create simple files of their own devising.

12.C.3 Read analog and digital meters on instruments used to make direct measurements of length, volume, weight, elapsed time, rates, and temperature, and choose appropriate units for reporting various magnitudes.

12.C.4 Use cameras and tape recorders for capturing information.

12.C.5 Inspect, disassemble, and reassemble simple mechanical devices and describe what the various parts are for; estimate what the effect that making a change in one part of a system is likely to have on the system as a whole.

12.D Communication Skills

12.D.1 Organize information in simple tables and graphs and identify relationships they reveal.

12.D.2 Read simple tables and graphs produced by others and describe in words what they show.

12.D.3 Locate information in reference books, back issues of newspapers and magazines, compact disks, and computer databases.

12.D.4 Understand writing that incorporates circle charts, bar and line graphs, two-way data tables, diagrams, and symbols.

12.D.5 Find and describe locations on maps with rectangular and polar coordinates.

12.E Critical-Response Skills

12.E.1 Question claims based on vague attributions (such as "Leading doctors say...") or on statements made by celebrities or others outside the area of their particular expertise.

12.E.2 Compare consumer products and consider reasonable personal trade-offs among them on the basis of features, performance, durability, and cost.

12.E.3 Be skeptical of arguments based on very small samples of data, biased samples, or samples for which there was no control sample.

12.E.4 Be aware that there may be more than one good way to interpret a given set of findings.

12.E.5 Notice and criticize the reasoning in arguments in which (1) fact and opinion are intermingled or the conclusions do not follow logically from the evidence given, (2) an analogy is not apt, (3) no mention is made of whether the control groups are very much like the experimental group, or (4) all members of a group (such as teenagers or chemists) are implied to have nearly identical characteristics that differ from those of other groups.